PRACTICAL
INVERTEBRATE ZOOLOGY

PRACTICAL INVERTEBRATE ZOOLOGY

A laboratory manual for the study of
the major groups of invertebrates,
excluding protochordates

Compiled and edited by

R. P. DALES

CONTRIBUTORS

F. E. G. COX, *Professor of
Zoology, King's College,
University of London*

R. P. DALES, *Professor of
Zoology, Bedford College,
University of London*

J. GREEN, *Professor of
Zoology, Westfield College,
University of London*

J. E. MORTON, *Professor of
Zoology, University of Auckland,
New Zealand*

D. NICHOLS, *Professor of
Biological Sciences,
University of Exeter*

D. WAKELIN, *Senior Lecturer
in Zoology,
University of Glasgow*

SECOND EDITION

BLACKWELL SCIENTIFIC PUBLICATIONS
OXFORD LONDON EDINBURGH
BOSTON MELBOURNE

© Michael Packard 1981
Published in conjunction with
Blackwell Scientific Publications
Editorial offices:
Osney Mead, Oxford, OX2 0EL
8 John Street, London, WC1N 2ES
9 Forrest Road, Edinburgh, EH1 2QH
52 Beacon Street, Boston, Massachusetts 02108,
　USA
214 Berkeley Street, Carlton, Victoria 3053,
　Australia

First published 1981

Typeset by Enset Ltd
Midsomer Norton, Bath and printed and bound
in Great Britain by Billing and Sons Limited
and Kemp Hall Bindery,
Guildford, London, Oxford, Worcester

British Library Cataloguing in Publication Data

Practical invertebrate zoology.—2nd ed.
　1. Invertebrates—Physiology—Laboratory manuals
　I. Dales, Rodney Phillips 592′.01′028　　QL364

ISBN 0-632-00755-9

AUTHORSHIP

F. E. G. Cox
   1, Protozoa; 2, Sponges

R. P. Dales
   3, Coelenterates; 4, Ctenophores; 7.4, Endoprocts;
   8, Annelids; 9, Echiuroids; 10, Sipunculids;
   12, Lophophorates; 19, Insects

J. Green
   7.1, Rotifers; 16, Crustacea

J. E. Morton
   11, Molluscs

D. Nichols
   13, Echinoderms

D. Wakelin
   5, Platyhelminths; 6, Nemertines; 7.2, Nematodes;
   7.3, Acanthocephala; 14, Chaetognaths; 15, Chelicerates;
   17, Onychophorans; 18, Myriapodous arthropods

CONTENTS

PREFACE

This book has been designed with a dual function: to meet the needs of the student in the laboratory, and to act as a sourcebook for the teacher.

Changing emphasis in the direction of biological work is tending to reduce the proportion of time available to undergraduates in zoology for study of animal morphology and phylogeny. This book has been compiled with the conviction that whatever the interests of a student of zoology may be and however specialized these may later become, there is no substitute for a thorough knowledge of animal construction and functional morphology.

An appreciation of invertebrate structure and evolution can only be obtained by personal practice in a laboratory. This is not a comprehensive textbook, it is a laboratory guide to be used in conjunction with a standard textbook such as that of Barnes's *Invertebrate Zoology*, 4th edition 1980, Meglitsch (1972) or Fretter and Graham (1976) or in conjunction with Laverack & Dando (1979).

INTRODUCTION TO FIRST EDITION

Ten years ago, Professor Morton and I discussed the possibility of compiling an invertebrate practical book. The task was formidable, and with the intervention of other duties and research the idea was laid aside, though never dismissed, from our minds. At that time there were two books in particular which served students needs: Bullough's *Practical Invertebrate Anatomy* (Macmillan, 1950), and F. A. Brown's *Selected Invertebrate Types* (Wiley) published in New York in the same year. Over thirty years have elapsed since these two works were published. Many books on invertebrates have appeared since then, not only books devoted to single phyla, but several major textbooks on invertebrates as a whole, including most of the volumes in the series by Hyman and that edited by Grassé.

The present book is the outcome of collaboration between six teachers each of whom is specially interested in one or other of the main groups of invertebrates. Our aims are the same in wishing to provide sufficient information for a student to make a start at getting to know the many kinds of invertebrate alive today, and to indicate the features of which each holds a special interest. Our aim was to provide a manual in which a coherent approach was achieved without the sacrifice of individual emphasis. We have tried to achieve uniformity of presentation for ease of use, but not at the cost of veiling personal enthusiasms.

This book is not complete in itself and is designed to be used with a standard textbook such as that of Barnes (1980) and within the context of lectures and practical classes. We hope that it will help the teacher and organizer of such classes as well as the student: it is this dual purpose which has dictated much of the form and arrangement of information. While some species of animal are obvious choices for study there may be places where they are difficult or impossible to get. Directions are often general, so that they may still be useful if other species are used. Suggestions, too, are given whenever possible for sources of further information and for ideas for experiments or demonstrations. Our choice of examples has been consciously guided by practical supply as well as by particular interest. We have tried to make the book useful throughout the world.

Each Phylum or major group comprises a Section (1–19). In major sections, there is an Introduction followed by a brief outline of general methods of particular usefulness in studying that group of animals, an outline classification (including brief diagnoses) intended for reference, followed by detailed directions for preparations, dissections and experiments. Illustrations are given where required. Most are original and have been drawn by the author of the relevant section. As already emphasized, this manual is designed to be used alongside a standard textbook, so that illustrations have not been given where such are available and accurate. While suggestions are made for experiments, these are such as might be done within the framework of a general invertebrate zoology course.

xi

Further details of experiments more appropriately included in a physiology course will be found in such books as those of Welsh and Smith (1949) and Clark (1966). Similarly, while reference is made to many histological methods, and details of many of them are given, it is expected that a student using this manual will have access to other books supplying such details as we may omit.

We all owe a debt to our many colleagues, and not least to the generations of students without whose perennial enthusiasm this book would never have been conceived. We are also grateful to the authors and editors of journals for permission to include figures redrawn from the *Quarterly Journal of Microscopical Science*, Volumes 92 and 95 (figs. 13, 43) the *Proceedings of the Zoological Society*, Volume 141 (fig. 40, A) the *Journal of the Marine Biological Association*, Volumes 34 and 42 (figs. 44, 52) *Journal of the Society for Experimental Biology*, Volume 14 (fig. 43) and to Messrs. John Wiley and Sons (fig. 16).

R.P.D.

Bedford College
1969

1. PROTOZOA

1.1 Introduction

The protozoa stand apart from the rest of the animal kingdom in that they consist of single cells and all the functions of an organism must be localized within the cell. Protozoa, being unicellular, are small and for practical purposes cannot be treated in the same way as other animals; for example, they cannot be dissected and the examination of serial sections is unrewarding. The techniques used in studying protozoa are basically those of the microbiologist. The microscope, therefore, is the basic tool of protozoology but even with the best light microscopes it is impossible to see anything of the structure of some of the smaller species. Much of what is known about the protozoa has been derived from studies at the electron microscope level and reference must be made to published electron micrographs in order that the structure of particular organelles can be elucidated. The small size may make the examination of protozoa difficult but it does permit the examination of living specimens, and in this respect the protozoa have an advantage over the rest of the animal kingdom. Practically all the organelles can be seen in living specimens and as protozoa are relatively easy to obtain and cheap the study of the group can be based almost entirely on living material. The protozoa have one final advantage, and this is that they can be used in a number of experiments to illustrate basic principles of cell biology and zoology.

Protozoology is a field in its own right and the study of the protozoa involves techniques either unique to or modified for this group. In general, protozoa must be collected, slowed down for microscopical study, examined under a microscope alive and observations supplemented by vitally stained and permanently stained material as well as by electron micrographs. Protozoa can be collected from a variety of sources, and in fact occur in almost any body of water containing sufficient organic material. The most convenient sources of protozoa are commercial organizations which provide cultures of particular species and certain hosts which are almost invariably parasitized by protozoa. Details of how to collect free-living protozoa are given by Mackinnon and Hawes (1961) and a list of easily obtained parasitic protozoa is given on p. 18. Specific collection and culture methods are given in Mackinnon and Hawes (1961) and additional culture methods by Manwell (1961). By far the most comprehensive list of culture methods for protozoa is given in the Journal of Protozoology, volume 5, pp. 1–38 (1958). Some of the better known culture methods are given in the 'List of Strains' available from the Culture Centre of Algae and Protozoa, 36 Storey's Way, Cambridge CB3 0DT. Details Cambridge. Details of methods for the cultivation of parasitic protozoa are given by Taylor and Baker (1968).

Each student should have a good, well-aligned microscope with a high power objective. Ideally the microscope should have a binocular head, phase contrast objectives and condenser, built-in light source and mechanical stage. An oil immersion objective is essential for the examination of some protozoa. If it is not possible for each student to have such a microscope then as many as possible should be available for demonstrations. The most useful microscope for general protozoology is one of the larger Leitz models fitted with phase contrast objectives and a Heine condenser which gives bright field, phase and dark ground illumination with the minimum of adjustment. With a limited number of phase contrast microscopes and oil immersion objectives it is advantageous to have the smaller protozoa available as demonstrations and slides, with cultures of the larger specimens available for individual study.

Many flagellates and ciliates have to be slowed down to observe them because they move across the microscope field so rapidly. Such organisms can be slowed down by making their medium viscous or by adding an anaesthetic to it. Amoebae and Sporozoa do not have to be slowed down, nor do sedentary ciliates or the majority of parasites. Most features of a protozoan cell can be seen in the living organism, but it is sometimes necessary to pinpoint certain structures such as food vacuoles or mitochondria. This can be done by adding the so-called "vital dyes" to the medium in which the organisms are being examined. Permanent preparations of protozoa tend to be very disappointing but they should be studied to supplement observations on living material. The most widely used staining methods are iron haematoxylin for general structures, leuco-basic fuchsin for the nucleus and silver for the ciliary and pellicular structures of ciliates and for the flagella and associated organelles of flagellates. Detailed accounts of the methods used in protozoology can be found in Mackinnon and Hawes (1961) and also in Manwell (1961).

Representatives of all the protozoan orders are briefly described. These representatives have been selected because they illustrate the diagnostic characteristics of the order they represent and because they are easy to obtain. Experiments have been selected because they illustrate certain basic principles and because they are easy to perform. References are given where appropriate and in general the majority of these have been taken from a few books. The most useful books are: Mackinnon and Hawes, *An Introduction to the Study of Protozoa*, Jahn and Jahn, *How to know the Protozoa* and Vickerman and Cox, *The Protozoa*. Other useful books include Kudo, *Protozoology*, 5th edition, which is full of illustrations, and Hall, *Protozoology*, which is also well illustrated. In order to identify protozoa using keys the best book is Calkins, *The Biology of the Protozoa*, which is unfortunately now sadly out of date.

1.2. General Methods

1.2.1. MEASUREMENT

It is essential to know how large a protozoan is, for two main reasons. Firstly the size is frequently an important diagnostic characteristic and secondly it is necessary to know what size of organism to look for. For these reasons the microscope must be calibrated. This is best done by using a permanent eyepiece micrometer and a stage micrometer slide. Match the units of the eyepiece micrometer with the known units on the slide (these differ from slide to slide) and record the eyepiece units in micrometres (μm). Do this for each objective. It may be helpful to prepare a graph which permits an actual size to be read off for a given number of ocular units. The actual size of any protozoan (or other object) can be calculated from these measurements. An alternative method is to calculate the diameter of the microscope field, for each objective, using a stage micrometer slide. When looking for a particular protozoan find out how large it is and form a mental picture of how much of the microscope field it ought to occupy. Select an appropriate objective and search for the organism. It is possible to waste a lot of time by missing organisms which are larger or smaller than an incorrect preconceived size. Always record the size of a protozoan in micrometres on any drawing.

1.2.2. SLOWING PROTOZOA

Many living protozoa move too fast to be examined properly and must be slowed down. There are three main ways in which this can be done.

(a) 10% *methyl cellulose*
Add 10 g methyl cellulose to 50 ml water and bring to the boil. Allow to cool and leave for 30 mins. Make up to 100 ml with cold distilled water stirring until thoroughly mixed. Dilutions of methyl cellulose between 1–10% are also efficient but take longer to slow the organisms.

(b) 2% *sodium carboxymethylcellulose*
Add 2 g sodium carboxymethylcellulose slowly to 100 ml boiling distilled water.
With either of the above preparations make a small ring on a slide and add a drop of water containing the protozoa to the centre of the ring and cover gently with a coverslip. The protozoa will gradually slow down.

(c) *Nickel sulphate*
Nickel sulphate acts as an anaesthetic but different protozoa respond to different concentrations. 0.01% $NiSO_4.6H_2O$ is a useful concentration from which dilutions can be made. Exposure to the nickel sulphate for 15 minutes should suffice.

1.2.3. EXAMINATION IN HANGING DROP

The most convenient way to examine protozoa is on a slide without a coverslip but this is only suitable for low power examination. In most cases the material containing the protozoa can be covered with a coverslip and sealed with very hot vaseline to prevent evaporation. With large organisms it is necessary to support the coverslip on broken pieces of glass, to prevent crushing. The best way to examine protozoa is to place a small drop of water containing the organisms on a coverslip and then to invert this on a cavity slide. If this preparation is then sealed the protozoa will survive for a considerable time in the hanging drop.

1.2.4. VITAL STAINING

Most of the organelles of a protozoan can be seen by means of phase contrast microscopy. It is sometimes useful to be able to stain certain structures in living protozoa and this can be done by using vital dyes (which actually kill the organisms after a while). These dyes can be added in 0.01% aqueous solutions to equal amounts of fluid containing the protozoa, or alternatively 0.1% solutions in alcohol can be used to coat slides on which drops of the protozoan suspension are later mixed. Useful vital stains include Neutral red which colours food vacuoles; Janus green B which dyes mitochondria; Brilliant cresyl blue which dyes a number of structures and Sudan black B and Sudan IV which colour lipids. Indian ink or carmine colour the food vacuoles when fed to most ciliates.

1.2.5. PERMANENT PREPARATIONS

From the point of view of permanent preparations protozoa fall into two categories; parasites in the blood of vertebrates and the rest. Protozoa in the blood are fixed and stained in the same way as ordinary blood films. Thin films fixed in methyl alcohol and stained in Giemsa's stain are usually regarded as being the most satisfactory. Other protozoa must be processed in four stages, attachment to the coverslip or slide, fixation, staining and mounting. Most protozoa are best stained on No. 1 coverslips in Columbia jars; specimens tend to get lost in coplin jars and watch glasses. The secret of successful attachment is to trap the organisms in a thin moist layer prior to fixation. A number of methods are given in Mackinnon and Hawes (1961). The most suitable method for attachment prior to fixation is to coat the coverslip with filtered 1:1, egg albumin:glycerol, add the protozoa and fix when nearly dry. If the protozoa have already been fixed they can be mixed with 1% celloidin in 1:1, ether:alcohol mixture. They then attach to a coverslip. Full details are given by Mackinnon and Hawes (1961).

(a) *Preparation of wet smears*

Place a small amount of the material containing the protozoa on a coverslip and, working rapidly to prevent drying, spread it out into a layer sufficiently thin and uniform to permit microscopical examination. Fix by plunging face downwards into a fixative while still wet. Do not use diluted material as this tends to float off the glass. This method is used for the preparation of such material as gut contents, faeces, marrow, brain etc.

(b) *Fixation*

Smears may be fixed in a variety of fixatives, the most commonly used are Bouin's and Schaudinn's mixtures (see p. 338).
The wet smear on a coverslip is placed face downwards into the fixative in a watch glass. The period of fixation varies with the material used: as a general rule 10–30 minutes is satisfactory. *After fixation the fixative must be washed out of the preparation*, otherwise the remainder of the process will be ruined.

(c) *Staining*

1. *Heidenhain's iron haematoxylin-alcoholic method*

1. Fix in Schaudinn's solution for 20 minutes
2. 70% alcohol for 10 minutes
3. 70% alcohol to which has been added sufficient Lugol's iodine to produce a light mahogany colour, for 10 minutes
4. 70% alcohol for 10 minutes
5. Mordant for 12–24 hours in 1 part 4% iron alum to 10 parts 50% alcohol
6. Wash for 5 minutes in 70% alcohol to remove the excess of mordant solution

7. Stain for 24 hours in 1 part 1% Heidenhain's haematoxylin to 10 parts 70% alcohol. Care should be taken to avoid precipitation of stain on the preparation, and films etc., should be kept face downwards during this stage
8. Differentiate in 1 part 4% iron alum to 10 parts 50% alcohol. No definite time can be given for this process. The section or film should be examined from time to time under a lower power microscope, so that decolorization can be controlled accurately
9. Wash for 2 hours in several changes of 70% alcohol to ensure that the last trace of iron alum has been removed from the preparation
10. Dehydrate in absolute alcohol
11. Clear in clove oil
12. Remove clove oil with xylol
13. Mount in Canada Balsam

2. *Heidenhain's iron haematoxylin–rapid method*

1. Wash out the fixative (Schaudinn's in this case) in 30% alcohol
2. Distilled water for 5 minutes
3. Mordant in 4% aqueous iron alum for $\frac{1}{2}$ to 1 hour
4. Rinse in tap water
5. Stain in haematoxylin for $\frac{1}{2}$ to 1 hour (a period equal to that in the mordant)
6. Rinse in tap water
7. Differentiate in 2% iron alum and return to tap water and examine under a microscope. Do this until the cytoplasm has lost its stain and stop the process before any stain comes out of the nuclei
8. Blue in tap water
9. Bring to 96% alcohol (via 50%)
10. Dehydrate, clear and mount
[*N.B.* Most protozoologists use the alcoholic method]

3. *Phosphotungstic acid-haematoxylin for intestinal protozoa* (modification by W. Cooper)

Dissolve 1 g haematoxylin in a little distilled water, using heat, cool and make up to 80 ml. Add 20 ml 10% phosphotungstic acid (Analar) and allow to ripen for several months. This process can be speeded up by adding 10 ml 0.25% aqueous potassium permanganate.
1. Prepare coverslip preparations of the specimen and fix in Schaudinn's solution for 20–30 minutes
2. Wash briefly in 70% alcohol
3. Treat for 5–10 minutes in 70% alcohol to which 3–5% Lugol's iodine has been added
4. Wash in 70% alcohol containing 3–5 ml 5% sodium hyposulphite
5. Wash in 70% alcohol
6. Place in absolute alcohol for 10 minutes to harden the protozoa
7. Bring, through 70%, 50% and 30% alcohol, to water
8. Immerse face downwards in covered dish of stain for 12 hours, or overnight. (If freshly made stain is used, place the dish in a 37°C incubator)
9. Wash briefly in tap water and dehydrate rapidly by passing the film through 25%, 50%, 70% and absolute alcohol, clove oil and xylol
10. Mount in Canada Balsam
Differentiation is not necessary, as this method does not overstain.

4. *Feulgen reaction for DNA*

1. Fix in Schaudinn's fixative
2. Wash in 70% alcohol and bring to water
3. Place in Normal HCl for 1 minute
4. Place in Normal HCl at 60°C for 5 minutes
5. Wash in HCl
6. Wash in distilled water
7. Place in Schiff's reagent (Leuco-basic fuchsin) for 1–2 hours
8. Wash in saturated *sulphurous* acid for 1 minute 3 times
9. Wash in tap water for 15 minutes
10. Bring to 96% alcohol
11. Stain in 0.1% fast green FCF for 1–2 minutes
12. Wash briefly in 96% alcohol
13. Dehydrate
14. Clear in xylol
15. Mount

5. *Impregnation with silver*

In order to demonstrate the silver-line-system of ciliates, it is necessary to impregnate them with silver. The original method devised by Klein in 1927 is crude in comparison with more recent techniques, but good results can be obtained and it is easily performed in undergraduate classes. Improved silver impregnation techniques are given by Mackinnon and Hawes (1961). Klein's method, which is not suitable for ciliates in seawater or water with a high organic content, is given below.

1. Make a ring with a diamond pencil near the centre of a slide and allow a drop of water containing ciliates to dry within the circle. Try to dry the ciliates as quickly as possible
2. Place the slide in 2% aqueous silver nitrate for 20 minutes. Keep away from bright light
3. Wash quickly with distilled water
4. Place slides ciliates upwards in distilled water in a white staining dish and expose to bright sunlight. Leave until the silver has been reduced and the ciliates are brown in colour
4. Wash thoroughly in distilled water
5. Air dry
6. Do not mount but examine under oil. Only a few specimens will be really well impregnated so search for these

6. *Impregnation with protargol*

This method gives a much better picture than the Klein dry silver method and can be used for flagellates. It is, however, expensive and time consuming. '

1. Fix smears on coverslips in aqueous Bouin's fixative
2. Wash in tap water and store in 70% alcohol to remove picric acid
3. Bring to water
4. Place in 0.25% potassium permanganate for 10 minutes
5. Wash in distilled water
6. Place in 2% oxalic acid for 5–10 minutes
7. Wash thoroughly in distilled water

8. Fill a columbia jar with 1% protargol (silver abumose) and add a coil of copper wire (0.5 g/10 ml protargol). Place the coverslip in the jar and leave at 30–37°C for 24 hours
9. Wash in distilled water
10. Place in 1% hydroquinone in 5% sodium sulphite for 5–10 minutes
11. Wash *very carefully* but thoroughly in distilled water
12. Place in 1% aqueous gold chloride for 5–10 minutes
13. *Rinse* in distilled water
14. Place in 2% oxalic acid until purple
15. Wash in distilled water
16. Place in 5% sodium thiosulphate for 5 minutes
17. Wash thoroughly
18. Dehydrate
19. Clear
20. Mount

1.2.6. PARASITIC PROTOZOA

In general the viscous nature of the body fluids of the host make it unnecessary to slow down parasitic protozoa. Blood parasites, such as trypanosomes, should be examined in drops of blood compressed under a coverslip. As the blood clots trypanosomes tend to appear at the edge of the clot. Intestinal contents are usually best diluted in an appropriate Ringer's solution or saline, 0.9% for parasites of mammals, 0.65% for poikilotherms and freshwater invertebrates and seawater for the parasites of marine invertebrates. Ringer's solution or saline for the parasites of mammals and birds should be kept at about 37°C otherwise the protozoa slow down and cannot be recognized.

(a) *Examination of blood for parasitic protozoa*

Only trypanosomes can be clearly seen in fresh blood. In order to identify, and even to recognize other protozoa, blood films must be made and stained. Thick films indicate the presence or absence of parasites while thin films are essential for diagnostic purposes.

Preparation of films

Thin films. Take a *small* drop of blood on a clean slide and then place the edge of another slide against the drop at an angle of 45° so that the blood runs out along the junction of the two slides. Push the second slide firmly along the first until a thin film of blood has been made. Dry the film by waving it rapidly in the air; do not apply heat.
Thick films. For a thick film the drop should be somewhat larger than for a thin film. Place a drop of blood on a clean slide and then spread it out with a needle so that it eventually covers an area approximately three times that of the original drop, a rough guide being that it should be so thick that it is *just possible* to read newsprint through it. Dry the slide slowly in a horizontal position and protect it from dust and flies.

Fixation

Thin blood films are fixed for one to three minutes in methyl alcohol (methanol), after which they are rapidly washed in tap water. Thick films must not be fixed.

Staining

> *Giemsa's stain*
>
> > *Method for thin films*
> > 1. Pour into a clean staining dish a measured quantity (e.g. 15 ml) of distilled water previously adjusted to pH 7.2
> > 2. Add 1 drop of stain for each ml distilled water
> > 3. Fix film by covering it with methyl alcohol (=methanol) for 1 minute
> > 4. Wash in tap water
> > 5. Place the slide *film downwards* in the stain while still wet
> > 6. Leave for 30–40 minutes
> > 7. Remove the slide from the dish and flush for an instant (*not longer*) in tap water and then stand the slide in an upright position to drain and dry
> >
> > *Method for thick films*
> > 1. Prepare dish of stain as for thin films, or stain both films in the same dish
> > 2. *Without any fixation*, place slide *film downwards* in the stain
> > 3. Leave for 5–20 minutes, depending on the quality of the stain. Agitate the dish gently from time to time to remove the haemoglobin from the area of the film
> > 4. Remove film carefully from the dish and flush for an instant (*not longer*) in a dish of tap water
> > 5. Place slide in a semi-upright position to drain and dry
>
> *Leishman's stain*
>
> > *Method for thin films*
> > 1. Place the slide film uppermost on a staining rack
> > 2. Cover the film with about 12 drops of Leishman's stain and leave for 30 seconds. The methyl alcohol in the stain fixes the film
> > 3. With a pipette add double the quantity (24 drops) of distilled water adjusted to pH 7.2
> > 4. Mix the solution thoroughly by very gently rocking the slide to and fro, taking care not to spill the stain
> > 5. Allow to stain for 15–20 minutes
> > 6. Flush for an instant (*not longer*) in a gentle flow of water
> > 7. Stand the slide in an upright position to drain and dry
> >
> > *Method for thick films*
> > 1. Place the slide film side downwards in a dish (or stand it upright in a cylinder) containing tap water
> > 2. Leave until all the haemoglobin has been removed (2 to 3 minutes) and the film has become completely white
> > 3. Stand the slide in an upright position to drain and dry
> > 4. Then proceed as for Leishman's stain for thin films

Methods of adjusting pH of water to 7.2

A pH of 7.2, as required for staining blood, can be obtained in several ways, the best of which is by preparing a buffered solution. This can be done by dissolving 3.0 g anhydrous disodium phosphate and 0.6 g anhydrous potassium dihydrogen phosphate in 1 litre distilled water. If crystals of $Na_2HPO_4.12H_2O$ have to be used, 7.5 g must be added. More convenient methods for obtaining the correct pH are (1) boiling the distilled water for 10 minutes in a hard glass flask, (2) using bromo-thymol-blue as an indicator, add drops of saturated lithium carbonate until the water takes on and retains a bluish-green colour.

Procedure for examining blood films

Blood films should not normally be covered with a coverslip. Place a drop of oil on the film and search systematically with the 1/12″ objective. For routine surveys it is usually sufficient to examine 250 microscope fields. Lower power objectives should be used after the examination under oil (the oil on the film improves resolution) to scan the film in case anything large has escaped the oil immersion scrutiny. Thin films must always be used for final diagnosis although thick films are more convenient for seeing if parasites are or are not present. If parasites are present and the film needs to be preserved, soak a cleaning tissue in xylol and draw it gently across the slide. When dry mount under a No. 1 coverslip in green euparal.

Thin blood films can be identified by writing on the films themselves with a soft pencil. Such marks do not come off after staining.

Examination of tissues for parasitic protozoa

The presence of parasitic protozoa in any tissue can be deduced from the pathological conditions they cause, but in order to be certain about the identity of the protozoa, sectioned material has to be used. In general, the methods employed are those of normal histology and material is fixed and embedded in the usual way. Sections are usually cut at 7–10m. Two commonly used staining schedules are given below, the first for material fixed in formalin, Bouin's fixative or Schaudinn's fixative and the second for material fixed in Carnoy's fixative.

    1.   *Ehrlich's haematoxylin and eosin*

        1. Wash out fixative in 30% and 50% alcohol if the fixative used was Schaudinn's and 70% alcohol if Bouin's (10 minutes)

        2. Bring to 70% alcohol (5 minutes)

        3. Stain until the specimen is dark blue-black in colour (20 minutes)

        4. Wash in 70% alcohol (5 minutes)

        5. Differentiate in acid alcohol. Watch this process under the microscope and stop it when the nuclei stand out clearly as the stain leaves the cytoplasm

        6. Wash in 70% alcohol (5 minutes)

        7. 50% alcohol (5 minutes)

        8. Wash in tap water until the specimen is blue in colour

9. Stain in 1% aqueous eosin for 1–2 minutes
10. Wash rapidly in tap water
11. Bring rapidly to 96% alcohol (via 50% and 70%)
12. Dehydrate in absolute alcohol, two changes (5 minutes each)
13. Clear in xylol
14. Mount in Canada Balsam

2. *Giemsa's stain for sections (Shortt and Cooper's method)*

This method has the advantage of staining cells and protozoan parasites in sectioned material in a manner similar to that obtained by staining blood smears in Giemsa. Parasite cytoplasm stains blue, parasite nuclei stain red and host tissue stains purple. Carnoy's fixative consists of absolute alcohol 60%, chloroform 30%, glacial acetic acid 10%. The stain consists of Giemsa stain 10 ml, acetone 10 ml, methyl alcohol 10ml, distilled water pH 7.2 100 ml.

1. Fix in Carnoy's fixative (6 hours), embed and cut sections
2. Bring wax sections to water
3. Stain in Giemsa prepared as above (1 hour)
4. Wash momentarily in tap water
5. Differentiate in 15% colophonium resin in acetone for 15 seconds or longer, checking under the microscope all the time. Replace the colophonium as a surface film forms.
6. Wash in 70% acetone/30% xylol several times
7. Clear in xylol
8. Mount in green euparal

1.3.   Classification

Phylum PROTOZOA

Class MASTIGOPHORA

One or more flagella typically present during vegetative stages or during the greater part of the life history; reproduction typically by symmetrogenic binary fission although sexual reproduction occurs in some groups; nutrition autotrophic, heterotrophic or mixotrophic; no spore formation; no conjugation.

Sub-class PHYTOMASTIGOPHORA

Chromatophores present or secondarily lost, the relationship to pigmented forms remaining obvious; majority free living.   *Chlamydomonas*

Many of the unicellular organisms which are regarded as phytoflagellates belong to groups which grade into the higher algae. The sub-class Phytomastigophora is an artificial one existing only in Zoology Departments during Protozoology courses. The zoological orders traditionally accepted do not have exact botanical counterparts and in this book only a few representative "phytoflagellates" will be considered. The most acceptable botanical classification is that of Christensen (1962) Botanik, Bind II, Alger.

Sub-class ZOOMASTIGOPHORA

Chromatophores absent; not obviously related to pigmented forms; one to many flagella; additional organelles may be present in mastigonts; sexual reproduction rare; free living or parasitic.

Order RHIZOMASTIGIDA

One to four or many flagella; pseudopodia present either simultaneously with flagella or alone during the lesser part of the life history; free living or parasitic.   *Tetramitus*

Order PROTOMONADIDA

One to four flagella; kinetoplast present but may be secondarily lost; free living or parasitic
Trypanosoma, Leishmania

Order POLYMASTIGIDA

Typically three to eight flagella; one, two or many nuclei; free living or parasitic.
Chilosnastix

Order TRICHOMONADIDA

Typically four to six flagella, one of which is recurrent; axostyle and parabasal body present; cysts unknown; all parasitic.   *Trichomonas*

Order HYPERMASTIGIDA

Many flagella; single nucleus; all occur in wood-eating insects.   *Trichonympha*

Order OPALINIDA

Many cilia-like flagella in oblique rows over entire body surface; two or many identical nuclei; parasitic in cold blooded vertebrates.   *Opalina*

Class SARCODINA

One or more pseudopodia typically present; amoeboid during most of life cycle; cytoplasm usually differentiated into endoplasm and ectoplasm; naked or with internal skeletons or external tests formed from various materials; asexual reproduction by means of fission; sexual reproduction with flagellate or amoeboid gametes; no spore formation; no conjugation.

Sub-class RHIZOPODA

Pseudopodia lobose, filose or reticulose; without internal skeletons.

Order PROTEOMYXIDIIDA

No test or shell; filose or reticulose pseudopodia may be formed; frequently with flagellate stages in the life history; one to many nuclei.

Order MYCETOZOIDA

Large multinucleate plasmodium; resting cysts or sporangia may be formed; complex life cycle involving sexual reproduction.

Order AMOEBIDA

Pseudopodia typically lobose; rarely filiform or anastomosing; no test or shell; encystment common; no sexuality known.   *Amoeba*

Order TESTACIDA

Pseudopodia typically lobose; simple single chambered test or rigid external membrane; sexuality unknown.   *Anella*

Order FORAMINIFERIDA

Pseudopodia typically reticulose; simple or multilocular test, perforate or imperforate; reproduction with alternation of sexual and asexual generations, one of which may be secondarily repressed.   *Elphidium*

Sub-class ACTINOPODA

Pseudopodia typically radiating axopodia; body usually spherical; distinct inner and outer zones of cytoplasm; often with internal skeleton.

Order HELIOZOIDA

Cytoplasm divided into outer vacuolated and outer dense zones; body naked or with test; one of many nuclei; skeleton, when present, siliceous; sexual reproduction common; mainly freshwater.   *Actinosphaerium*

Order RADIOLARIDA

Inner and outer zones of cytoplasm separated by a central capsule; skeleton siliceous or composed of strontium sulphate; sexual reproduction rare; all marine.   *Acanthometra*

Class SPOROZOA

Without obvious means of locomotion; internal parasites of other organisms; typically with resistant stage in life history; complex life cycle.

Sub-class TELOSPOREA
Infective body sporozoite; polar filaments absent; sexual reproduction occurs in all species.

Order GREGARINIDA
Typically parasites of digestive tract and body cavity of invertebrates; mature trophozoites large and extracellular. *Monocystis*

Order COCCIDIIDA
Typically parasites of epithelial cells of invertebrates and vertebrates; mature trophozoites small and usually intracellular; sporozoites typically enclosed. *Eimeria*

Order HAEMOSPORIDIIDA
Parasites of red blood corpuscles of vertebrates usually forming pigment from host cell haemoglobin; sporogony in invertebrate and schizogony in vertebrate hosts; sporozoites naked. *Plasmodium*

Sub-class PIROPLASMEA
Small parasites of vertebrate erythrocytes; known vectors are ticks; life cycles obscure.

Order PIROPLASMIDA
Babesia

Sub-class TOXOPLASMEA
No sexual reproduction; no polar filaments; parasitic in vertebrates.

Order TOXOPLASMIDA
Toxoplasma

Class CNIDOSPORA

Spores contain one or more polar filaments; infective body sporoplasm.

Order MYXOSPORIDIIDA
Spore with one (rarely two) sporoplasm and with one to six polar capsules; each capsule contains a coiled polar filament; spore membrane composed of two to six valves; coelozoic or histozoic in fishes and occasionally in amphibians and reptiles. *Myxobolus*

Order ACTINOMYXIDIIDA
Spore with three polar capsules, each enclosing polar filament; membrane composed of three valves; several to many sporoplasms; parasitic in invertebrates, especially annelids. *Triactinomyxon*

Order HELICOSPORIDIIDA
Spore with three sporoplasms; sporoplasms surrounded by spirally coiled thick filament; the whole covered by membrane; histozoic. *Helicosporidium*

Order MICROSPORIDIIDA
Spore with one sporoplasm and one or two relatively long polar filaments; cytozoic in invertebrates and some vertebrates. *Nosema*

Class CILIATA

Simple cilia or compound ciliary organelles in at least one stage of life cycle; subpellicular infraciliature; dimorphic nuclei; binary fission basically homothetogenic; sexual phenomena involving conjugation and autogamy; heterotrophic; mostly free living, some parasitic.

Sub-class HOLOTRICHA
Somatic ciliature simple and uniform; buccal ciliature, if present, typically of tetrahymenal type and inconspicuous, except in one order (Peritrichida).

Order GYMNOSTOMATIDA
Essentially no oral ciliature; cytostome-cytopharyngeal complex with trichites, opens directly to outside; morphology and ciliation usually simple; mostly large. *Prorodon*

Order TRICHOSTOMATIDA
Somatic ciliature typically uniform, but highly asymmetrical in some forms; vestibular but no buccal ciliature in oral area. *Balantidium*

Order CHONOTRICHIDA

Somatic ciliature absent in mature individuals; vestibular ciliature in apical "funnel" derived from field of ventral cilia present on migratory larval forms; adults vase-shaped, attached to crustaceans by non-contractile, non-scopula-produced stalk; reproduction by budding.   *Spirochona*

Order APOSTOMATIDA

Somatic ciliature of mature forms spirally arranged; typically with unique rosette near inconspicuous cytostome; polymorphic life cycles with marine crustaceans usually involved as hosts.   *Gymnodinioides*

Order ASTOMATIDA

(a somewhat polyphyletic group, all the species of which should possibly be re-located in other orders of Holotricha).

Somatic ciliature typically uniform; cytostome absent; often large; some species with endoskeletons and hold-fast organelles; catenoid "colonies" (chain formation) typical of some groups; mostly parasitic in oligochaete annelids.   *Anoplophrya*

Order HYMENOSTOMATIDA

Somatic ciliature typically uniform; buccal cavity ventral, with ciliature fundamentally composed of one undulating membrane on the right and adoral zone of three membranelles on the left; often small.   *Paramecium*

Order THIGMOTRICHIDA

Tuft of thigmotactic somatic cilia typically present near anterior end of body; buccal ciliature, if present, located sub-equatorially on ventral surface or at posterior pole; usually parasitic in or on bivalve molluscs.   *Ancistrum*

Order PERITRICHIDA

Somatic ciliature typically absent in mature form; oral ciliature conspicuous, located predominantly at apical pole of body; often attached to substrate by contractile, scopula-produced stalk; colonial organization common; migratory larval form (telotroch) produced by unequal fission, possesses aborally located ciliary girdle.   *Vorticella*

Order SUCTORIDA

Mature stage without cilia, typically sessile, attached to substrate by non-contractile scopula-produced stalk; with few to many suctorial tentacles which serve for ingestion; somatic ciliature present in astomatous migratory larval stage; reproduction by budding.   *Discophrya*

Sub-class SPIROTRICHA

Somatic ciliature sparse in all but one order; cirri present in one group; buccal ciliature conspicuously developed, with adoral zone typically composed of many membranelles on left side of peristomial area.

Order HETEROTRICHIDA

Somatic ciliature, when present, usually uniform; frequently large; some species pigmented; a few species loricate with migratory larval forms.   *Stentor*

Order OLIGOTRICHIDA

Somatic ciliature sparse or absent; buccal membranelles conspicuous, often extending around apical end of body; typically small, mostly marine.   *Halteria*

Order TINTINNIDA

All loricate, but motile; lorica exhibiting variety of shapes, sizes and composition; oral membranelles conspicuous when extended from lorica; typically marine, pelagic, one species widespread in large freshwater bodies.   *Tintinnopsis*

Order ENTODINIOMORPHIDA

Simple somatic ciliature absent; oral membranelles, functional in feeding, restricted to small area; dorsal zone of membranelles and other membranellar tufts present in many species; with firm pellicle, often drawn out posteriorly into spines; parasitic in herbivores.

                                                                                        Epidinium

Order ODONTOSTOMATIDA

Somatic ciliature usually sparse; oral ciliature reduced to eight membranelles; body small, wedge-shaped, laterally compressed; pellicle, or carapace, sometimes with spines.　　*Epalxis*

Order HYPOTRICHIDA

Cirri (compound somatic ciliature) arranged in various patterns on ventral body surface; adoral zone of membranelles prominent; body dorso-ventrally flattened; pellicle quite rigid in some species.　　*Euplotes*

Comments on the classification of the Protozoa

The classification of any group of organisms as large and as varied as the Protozoa is obviously a subject for considerable differences of opinion, and the literature tends to become confused if a number of different classifications are in use at any one time. These notes are intended to clarify the present position and to justify the scheme of classification outlined above. Until the 1950's only one scheme of classification was in general use and this was that of Doflein and Reichenow (1927–29) which gained practically universal acceptance with the publication of Hyman's first volume of her great work on invertebrates in 1940. In 1952–3 the French work dealing with the whole animal kingdom, Traité de Zoologie began publication, together with Grassé's classification of the Protozoa, which was very different from the generally accepted one. Many parts of Grassé's classification were adopted by protozoologists working in a variety of fields. In 1961 a definitive work on the ciliates by J. O. Corliss was published, and his classification was widely accepted *in toto*. From this point onwards the situation became complex. At the first International Conference on Protozoology at Prague all participants received a duplicated scheme of classification prepared by a committee of which the chairman was B. M. Honigberg, and the members mainly American. It was obvious that this complex scheme of classification was not acceptable as a whole to any members of the committee. In 1964 this scheme was published and almost simultaneously alternative and quite different schemes appeared from Russian (1963) and Polish (1964) groups. In 1965 the second International Conference on Protozoology was held and the discussion which took place, and was subsequently published in 1966, leave no doubt that none of the published schemes is anything like satisfactory to the majority of protozoologists.

The three most recent classifications called, mainly for convenience, the American, Russian and Polish schemes, are outlined below. These have been summarized so much that they will be difficult to understand without reference to the classification given above which, for convenience, will be called the conventional one. Explanatory notes have been added in brackets.

1. AMERICAN [J. Protozool., **11**, 7 (1964)]

Phylum PROTOZOA (equivalent to Algae and Fungi)

Sub-phylum 1. *SARCOMASTIGOPHORA* (= Sarcodina + Mastigophora + some Sporozoa)

Super-class 1.　Mastigophora
,,　　　　2.　Opalinata
,,　　　　3.　Sarcodina (including PIROPLASMIDA)

Sub-phylum   2.   *SPOROZOA*
   „   3.   *CNIDOSPORA* (= Cnidosporidia)
   „   4.   *CILIOPHORA*

2.   RUSSIAN [Acta Protozoologica, **1,** 327 (1963)]

Phylum PROTOZOA

Sub-phylum 1.   *PLASMODROMA* [HOMOCARYOTA] (nuclei similar)
   Class 1.   Sarcodina (including PIROPLASMIDA)
    „   2.   Cnidosporidia
    „   3.   Toxoplasmatea
    „   4.   Mastiogophora
    „   5.   Sporozoa

Sub-phylum 2.   *CILIOPHORA* [HETEROCARYOTA] (nuclei dissimilar)
   Class 1.   Ciliata

3.   POLISH [Acta Protozoologica, **2,** 1 (1964)]

Sub-regnum PROTOZOA

   Phylum 1.   *Mastigota*
   Sub-phylum 1.   *MASTIGOPHORA*
      „   2.   *TELOSPORIDIA*

   Phylum 2.   Sarcodina
   Sub-phylum 1.   *RHIZOPODA*
                  ACTINOPODA
                  CNIDOSPORIDIA

   Phylum 3.   Ciliophora
   Sub-phylum 1.   *CILIATA*

The conventional scheme of classification has much to commend it. Firstly, it is much simpler than the alternative schemes. Secondly it is directly comparable with the schemes which appear in the standard protozoological text books, e.g. Wenyon (1926), Calkins (1933), Hyman Vol. 1 (1940), Hoare (1949), Doflein and Reichenow (6th ed., 1949–53), Mackinnon and Hawes (1961), Manwell (1961), Sandon (1963), Kudo (1966), Vickerman and Cox (1967) and the invertebrate text books by Barrington (1967) and Barnes (1980). Thirdly, it has stood the test of time and advances in our knowledge have been assimilated into the classification without the need for a major overhaul. One example will suffice to illustrate this point. The piroplasms, important pathogens of cattle, have traditionally been placed near the malaria parasites. In the 'reformation' of the 1960's they were placed within the Sarcodina, causing a reappraisal of this group, on purely negative grounds. Recently the electron microscope has shown that the piroplasms are remarkably like the malaria parasites in structure, so that their traditional position is justified by this new knowledge.

1.4. Sources of Protozoa

1.4.1. FREE-LIVING PROTOZOA

The most convenient sources of protozoa are samples of water from clean streams or ponds containing organic matter. The water should be collected together with a certain amount of vegetation, and allowed to stand away from heat and direct light for a few days before use. Protozoa then tend to sort themselves out so that different species occupy different parts of the habitat and samples for examination should be taken from the layer of detritus and the surface scum as well as the open water. Aquarium tanks are often good sources of protozoa. Soil protozoa can usually be found in plant pots. A brief account of the collection of protozoa is given in Vickerman and Cox, and a much more comprehensive coverage is given by Mackinnon and Hawes. A useful source of protozoa is available commercially under the tradename "Microzoo", which is manufactured by Tasco Sales Inc., Miami, Florida. The kit consists of protozoan cysts on vermiculite, a nutrient pad and methyl cellulose. The collection of protozoa is a good one and can be dried down and used again. Most of the ciliates are hypotrichs, but samples vary. The best source of known free-living protozoa is the Cambridge Culture Collection which specializes in green flagellates but also provides a number of amoebae and ciliates. A list can be obtained from The Botany School, The University of Cambridge, and this list also contains some details of culture methods used to maintain the collection. The American equivalent is the Culture Collection of Algae, Botany Department, The University of Indiana. A comprehensive list of protozoa, methods for their cultivation and selected references is given in the Journal of Protozoology, 5, 1–38 (1958).

Prepared slides of free-living and parasitic protozoa can be obtained from Turtox Biological Supplies, Chicago, who issue lists of parasites available.

1.4.2. ENTOZOIC AND EPIZOOTIC PROTOZOA

Certain invertebrates and vertebrates frequently serve as hosts to a variety of protozoa. Twelve common hosts between them harbour representatives of nearly all the main groups of parasitic protozoa. These animals are: earthworms, *Lumbricus terrestris*; polychaete worms, *Cirratulus cirratus* or *Cirriformia tentaculata*; the freshwater hog louse *Asellus aquaticus*; larvae of the beetle *Tenebrio molitor* (mealworms); cockroaches *Blatta orientalis* or *Periplaneta americana*; termites *Zootermopsis angusticollis*; mussels, *Mytilus edulis*; herrings, *Clupea harengus*; frogs, *Rana temporaria* or toads, *Bufo bufo*; bank voles, *Clethrionomys glareolus*; laboratory mice, *Mus musculus* and the rumen of domesticated sheep. The species mentioned are easily available in Great Britain but in other parts of the world closely related hosts harbour similar faunas. This is not intended to be a complete parasite list for each host. Further details of most of these species are given by Mackinnon and Hawes (1961), which should also be consulted for information on similar parasites and methods for permanent preparations.

MASTIGOPHORA

Protomonadida	*Trypanosoma*	Blood	Voles
Polymastigida	*Hexamita muris**	Caecum	Voles and mice
	*Hexamita intestinalis**	Large intestine	Frogs and toads
	*Giardia muris**	Small intestine	Voles and mice
Trichomonadida	*Trichomonas batrachorum**	Large intestine	Frogs and toads
	*Trichomonas muris**	Caecum	Voles and mice
Hypermastigida	*Trichonympha campanula**	Hind gut	Termites
Opalinida	*Opalina ranarum**	Large intestine	Frogs and toads

SARCODINA

Amoebida	*Entamoeba thomsoni*	Hind gut	Cockroaches
	Entamoeba ranarum	Large intestine	Frogs and toads
	Entamoeba muris	Caecum	Voles and mice

SPOROZOA

Gregarinida	*Monocystis spp**	Seminal vesicles	Earthworms
	*Gregarina spp**	Mid and hind gut	Mealworms
	Gregarina blattarum	Mid gut	Cockroaches
	*Selenidium spp**	Intestine	*Cirratulus* and *Cirriformia*
Coccidiida	*Eimeria clupearum**	Liver	Herring
	*Eimeria sardinae**	Testis	Herring
	Hepatozoon muris	Blood	Voles
Piroplasmida	*Babesia microti*	Blood	Voles

CILIATA

Trichostomatida	*Balantidium* spp*	Large intestine	Frogs and toads
Hymenostomatida	*Dasytricha**	Rumen	Sheep
	*Isotricha**	Rumen	Sheep
Astomatida	*Anoplophrya* spp	Intestine	Earthworms
	Durchoniella spp	Intestine	*Cirratulus* and *Cirriformia*
Thigmotrichida	*Ancistrum mytili*	Mantle cavity	Mussel
Peritrichida	*Carchesium*	On legs	*Asellus*
Heterotrichida	*Nyctotherus ovalis*	Hind gut	Cockroaches
	*Nyctotherus cordiformis**	Large intestine	Frogs and toads
Entodiniomorphida	*Entodinium* spp*	Rumen	Sheep
	Diplodinium spp*	Rumen	Sheep

* Almost invariably present.

1.5. Representative Protozoa

In this section, individual species have been selected as representatives of the main groups of protozoa. In most cases only one species in an order is described, but several species are mentioned when the group contains important members such as the malaria parasites. The types which are described in this section have been chosen because they illustrate the diagnostic characteristics of their group and also because they are easily available.

1.5.1. MASTIGOPHORA

Phytomastigophora

Chlamydomonas reinhardii (Fig. 1,A). To examine living specimens, they should be slowed down (see p. 4). Note the plant-like characteristics, the cell wall which is composed of cellulose, and the single cup-shaped chloroplast containing a pyrenoid. Nutrition is

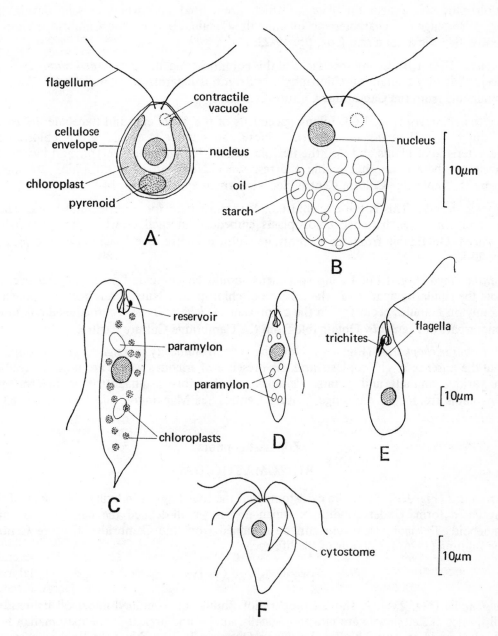

Figure 1

A, *Chlamydomonas reinhardii*, living specimen; B, *Polytoma* sp., living specimen; C, *Euglena sporigyra*, living specimen; D, *Astasia longa*, living specimen; E, *Peranema trichophorum*, living specimen; F, *Tetramitus rostratus*, living specimen

(A, B—same scale; C, D, E—same scale)

autotrophic. Note also the flagella which are paired and when present during the vegetative stage are characteristic of animals. Obtainable from the Cambridge Culture Centre. See Vickerman and Cox, pp. 19-23.

Polytoma (Fig. 1,B). Living specimens of this colourless flagellate resemble *Chlamydomonas* except that they contain no chlorophyll. Nutrition is heterotrophic but starch is formed. Obtainable from the Cambridge Culture Centre.

Euglena spirogyra (Fig. 1,C). Living specimens of this large euglenid flagellate are easier to study than *E.viridis* or *E.gracilis*. Note the numerous round green chloroplasts characteristic of plants and also the flexible pellicle, the anterior reservoir and locomotor flagellum (there is another within the reservoir). Nutrition is autotrophic. Obtainable from the Cambridge Culture Centre. See Vickerman and Cox, pp. 14-8.

Euglena gracilis. This is the euglenid flagellate most used for experimental work. There are about a dozen flattened chloroplasts embedded in each of which are paramylum granules. Obtainable from the Cambridge Culture Centre. See Mackinnon and Hawes, pp. 76-82.

Astasia longa (Fig. 1,D). Living specimens should be compared with *Euglena gracilis*. Note the similarity apart from the absence of chloroplasts. Nutrition is heterotrophic and paramylon granules occur free in the cytoplasm. *A. longa* can be regarded as a colourless counterpart of *E.gracilis*. Obtainable from the Cambridge Culture Centre.

Peranema trichophorum (Fig. 1,E). Note the general similarity to *Euglena gracilis* but also note the absence of chloroplasts and the presence of trichites which are used for feeding on various animals and plants. *Peranema* is therefore a phagotrophic heterotroph. Obtainable from the Cambridge Culture Centre. See Mackinnon and Hawes, pp. 83-7.

Zoomastigophora

RHIZOMASTIGIDA

Tetramitus (Fig. 1,F). Note the cytostome and the four flagella in living specimens of the flagellated form. Under conditions which are as yet ill-defined the flagellate becomes amoeboid. *Tetramitus rostratus* can be obtained from the Cambridge Culture Centre. See Bunting (1926) for details of the life cycle.

PROTOMONADIDA

Leishmania (Fig. 2,A). A stained preparation should be examined under oil immersion. All species of *Leishmania* are morphologically similar and appear in the mammalian host as small ovoid or round bodies (Leishman-Donovan bodies). Note the large numbers in the cells of the reticulo-endothelial system. The only internal structures visible are the nucleus and the kinetoplast. Giemsa is the most useful stain.

Leptomonas (Fig. 2,B). Living or stained specimens should be examined under oil immersion. In living specimens note the single flagellum. In preparations stained in Giemsa's stain note the single nucleus and the single flagellum arising close to the

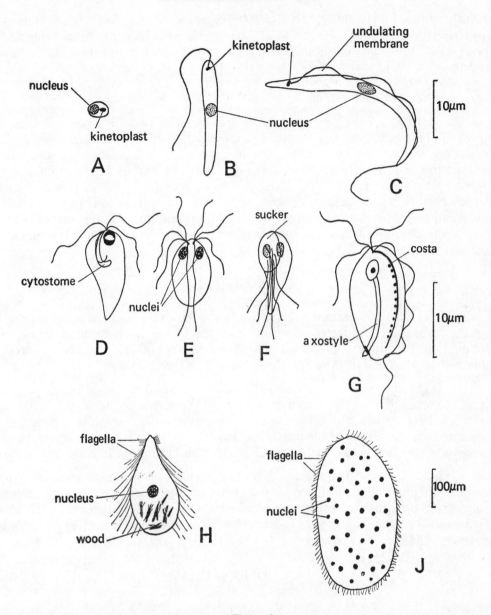

Figure 2

A, *Leishmania* sp. specimen stained in Giemsa's stain; B, *Leptomonas jaculum*, specimen stained in iron haematoxylin; C, *Trypanosoma lewisi*, specimen stained in Giemsa's stain; D, *Chilomastix bettencourti*, specimen stained in iron haematoxylin; E, *Hexamita muris*, specimen stained in iron haematoxylin; F, *Giardia muris*, specimen stained in iron haematoxylin; G, *Trichomonas muris*, specimen stained in iron haematoxylin; H, *Trichonympha campanula*, specimen stained in iron haematoxylin; J, *Opalina ranarum*, specimen stained in iron haematoxylin

(A, B, C—same scale; D, E, F, G—same scale; H, J—same scale)

anteriorly situated kinetoplast. This can be regarded as the basic form among the trypanosomatid flagellates. The stage of *Leishmania* found in the insect vector is the 'leptomonad' form and this is the form which occurs in cultures. In addition various invertebrates harbour species of *Leptomonas*—for example, *L.jaculum* occurs in the water scorpion *Nepa cinerea* (see Mackinnon and Hawes, pp. 112–5). Giemsa or iron haematoxylin are the most useful stains.

Trypanosoma (Fig. 2,C). Living or stained specimens should be examined under oil immersion. In living specimens note the movement of the undulating membrane. In preparations stained in Giemsa note the kinetoplast, which is posterior in position, the flagellum, forming the edge of the undulating membrane, and its free continuation. See Vickerman and Cox, pp. 24–7. *T.lewisi* can sometimes be found in rats and similar trypanosomes occur in field mice and voles. The diagnostic characteristics of the mammalian trypanosomes are summarized in Mackinnon and Hawes, pp. 117–23 and a more up to date classification at the sub-generic level is given by Hoare (1966). Giemsa is the best stain.

POLYMASTIGIDA

Chilomastix (Fig. 2,D). Stained specimens should be examined under oil immersion. Note the three flagella directed anteriorly and the fourth lying within a conspicuous cytostome which is supported by fibrils. *C.bettencourti* is common in the caecum of rodents and resembles *C.mesnili* of man. Iron haematoxylin is the best stain.

Hexamita (Fig. 2,E). Stained specimens should be examined under oil immersion. Note that all the organelles are duplicated and arranged symmetrically about the longitudinal axis. Six flagella are directed anteriorly and two posteriorly. *H.muris* is common in the caecum and small intestine of rodents and *H.intestinalis* in the large intestine and rectum of frogs (see Mackinnon and Hawes, pp. 133–5). Iron haematoxylin is the best stain.

Giardia (Fig. 2,F). Stained specimens should be examined under oil immersion. All the organelles are duplicated and there are four pairs of flagella arranged in a characteristic pattern. Note the broad sucker by means of which the flagellate attaches itself. *G.muris* is common in the small intestine of rodents (see Mackinnon and Hawes, pp. 135–7). Iron haematoxylin is the best stain.

TRICHOMONADIDA

Trichomonas (Fig. 2,G). Living or stained specimens should be examined under oil immersion. In the living animals note the movement of the undulating membrane. In coloured specimens note the axostyle or supporting rod which runs the length of the body. From an anterior complex of basal bodies arise three anterior flagella and one recurrent one which runs along the edge of the undulating membrane which is itself supported by a costa. *T.muris* is common in the caecum and small intestine of rodents (see Mackinnon and Hawes, pp. 124–30) and *T.batrachorum* is equally common in the large intestine of frogs and toads (Mackinnon and Hawes, pp. 130–2). Iron haematoxylin is the best stain. Protargol is also very useful.

HYPERMASTIGIDA

Trichonympha (Fig. 2,H). Living and stained specimens should be examined. In the living flagellates note the numerous flagella and contrast their length and movement with those of *Opalina* and the ciliates. In stained specimens note that the flagella are arranged in longitudinal rows and that there is only one nucleus. *Trichonympha campanula* is almost invariably found in the hind guts of termites belonging to the genus *Zootermopsis* (see Mackinnon and Hawes, pp. 142–6). staining may be carried out in haematoxylin, iron haematoxylin or protargol.

OPALINIDA

Opalina (Fig. 2,J). Living and coloured specimens should be examined. In living specimens note the numerous flagella and the similarity to ciliates. In coloured specimens note the many monomorphic nuclei and the fact that the rows of flagella do not run longitudinally.

Opalina ranarum is common in the hind guts of frogs and toads. See Mackinnon and Hawes, pp. 150–4. Haematoxylin or iron haematoxylin are good general dyes but protargol is required to show the details of the flagellar arrangements.

1.5.2. SARCODINA

AMOEBIDA

Naegleria (Fig. 3,A). Living amoeboid forms should be examined in a hanging drop. Note the formation of a single pseudopodium with clear ectoplasm at the leading edge contrasting with the granular endoplasm. Note also the nucleus containing a large nucleolus and one or more contractile vacuoles. *Naegleria* is an amoeboflagellate *i.e.*, it has amoeboid and flagellated stages in its life history (see p. 39). *N.gruberi* can be isolated from soil but is also available from the Cambridge Culture Collection. See Mackinnon and Hawes, pp. 26–30.

Amoeba proteus (Fig. 3,B). Living amoebae may be examined in a petri dish with the aid of a binocular microscope using daylight as the light source. Note the single nucleus, the contractile vacuole and the uroid at the posterior end. Observe the formation of pseudopodia and the ectoplasm which is visible only at the tips of the advancing pseudopodia. Watch the amoebae change direction. Examine some amoebae under a coverslip at a higher power and note the shape of the triuret crystals in the cytoplasm. Run a drop of 1% methyl green in 0.5% glacial acetic acid under the coverslip. This kills and fixes the amoebae revealing the nucleus clearly, the dense mass within the nucleus is not a nucleolus but is an artefact of fixation. See Mackinnon and Hawes, pp. 6–18 and Vickerman and Cox, pp. 28–33.

Amoeba proteus is only one of a number of large freshwater amoebae which have the same general form. *A.proteus* is ridged longitudinally and flows through a single pseudopodium. *A.discoides* is similar but not ridged. *A.dubia* is not ridged but flows through several pseudopodia at the same time. *Pelomyxa carolinensis* is multinucleate and forms pseudopodia which are longitudinally ridged. *Pelomyxa palustris* is also multinucleate but seldom forms pseudopodia. See Jahn and Jahn, pp. 120–4.

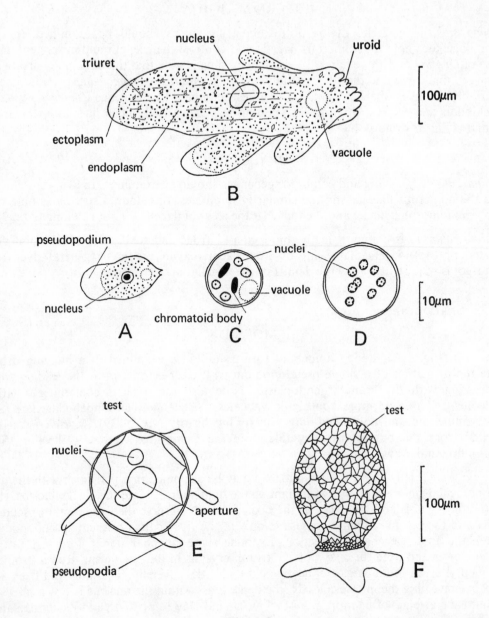

Figure 3

A, *Naegleria gruberi*, specimen stained in iron haematoxylin; B, *Amoeba proteus*, living specimen; C, *Entamoeba histolytica*, cyst stained in iron haematoxylin; D, *Entamoeba coli*, cyst stained in iron haematoxylin; E, *Arcella vulgaris*, living specimen; F, *Difflugia* sp., living specimen

(A, C, D—same scale; E, F—same scale)

Entamoeba histolytica and *E.coli* (Figs. 3,C,D). Cysts of *E.histolytica* should be stained and examined ùnder oil immersion. Compare them with the cysts of *E.coli*. In *E.histolytica* note the four nuclei each with a central endosome, fine peripheral chromatin within the nuclear membrane and rod-shaped chromatoid bodies. The cysts of *E.coli* possess eight nuclei, the endosome is eccentric, the peripheral chromatin is clumped and the chromatoid bodies are splinterlike or absent. Both amoebae occur in the intestine of man. *E.histolytica* may be pathogenic but *E.coli* is always a harmless commensal.

Entamoeba invadens from snakes and *E.ranarum* from frogs are morphologically similar to *E.histolytica* and easier to obtain. See Mackinnon and Hawes, pp. 32–7. *Entamoeba muris*, from the caecum of rats, mice and voles, is morphologically similar to *E.coli*. See Mackinnon and Hawes, pp. 38–9. Cysts and trophozoites are best stained in iron haematoxylin.

TESTACIDA

Arcella (Fig. 3,E). Living or stained specimens may be examined. Note the pseudo-chitinous test with a central aperture through which the pseudopodia are thrust. *A.vulgaris* is found in water rich in organic material. See Mackinnon and Hawes, pp. 44–6. *Difflugia* (Fig. 3,F). Living or stained specimens may be examined. Note the vase-like test which is formed from foreign particles cemented onto an organic base. Species of *Difflugia* are common in stagnant water. See Mackinnon and Hawes, pp. 46–7.

FORAMINIFERIDA

Elphidium (Fig. 4,A). Living, and stained and decalcified specimens should be examined. In the living animals observe the pseudopodia which pass out through the foramina in the test to form a feeding net. In prepared specimens distinguish between the microspheric form which develops from a zygote and the megalospheric form which develops from an asexually produced amoebula. In the former the first chamber, or proloculum, is small whereas in the latter it is large. See Mackinnon and Hawes, pp. 51–5.

Foraminiferan tests. A range of foraminiferan tests may be examined to see the various ways in which the chambers are arranged.

Actinopoda

HELIOZOIDA

Actinosphaerium (Fig. 4,B). Living or stained specimens should be examined. In the living specimens observe the ectoplasm with large vacuoles and the endoplasm with smaller ones. The long thin axopodia radiate from the endoplasm which contains many nuclei and these can be demonstrated by flooding the specimens with 1% methyl green in 0.5% glacial acetic acid. In stained specimens note that the ectoplasm and endoplasm are not separated by a distinct membrane. *Actinosphaerium eichorni* occurs in well oxygenated freshwater. See Mackinnon and Hawes, pp. 56–9.

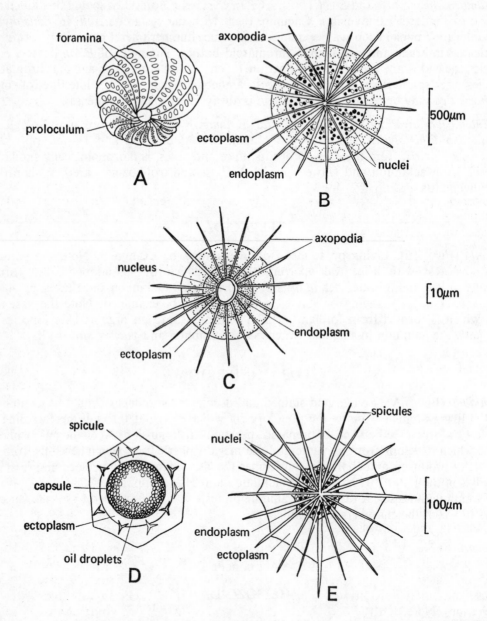

Figure 4

A, *Elphidium crispum*, test; B, *Actinosphaerium eichorni*, living specimen; C, *Actinophrys sol*, living specimen; D, *Sphaerozoum* sp., from a section stained in Ehrlich's haematoxylin; E, *Acanthometra* sp., living specimen

(A, B—same scale; D, E—same scale

Actinophrys (Fig. 4,C). Living or stained specimens should be examined. Note that the ectoplasm and endoplasm are not clearly demarcated and that there is only one nucleus. *Actinophrys sol* is common among vegetation from freshwater containing abundant green algae. See Mackinnon and Hawes, pp. 59–62.

RADIOLARIDA

Sphaerozoum (Fig. 4,D). Stained specimens of this colonial radiolarian should be examined. Note the central capsule enclosed by ectoplasm which in its turn is surrounded by siliceous spicules which represent the skeleton. See Mackinnon and Hawes, pp. 64–5. *Sphaerozoum* cannot really be regarded as a typical radiolarian but better examples are difficult to obtain.

Acanthometra (Fig. 4,E). Examine living or preserved material. Note the central capsule enclosing the endoplasm and surrounded by the ectoplasm. Note also the skeleton composed of spicules of strontium sulphate radiating from the centre of the organism. A species of *Acanthometra* occurs in the plankton in British seas. See Mackinnon and Hawes, pp. 66–8.

The variety of forms of radiolarian skeletons may be seen in radiolarian ooze.

1.5.3. SPOROZOA

Telosporea

GREGARINIDA

Monocystis (Fig. 5,A). Several genera of acephaline gregarines occur in the coelom and seminal vesicles of earthworms and although these are usually grouped together as monocystids they are not all members of the genus *Monocystis*. Remove the seminal vesicles from an earthworm and examine a portion in Ringer's solution (0.65%) to observe the feeble movements of the trophozoites. Young trophozoites are frequently covered with adherent earthworm sperms and may be confused with normal sperm masses. Note that the body of the trophozoite is undivided (monocystid). Look for associated gametocytes in gametocysts, gamete formation, zygotes and sporocysts. The sporocysts are usually very common and may be recognized by their characteristic boat-shaped appearance. Permanent preparations are best fixed in Bouin's or Schaudinn's fixatives and stained in haematoxylin or iron haematoxylin. See Mackinnon and Hawes, pp. 170-9 and Vickerman and Cox, pp 34–8.

Gregarina (Fig. 5,B). Cephaline gregarines may be found in mealworm guts. Cut off the head and tail of a living mealworm (larva of the beetle *Tenebrio molitor*) and draw out the gut. Place this in a watch glass with a little invertebrate Ringer's to keep it moist. Tease out small portions of the gut and examine in Ringers solution for cephaline gregarines. Note that the body of the trophozoite is divided into three regions by septa, the epimerite (which is frequently broken off), the protomerite and the deutomerite containing the nucleus. The movement of the trophozoites is characteristic and is known as gregarine movement. Look for gametocysts. The life cycle is similar to that of the monocystid gregarines. To make permanent preparations material is best fixed in Bouin's fixative and stained in haematoxylin. Other genera besides *Gregarina* occur in the gut of mealworms, and gregarines are common parasites of many terrestrial arthropods. See Mackinnon and Hawes, pp. 183–92.

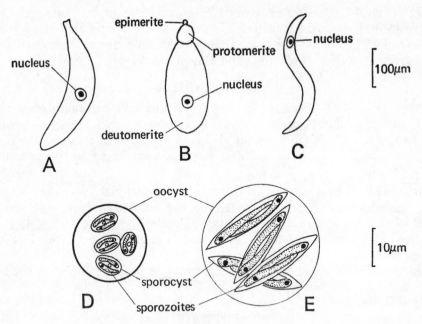

Figure 5

A, *Monocystis agilis*, stained in iron haematoxylin; B, *Gregarina* sp., stained in iron haematoxylin; C, *Selenidium* sp., stained in iron haematoxylin; D, *Eimeria clupearum*, oocyst, living specimen; E, *Eimeria sardinae*, oocyst, living specimen

(A, B, C—same scale; D, E—same scale)

Selenidium (Fig. 5,C). Trophozoites occur in the gut of the polychaete worms *Cirriformia tentaculata* and *Cirratulus cirratus*. Examine the contents in seawater and search for the trophozoites of *Selenidium* which resemble small nematodes. Note the elongate shape and the surface striations. The life cycle of *Selenidium* resembles that of *Monocystis* except that there is a period of multiplication between the sporozoite and the trophozoite stage. These schizogonic stages may be seen in transverse sections of heavily infected worms. Permanent preparations of trophozoites are best made from material fixed in Bouin's fixative and stained in iron haematoxylin. Polychaetes from Plymouth are almost invariably infected with *Selenidium* (and other gregarines). See Mackinnon and Hawes, pp. 167–9.

COCCIDIIDA

Eimeria (Fig. 5,D,E). Stained sections of the liver of a rabbit infected with *Eimeria stiedae* should be examined. Sporozoites enter epithelial cells of the bile ducts and a schizogonic phase precedes sporogony. Note the multinucleate schizonts, the bunches of banana-like merozoites, the comma-shaped microgametes, the oval macrogametes and the thick-walled oocysts. It is unlikely that all these stages will occur in the same section. See Mackinnon and Hawes, pp. 201–6. Sections of chick caeca infected with *Eimeria tenella* are usually better for the study of the life-cycle of *Eimeria* than those from rabbits, but it is necessary to examine a series taken at intervals between $2\frac{1}{2}$ and 7 days. See Mackinnon and Hawes, pp. 195–201. *Eimeria schubergi* is fairly common in the intestine of the centipede *Lithobius forficatus*. See Wenyon (1926).

Oocysts of *Eimeria* may be obtained from the faeces of infected rabbits or chickens but the most convenient source is the soft roe or liver of a herring. In this fish oocysts are common and are recognizable after freezing or canning. Squash a portion of liver under a coverslip and look for the round oocysts containing four sporocysts. Each sporocyst contains two sporozoites. This is *Eimeria clupearum*. *Eimeria sardinae* may be identified in the testis in the same way; the oocyst is much larger than that of *E.clupearum* and the sporocysts are cigar-shaped. Permanent preparations are not successful.

Hepatozoon (Fig. 6, A). In the coccidian *Hepatozoon* gametocytes enter the blood cells of their vertebrate host and transmission is effected by biting arthropods. *Hepatozoon muris* occurs in white blood cells of mice and voles as a crescentric uninucleate body in the cytoplasm. This parasite can best be seen in thin blood films stained in Giemsa's stain.

HAEMOSPORIDIIDA

Haemoproteus (Fig. 6,B). Thin blood films stained in Giemsa's stain should be examined under oil immersion. *Haemoproteus* is common in the red blood cells of birds where it appears as an elongate body lying alongside the nucleus and curving round it at the ends. Note the single nucleus and the presence of pigment granules. The stages in the blood are gametocytes and there are no schizogonic stages in the blood. Contrast this with the situation in *Plasmodium*.

Plasmodium. Various species of *Plasmodium* occur in reptiles, birds and mammals. The diagnostic stages are those in the blood and these can best be seen in thin smears stained in Giemsa's stain. See Garnham (1966).

Plasmodium gallinaceum (Fig. 6,C). Thin blood films stained with Giemsa's stain should be examined under oil immersion. The gametocytes, if present, are large uninucleate bodies. Look for young vacuolated ring stages and then for schizonts with increasing numbers of nuclei up to about 16–24. Note the appearance of pigment in the early trophozoites and its clumped appearance in the late schizont. Contrast this situation with that in birds infected with *Haemoproteus*. The malaria parasites of birds provide a better introduction to the study of malaria than those of mammals because of their larger size. See Garnham (1966).

Plasmodium berghei. Thin blood films stained with Giemsa's stain should be examined under oil immersion. This malaria parasite of rodents is commonly used in research, as it is fairly easily obtainable. Note the intensity of the infection and the fact that individual red blood cells may be infected with a number of parasites. Look for all stages of schizogony from the small uninucleate rings to the mature schizonts with 8–16 nuclei. See Garnham (1966).

Malaria parasites of man. The diagnostic characters of the four malaria parasites found in man are summarized below (Figs. 6,D,E,F,G).

> *P.vivax*. The infected red blood cell is larger than normal, is stippled and the gametocytes do not fill the cell.

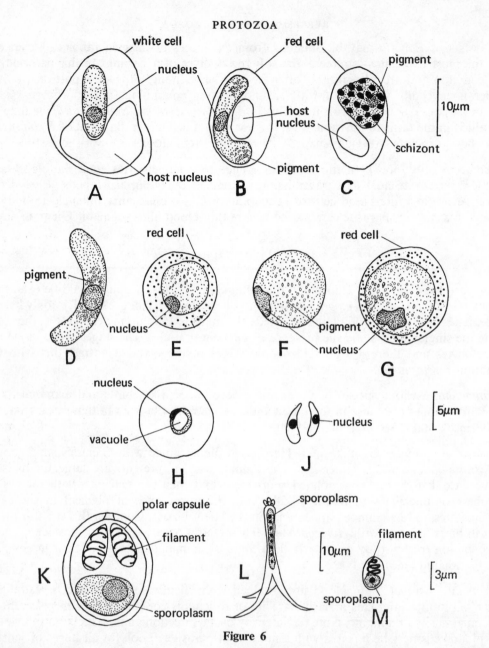

Figure 6

A, *Hepatozoon muris*, gametocyte, stained in Giemsa's stain; B, *Haemoproteus columbae*, gametocyte, stained in Giemsa's stain; C, *Plasmodium gallinaceum*, schizont, stained in Giemsa's stain; D, *Plasmodium falciparum*, macrogametocyte, stained in Giemsa's stain; E, *Plasmodium ovale*, macrogametocyte, stained in Giemsa's stain; F, *Plasmodium malariae*, macrogametocyte, stained in Giemsa's stain; G, *Plasmodium vivax*, macrogametocyte, stained in Giemsa's stain; H, *Babesia microti*, stained in Giemsa's stain; J, *Toxoplasma gondii*, stained in Giemsa's stain; K, *Myxobolus pfeifferi*, from a section stained in Giemsa's stain; L, *Triactinomyxon* sp., from a section in iron haematoxylin; M, *Nosema bombycis*, from a section stained in iron haematoxylin

(A, B, C—same scale; D, E, F, G, H, J—same scale; K, L—same scale)

P.ovale. Resembles *P.vivax* but the stippling is more pronounced and somewhat violet. The oval shape of the corpuscles is an artefact and depends on the time taken for the blood film to dry.

P.malariae. The host cell is normal in size and is not stippled. The gametocytes fill the cell.

P.falciparum. Only ring stages and characteristic crescentric shaped gametocytes are present in the blood.

Exoerythrocytic stages. Sections of mammalian and avian schizonts stained in Giemsa-colophonium should be examined. The exoerythrocytic stages of bird malarias occur in a variety of tissues and possess usually less than 1000 merozoites. Mammalian parasites occur in the liver and the mature schizont possesses more than 1000 merozoites.

Sporogonic stages. Whole mounts and sections of infected mosquito stomachs should be examined. Note the large oocysts. Stained preparations of sporozoites should be examined under oil immersion. Sporozoites are best stained in Giemsa's stain. Note the elongate uninuclear appearance.

Garnham's (1966) book on the malaria parasites is an invaluable and beautifully illustrated guide to these organisms.

Piroplasmea

PIROPLASMIDA

Babesia (Fig. 6,H). Thin blood films, stained in Giemsa's stain should be examined under oil immersion. In red blood cells all species of *Babesia* are morphologically similar, the vacuolated ring stage dividing into two or possibly four. Note the absence of pigment. The vacuoles tend to be much more clearly defined in *Babesia* than in *Plasmodium*. *Babesia microti* has been recorded from nearly all the small mammals of Britain, it may be quite common in certain localities. Other species cause diseases in cattle.

Toxoplasmea

TOXOPLASMIDA

Toxoplasma (Fig. 6,J). Spores from the peritoneal exudate of an experimentally infected mouse, should be dyed in Giemsa's stain, and examined under oil immersion. Note the characteristic crescentic shape with one end more pointed than the other. This spore is the only stage in the life cycle and groups of such spores may be enclosed in cysts or pseudocysts. The only species is *Toxoplasma gondii* and this occurs in practically all warm-blooded animals including man.

1.5.4. CNIDOSPORA

MYXOSPORIDIIDA

Myxobolus (Fig. 6,K). Living or stained spores, may be examined entire or in sections, under oil immersion. Note that the spore has two valves and two polar capsules each containing a coiled filament. Behind the polar capsules is the sporoplasm or infective body. *Myxobolus* is a common parasite of fish in which it invades various tissues often causing large tumours. See Jahn and Jahn, pp. 161–2.

ACTINOMYXIDIIDA

Triactinomyxon (Fig. 6,L). Living or stained spores should be examined under oil immersion. Note that the spore consists of three valves and contains three polar capsules. *Triactinomyxon legeri* occurs in *Tubifex* from the Thames. It is rare.

HELICOSPORIDIIDA

Helicosporidium parasiticum. This is the only species in this order. The minute spores possess three sporoplasms surrounded by a spirally coiled filament. Parasitic in dipterans and mites.

MICROSPORIDIIDA

Nosema (Fig. 6,M). Living or stained spores, should be examined entire or in sections, under oil immersion. Note the extremely small size and that the spore consists of a single valve and a single sporoplasm. If living spores are ruptured by a sharp blow on the coverslip (preferably under a layer of filter paper) the filament may be extruded. The infective sporoplasm passes up this hollow filament. The majority of microsporidia infect insects in which they are fairly common.

1.5.5. CILIATA

Holotricha
GYMNOSTOMATIDA

Prorodon teres (Fig. 7,A). Living specimens of this primitive ciliate should be studied. Note the slit-like cytostome and otherwise unspecialized body with a uniform covering of cilia. Silver impregnation reveals the pattern of the infraciliature. See Mackinnon and Hawes, pp. 238–41. *Coleps*, a gymnostome covered with plates, is available from the Cambridge Culture Centre.

TRICHOSTOMATIDA

Balantidium (Fig. 7,B). Living and stained specimens of *Balantidium* spp. from the rectum of a frog should be examined. Note that the cytostome does not open directly onto the surface of the animal but that it is depressed into a vestibulum. The cilia lining the vestibulum are continuous with those of the body. In specimens stained in haematoxylin note the bean-shaped macronucleus and the small round-micronucleus. The species in the frog are *B.elongatum* and *B.entozoon* in the rectum and *B.duodeni* in the small intestine. See Mackinnon and Hawes, pp. 262–7. *Balantidium coli*, which infects man under certain circumstances, is common in pigs.

CHONOTRICHIDA

Spirochona (Fig. 7,C). Living and stained specimens should be examined. Note that the ciliate is sessile (on the branchial lamellae of *Gammarus*) and that the anterior end is expanded into a vase-like feeding apparatus. Note also that the body is not ciliated but that in silver-impregnated specimens the kinetosomes are present. *Spirochona gemmipara* is found on *Gammarus pulex*. See Mackinnon and Hawes, pp. 258–60. Other chonotrichs are marine.

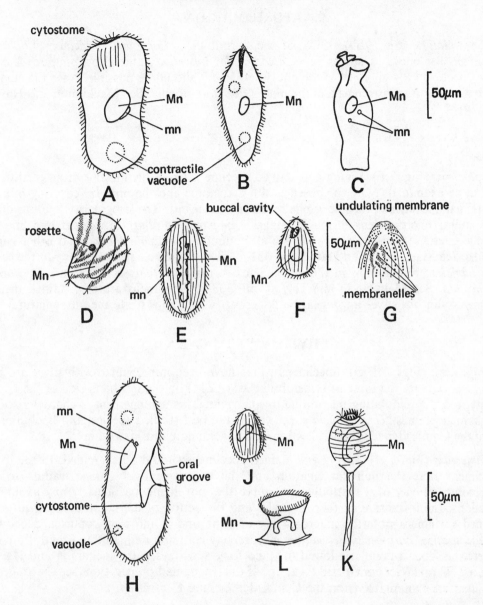

Figure 7

A, *Prorodon teres*, living specimen; B, *Balantidium elongatum*, living specimen; C, *Spirochona gemmipara*, living specimen; D, *Gymnodinioides inkystans*, living specimen; E, *Anophlophrya alluri*, stained in iron haematoxylin; F, *Tetrahymena pyriformis*, living specimen; G, *Tetrahymena pyriformis*, specimen impregnated with silver; H, *Paramecium caudatum*, living specimen; J, *Ancistrum mytili*, living specimen; K. *Vorticella microstoma*, living specimen; L, *Trichodina pediculus*, stained in Ehrlich's haematoxylin; Mn=macronucleus, mn=micronucleus

(All except G same scale)

APOSTOMATIDA

Gymnodinioides (Fig. 7,D). Gills of the hermit crab *Eupagurus prideauxi* should be examined for encysted stages of *Gymnodiniodes inkystans* (and also *Polyspira delagei*) which occur on the gill axes. If the gill is flattened it should be possible to see the ciliates. Note the clear macronucleus and the nine spiral rows of cilia. See Mackinnon and Hawes, pp. 341–4.

ASTOMATIDA

Anoplophrya (Fig. 7,E). Living and stained specimens from the intestine of an earthworm should be examined. Note the absence of a cytostome and the uniform covering of cilia. In stained specimens note the irregular elongate nucleus. The irregularity is an artefact. The lateral micronucleus is much smaller. *Anoplophrya alluri*, *A.vulgaris* and another astome ciliate, *Maupasella cepedei*, are all common in *Eisenia foetida* and other earthworms. See Mackinnon and Hawes, pp. 331–8. Two astome ciliates belonging to the genus *Durchoniella* are common in the polychaete worms *Cirratulus cirratus* and *Cirriformia tentaculata*. See Mackinnon and Hawes, pp. 338–9. Specimens are best stained in iron haematoxylin but silver impregnation is necessary to demonstrate the infraciliature.

HYMENOSTOMATIDA

Tetrahymena (Figs. 7,F,G). Specimens which have been impregnated with silver are best to examine. Note the three oral membranelles (AZM) on the left hand side of the buccal cavity and the undulating membrane on the right. This arrangement is characteristic of the hymenostomes. *Tetrahymena pyriformis* is the best studied species, and is obtainable from the Cambridge Culture Collection. See Mackinnon and Hawes, pp. 275–80.

Paramecium (Fig. 7,H). Living and stained specimens should be examined. In the living specimens note the uniform ciliation and the expansive oval groove leading to the cytostome by way of a vestibulum. Observe the movement, particularly any avoidance reactions, the formation of food vacuoles and the action of the contractile vacuoles. In stained specimens note the macronucleus and the one or more micronuclei, depending on the species. *Paramecium caudatum* is extremely common and is often found if a jar of vegetation from a pond is allowed to stand for a few days. See Mackinnon and Hawes, pp. 280–92 and Vickerman and Cox, pp. 45–50. In stained preparations see the stages of conjugation. Obtainable from the Cambridge Culture Centre.

THIGMOTRICHIDA

Ancistrum (Fig. 7,J). Living or stained specimens should be examined. Note that the rows of cilia are more tightly crowded on the ventral than on the dorsal surface; these cilia are also used for attachment. Note also the extensive buccal cavity on the right hand side leading to the posteriorly situated cytostome. *Ancistrum mytili*, together with other thigmotrichs, is found in the mantle cavity of the mussel *Mytilus edulis*. See Mackinnon and Hawes, pp. 307–9.

PERITRICHIDA

Vorticella (Fig. 7,K). Living and stained specimens should be examined. Note the contractile stalk, the extensive oral ciliature and the absence of cilia from the remainder of the body. The macronuclues can be seen in living specimens. *Vorticella microstoma* is common in freshwater. See Mackinnon and Hawes, pp. 318–24. *Vorticella* is solitary. *Carchesium* is a colonial peritrich commonly found attached to the legs of the freshwater crustacean *Asellus*.

Trichodina (Fig. 7,L). Examine living and stained specimens of this ectoparasitic peritrich. Note the absence of a stalk and the presence of an elaborate aboral disc. See Mackinnon and Hawes, pp. 324–8. *Trichodina pediculus* is common on *Hydra*.

SUCTORIDA

Discophrya. (Fig. 8,A). Living specimens should be examined. Note the general appearance which is quite unlike any other ciliate, the presence of suctorial tentacles and the absence of cilia. The larvae, which are ciliated, may be found in flourishing cultures. See Mackinnon and Hawes, pp. 250-5. *Discophrya* spp. are obtainable from the Cambridge Culture Centre.

Spirotricha

HETEROTRICHIDA

Stentor (Fig. 8,B). Living specimens of *Stentor coeruleus* should be examined. Note the conspicuous oral region, the uniformly ciliated body and the macronucleus which resembles a string of beads. *Stentor* is normally attached to the substratum by its aboral end but can swim free and is highly contractile. Observe the alternate rows of striped and clear bands running the length of the organism. See Mackinnon and Hawes, pp. 346–9 and Vickerman and Cox, p. 10. Obtainable from the Cambridge Culture Centre.

Nyctotherus (Fig. 8,C). Living and stained specimens should be examined. Note the uniform ciliation of the body and the extensive oral region which is ventral in position and leads *via* a spiral peristomal passage to the deeply situated cytostome. *Nyctotherus cordiformis* is common in frogs. See Mackinnon and Hawes, pp. 354–7. Staining is best carried out in haematoxylin.

OLIGOTRICHIDA

Halteria (Fig. 8,D). Living specimens should be examined. Note the spherical appearance, the extensive system of membranellae encircling the anterior end of the body and the absence of somatic cilia apart from a few bristle-like cirri. *Halteria geleiana* is common in freshwater. See Mackinnon and Hawes, pp. 359–60.

TINTINNIDA

Tintinnopsis (Fig. 8,E). Living specimens should be examined. Note the lorica, coated with foreign particles, within which the ciliate lives, and the extensive oral membranelles extending into the buccal cavity. These are used for feeding and swimming. Most tintinnids occur in the marine plankton in which they are most abundant in the summer. See Mackinnon and Hawes, pp. 361–3.

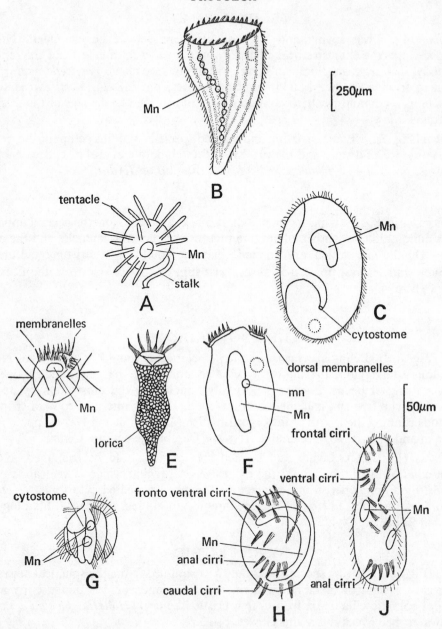

Figure 8

A, *Discophrya piriformis*, living specimen; B, *Stentor coeruleus*, living specimen; C, *Nyctotherus cordiformis*, stained in Ehrlich's haematoxylin; D, *Halteria geleiana*, living specimen; E, *Tintinnopsis* sp., living specimen; F, *Epidinium ecaudatum*, stained in iron haematoxylin; G, *Epalxis* sp., living specimen H, *Euplotes* sp., living specimen; J, *Oxytricha* sp., living specimen; Mn=macronucleus, mn=micronucleus

(All except B same scale)

ENTODINIOMORPHIDA

Epidinium (Fig. 8,F). Living and stained specimens should be examined. Note the absence of cilia from the body and the presence of two zones of membranelles, one associated with the cytostome and the other dorsal. The adoral zone of membranelles function in feeding and locomotion. See Mackinnon and Hawes, pp. 365–9. *Epidinium*, accompanied by other entodiniomorphids, is found in the rumen of herbivores such as sheep and cattle. The ciliates become unrecognizable when cooled. Permanent preparations are best stained in haematoxylin.

ODONTOSTOMATIDA

Epalxis (Fig. 8,G). Living specimens should be examined. Note the flattened body, the rigid pellicle and the reduced somatic ciliature. See Mackinnon and Hawes, pp. 381–3. These ciliates are not common and occur in water lacking in oxygen.

HYPOTRICHIDA

Euplotes (Fig. 8,H). Living specimens should be examined. Note the dorso-ventrally flattened body, the oral membranelles and the ventral cirri. These cirri are arranged in groups and the various hypotrichs are differentiated on their arrangement. Recognize the fronto-ventral, anal and caudal cirri. Observe the cirri by means of which *Euplotes* moves over the substratum. See Mackinnon and Hawes, pp. 370–80. A species of *Euplotes* is obtainable from the Cambridge Culture Centre.

Oxytricha (Fig. 8,J). Living specimens should be examined. Note the arrangement of the cirri.

1.6. Experiments with Protozoa

Twelve experiments are recommended to illustrate a number of principles of proto-zoology. Each is introduced by a reference to an elementary account of the subject of the experiment. The experimental details are brief but suffice to allow the instructor to decide whether the experiment is worth performing or not. Reference to a detailed account of the experiment is given. These experiments serve equally well as class experiments, class demonstrations or sources of prepared teaching material. The first three experiments are concerned with the nutrition of flagellates and illustrate the difficulties in drawing a line between the plant and animal kingdoms. Similarly the fourth experiment draws attention to the borderline between flagellates and amoebae. The fifth and sixth experiments are concerned with particulate feeding by ciliates and illustrate qualitative and quantitative aspects of a similar process. The seventh experiment also deals with ciliates but in this case the phenomenon is the general one of osmoregulation which happens to be easy to demonstrate in protozoa. The eighth and ninth experiments deal with repro-duction and illustrate the behaviour of mating types among flagellates and ciliates. The tenth experiment is concerned with the induction of mutations in protozoa. The last two experiments use *Paramecium*, the eleventh to demonstrate the effect of the peculiar symbiotic kappa particles and the last to illustrate behaviour. None of these experiments is intended to be complete, and it is up to the individual instructor to decide on what variations to use and what quantitive experiments to perform.

1.6.1. The Loss of Chlorophyll by *EUGLENA*

Euglena gracilis is a phototrophic autotroph, but if kept in the dark for about two weeks loses its chlorophyll and, in suitable culture media, behaves as an osmotrophic heterotroph. On being returned to light the chlorophyll reappears. If, however, the flagellate is subjected to heat, ultra-violet light or streptomycin the chlorophyll is again lost but is not reformed on return to normal conditions. Such euglenas become permanently bleached heterotrophs and comparable with *Astasia longa*. Thus the classical borderline between animals and plants is crossed. See Vickerman and Cox, pp. 9–11 and 17–8.

Prepare tubes of a suitable medium and inoculate with *Euglena gracilis*. When the cultures are growing well (and the numbers of flagellates still increasing) place a culture tube in a water bath at 35°C and transfer samples at intervals to fresh tubes at room temperature. Keep in the dark for 2 days and thereafter in the light. When growth has reached a satisfactory level (3 days) compare the sub-cultures. After 30 minutes at 33–37°C *Euglena* loses the ability to produce chlorophyll and this defect is inherited. The absence of chlorophyll can be seen in the flagellates which are yellow or colourless especially if grown on agar. Compare *E.gracilis* normal strains, strains grown in darkness for 14 days and bleached strains with *Astasia longa*.

Cultures of *Euglena gracilis* and *Astasia longa* can be obtained from the Cambridge Culture Centre. A suitable culture medium is:

Sodium acetate hydrate	0.1 g
Beef extract	0.1 g
Bacto tryptone	0.2 g
Yeast extract	0.2 g
$CaCl_2$	0.001 g
Distilled water	100 ml

Use in 5 ml quantities in test tubes observing sterile precautions.

Full details of the bleaching of *Euglena* are given by Schiff and Epstein (1965) and Blum, Sommer and Kahn (1965).

1.6.2. Nutrition in *OCHROMONAS*

Members of the genus *Ochromonas* are able to multiply in media rich in organic substances in either a soluble or a particulate form and in media devoid of organic compounds, except vitamins, in the presence of light. They are therefore both phototrophic autotrophs and phagotrophic heterotrophs. Both of these modes of nutrition co-exist in the same individuals. *Ochromonas danica* requires biotin and thiamine and *O.malhamensis* also requires vitamin B_{12}. There is no easily available elementary account of nutrition in *Ochromonas*.

Cultures of *O.danica* and *O.malhamensis* can be grown in defined media with or without bacteria or yeasts. In darkness the flagellates tend to become yellow and to take up the particulate food. In light, and in the absence of bacteria or yeasts, they remain green. Various experiments can be performed with these flagellates but the media used are complex. They are, however, fully described in the papers by Pringsheim (1955) and Aaronson and Baker (1959).

Cultures of *O.danica* and *Ochromonas* spp. can be obtained from the Cambridge Culture Centre.

1.6.3. FEEDING IN *PERANEMA*

Peranema trichophorum is a phagotrophic heterotroph despite its close affinities with *Euglena*. *Peranema* feeds on a variety of animals and plants including *Euglena*. A brief summary of food capture is given by Mackinnon and Hawes, pp. 86–7.

Starve a culture of *Peranema* overnight and add drops of the culture to drops of culture rich in *Euglena gracilis* in deep cavity slides. Observe over a period of time and note the process of prey capture and ingestion. *Peranema* sometimes feeds immediately, sometimes not at all.

Cultures of *Peranema trichophorum* and *Euglena gracilis* can be obtained from the Cambridge Culture Collection.

For a detailed account of food capture by *Peranema* see Chen (1950).

1.6.4. AMOEBO-FLAGELLATE TRANSFORMATION IN *NAEGLERIA*

The soil amoeba *Naegleria gruberi* has three phases in its life cycle: amoeboid, flagellated and encysted. The existence of a particular phase is determined by environmental conditions. The amoeboid stage occurs in damp conditions, the flagellated stage in flooded soil and the encysted stage in times of drought. Brief summaries and drawings of the various stages are given in Mackinnon and Hawes, pp. 25–30 and Vickerman and Cox, pp. 6–7.

Cultures of *Naegleria* on agar slopes are usually in the encysted state. Take a loopful of cysts and spread them on to an agar plate. Bacteria will grow out from the points of application and the amoebae will hatch and feed on them. Amoebae should be abundant about 48 hours after seeding the agar, and can be seen in the clear patches in the bacterial growth. Cysts form about 24 hours later, 72 hours after seeding. For the best results repeat this cycle twice. Enflagellation can be induced by flooding agar plates containing amoeboid forms with distilled water. This is best done soon after hatching, 12–24 hours after seeding the plate. Enflagellation begins about 30 minutes after the water is applied and should be complete in 4 hours. All these procedures should be carried out at about 25°C.

Cultures of *Naegleria gruberi* can be obtained from the Cambridge Culture Centre. Suitable nutrient agar plates and slopes can be prepared from:

Agar	1.5 g
Peptone	0.5 g
Distilled water	100 ml

For experimental details see Willmer (1956, 1963).

1.6.5. CYCLOSIS IN *PARAMECIUM*

Paramecium is a particulate feeder and particles are wafted towards the cytostome and into a food vacuole which forms in association with it. When full the vacuole becomes free and circulates within the ciliate, following a fairly regular path, for 1–3 hours before the contents are eliminated at the cytoproct. During its passage the contents of the vacuole become digested and this digestion is reflected in the pH of the vacuole. Further information is given by Mackinnon and Hawes, pp. 285–7.

Add to a drop of a rich culture of *Paramecium* on a slide a little yeast-congo red mixture. Seal a coverslip in position and observe the formation and colour of the food vacuoles. At $pH > 5$ the colour is orange and at $pH < 3$ it is blue.

Cultures of *Paramecium* may be obtained from the Cambridge Culture Centre. The yeast-congo red mixture is prepared by mixing 3 g compressed yeast with 30 mg congo red and boiling thoroughly for 10 minutes.

For further information see Buck (1943).

1.6.6. THE UPTAKE OF CARBON BY *TETRAHYMENA*

Tetrahymena is a particulate feeder and can take out of its medium suspended particles such as carmine and carbon. Carbon is concentrated in food vacuoles and later egested at the cytoproct and falls out of suspension. If the amount of carbon in suspension before the introduction of a known number of ciliates is measured and compared with the amount present a given time later the rate of filtration per ciliate per hour can be determined.

Dilute Indian ink to 1 part in 10,000 in distilled water and measure the optical density using a densitometer or colorimeter. Prepare several tubes of diluted Indian ink and add to each approximately one million ciliates. Examine the tubes at intervals for 24 hours. Centrifuge and estimate the optical density of the supernatant. Compare the readings with those obtained from various dilutions of the original diluted Indian ink and calculate the amount of carbon removed by the ciliates in unit time. Examine the ciliates themselves.

N.B. Not all commercial Indian ink samples are suitable as some tend to flocculate.

Cultures of *Tetrahymena pyriformis* can be obtained from the Cambridge Culture Centre. Carbon for injection manufactured by Gunther Wagner and sold by G. H. Smith and Partners (Colchester) is suitable.

These experiments are described in detail by Cox (1967).

1.6.7. OSMOREGULATION IN CILIATES

Water tends to enter freshwater ciliates by osmosis and is eliminated by means of contractile vacuoles. The activity of the contractile vacuoles is related to the osmotic pressure of the external medium. Ciliates tend to possess conspicuous contractile vacuoles and sessile ciliates are the easiest to handle.

The suctorian *Podophrya* attaches to silk threads in culture and the peritrich *Carchesium* lives attached to the legs of *Asellus*. All are therefore useful experimental animals. Place the ciliates in the cavity of a slide and irrigate with tap water, 0.1%, 0.2%, 0.4% and 0.8% sea water and finally tap water. Time the contractions at each dilution of sea water and in tap water. Ensure that the ciliates do not become too hot or dry out.

Discophrya can be obtained from the Cambridge Culture Centre.

Podophrya and *Discophrya* can be obtained from the Cambridge Culture Centre.

For further details see Clark (1966) and Kitching (1951).

1.6.8. MATING IN *CHLAMYDOMONAS*

Sexual reproduction in *Chlamydomonas* occurs between different mating types and is thought to be induced by the depletion of nitrogenous substances in the surrounding medium. The events which occur are summarized by Vickerman and Cox, pp. 22–3.

Take two cultures of different mating types on agar slopes and place in darkness for 24 hours. Flood each slope with 0.5 ml distilled water and stand in the light for 30–60 minutes. Take a sample of flagellates from each tube with a loop or fine pasteur pipette and mix on a slide. Clumping occurs after about 5 minutes and fusion occurs 6–8 hours later.

Suitable mating types of *Chlamydomonas* are obtainable from the Cambridge Culture Centre already on agar. These mating types are 11/16f and 11/16g.

For further details and information on the effect of nitrogen concentration on mating see Coleman (1962).

1.6.9. CONJUGATION IN *PARAMECIUM*

Each species of *Paramecium* exists in a number of varieties and within each variety mating occurs between different mating types. Conjugation is the most common sexual process in *Paramecium* and the process is summarized by Mackinnon and Hawes, pp. 292–300 and Vickerman and Cox, pp. 48–50.

Mix two mating types of *Paramecium*, take a small sample and observe under a microscope. Large clumps form and break down into pairs after about an hour. Conjugation occurs after about 24 hours. Various stages in conjugation can be obtained, and temporary or permanent preparations made by taking samples at varying periods after the formation of pairs. Temporary preparations can be made in 1% methyl green in 0.5% glacial acetic acid.

Suitable cultures of *Paramecium bursaria* (1660/1f and 1660/1g) can be obtained from the Cambridge Culture Centre.

Laboratory methods for investigating mating in *Paramecium* are given by Sonneborn (1950).

1.6.10 MUTANTS OF *CHLAMYDOMONAS*

Mutations can be induced in protozoa by exposing them to various substances. The mutants are usually difficult to detect but in *Chlamydomonas* it is possible to produce a flagellum mutation on exposure to proflavine or acridine orange. The mutant sinks to the bottom of the culture tube and forms dense plaques when plated onto agar. A brief note on this phenomenon is given in Vickerman and Cox, p. 23.

Chlamydomas reinhardii can be grown on agar plates containing acridine orange or proflavine. The flagellates are then transferred to a liquid medium in which the mutants sink to the bottom of the tube and can be found there after 3–5 days. Various mutants can be found. Some flagellates have no flagella; in others they are reduced or abnormal, while some with flagella are unable to swim or swim abnormally.

A detailed account of the procedures for obtaining flagellar mutants of *C.reinhardii* is given by Warr *et al* (1966) and the culture media used are described by Sagar and Granick (1954).

Cultures of *C.reinhardii* can be obtained from the Cambridge Culture Centre.

1.6.11. KAPPA IN *PARAMECIUM*

Paramecium aurelia containing the symbionts known as kappa kill sensitive paramecia either during conjugation or by the production of a substance called paramecin.

Examine "killer" and "sensitive" strains of *Paramecium aurelia*. Starve the ciliates and gently squash about 10 of them under a coverslip. Examine under a phase contrast microscope. The kappa bodies occur in the cytoplasm of the "killers".

Take some of the culture medium from a heavy culture of "killer" paramecia and add it to a small drop of fluid containing "sensitive" paramecia. The paramecin will adversely affect the "sensitives" and this can be seen under the microscope.

Centrifuge a culture of "killer" paramecia and homogenize the packed ciliates. Centrifuge again and retain the supernatant. Add this at various dilutions to small numbers of "sensitive" paramecia in order to determine the titre of the paramecin.

"Killer" and "sensitive" strains of *P.aurelia* can be obtained from the Cambridge Culture Centre. These are designated LB 1660/3 strain 51 sensitive and LB 1660/3 strain 51 killer.

The standard work on the inheritance of characters in *P.aurelia* is that of Beale (1954) and a review of kappa is given by Sonneborn (1959).

1.6.12. Behaviour of *PARAMECIUM*

The simple behaviour patterns of *Paramecium* brought about by the reversal of ciliary beat tend to keep the organism in a region of favourable conditions. These behaviour patterns are described by Mackinnon and Hawes, pp. 284–5 and Jahn and Jahn, pp. 191–2.

Prepare a wax trough about 1 mm deep on a 75×50 mm ($3'' \times 2''$) slide and place on a piece of black paper. Introduce into this trough a flourishing culture of *Paramecium*. Introduce various substances carefully into the liquid and cover tightly with a coverslip. The ciliates will collect in drops of 0.01% hydrochloric, acetic or sulphuric acid, around bubbles of carbon dioxide and also oxygen in the absence of dissolved oxygen. As the various substances diffuse throughout the medium the patterns of aggregation change.

Galvanotropism can be demonstrated by inserting $(+)$ and $(-)$ platinum electrodes into the two ends of the trough. At 1.5–3 volts the ciliates collect at the cathode, but with an increase in current the ciliates move towards the anode. Watch the pattern of ciliary beat as the ciliates near the cathode.

Cultures of *Paramecium* spp. can be obtained from the Cambridge Culture Centre. The most extensive work on the behaviour of *Paramecium* is that of Jennings (1931).

2. SPONGES

2.1. Introduction

The structure of a sponge lies somewhere between that of a colonial protozoan and that of a true metazoan. The sponges cannot conveniently be studied as cells, as can the protozoa, nor as animals which can be dissected and investigated experimentally, as can the higher invertebrates. The study of sponges must be based on living material, sectioned material and on structural elements which have been dissociated from the entire sponge. In order to understand anything about sponge morphology it is essential to know something about the ways in which the basic structure of these animals has become modified. Good accounts are given in Hyman Vol. 1, Burton (1967) and Barnes (1968). Basically a sponge consists of a sac, the walls of which are composed of mesenchyme cells lined on the outside by flattened pinacocytes and on the inside by flagellated choanocytes. The mesenchyme contains wandering amoebocytes and the wall is strengthened by the presence of some kind of skeleton. Linking the inside with the outside of the sponge are pore cells or porocytes which may be associated with muscle cells or myocytes. The flagella beat and draw a current of water into the sponge by way of the porocytes and out of the sponge by way of a single opening, the osculum. The sponge, therefore, consists of a mass of cells specifically modified for the passage of an inhalent current of water through the porocytes (or ostia) to the flagellated region and an exhalent current out through the osculum. The characteristic features of a sponge are the ostia and the osculum. This basic form has been modified so that the flagellated cells are enclosed in flagellated chambers and various passages lead the inhalent current into these chambers and the exhalent current away to the osculum. The water currents of sponges can be demonstrated by adding coloured particulate matter to the water in which the sponges are living (see p. 46). The complexes of passages must be examined in sectioned material (see p. 46). The skeletal elements may consist of a material called spongin or of calcareous or siliceous spicules. The spicules occur in a variety of forms and these are diagnostic of certain groups and can be studied by dissolving away the remainder of the sponge (see p. 47). The nomenclature of the spicules is complex but logical and adequately summarized by Hyman (Vol. 1) pp. 296–301 and Arndt (1934–5).

2.2. Classification

Because of the uniform structure of sponges it is not always easy to define the characters which separate species from species or genus from genus. This has led to a mass of synonyms, and in the account which follows the generic and specific names are those which appear in the Plymouth Marine Fauna, 3rd edition, 1957. It is realized that some of these have been changed but it is felt that the convenience of using a standard list out-weighs the fact that the names may no longer be zoologically acceptable, although widely used.

The classification given here is based on that of Hyman (Vol. 1) which is itself based on earlier classifications. The main disadvantage of this system of classification is that it is based essentially on the form of the spicules. This means that sponges cannot be identified adequately unless the spicules have been examined. The spicules themselves are difficult to classify and reference must be made continually to detailed illustrated accounts such as that of Hyman (1940). As far as undergraduate practical classes are concerned such detail is not necessary.

An alternative classification, which is somewhat simpler than the one below, but less generally accepted, is given by Burton (1967).

Phylum PORIFERA

Class CALCAREA
    Skeleton composed of calcareous spicules

        Order HOMOCOELA
        Inner layer of flagellated cells unbroken.   *Leucosolenia*

        Order HETEROCOELA
        Flagellated cells contained in chambers.   *Sycon, Grantia*

Class HEXACTINELLIDA
    Skeleton composed of six-rayed siliceous spicules (which may be modified).

        Order HEXASTEROPHORA
        Hexasters (spicules with branched rays) present; amphidiscs (spicules with a single axis with mushroom-like ends) absent.   *Euplectella*

        Order AMPHIDISCOPHORA
        Amphidiscs present; hexasters absent.   *Hyalonema*

Class DEMOSPONGIAE
    Skeleton composed of siliceous spicules which are not six-rayed (usually four-rayed) or spongin or both.

        Sub-class TETRACTINELLIDA
        Skeleton composed of four-rayed siliceous spicules or absent; no spongin.

        Order MYXOSPONGIDA
        Spicules absent.   *Oscarella*

        Order CARNOSA
        Spicules present; all similar in size.   *Plakina*

        Order CHORISTIDA
        Spicules present; large (megascleres) and small (microscleres).   *Stelletta*

Sub-class MONAXONIDA
Skeleton composed of two-rayed spicules.

Order HADROMERINA
Megascleres with broad end knobbed; microscleres, if present, star-like; spongin absent.
Spheciospongia

Order HALICHONDRINA
Megascleres of two or more kinds; little spongin.   *Halichondria, Hymeniacidon*

Order POECILOSCLERINA
Megascleres of two kinds united by spongin.   *Myxilla, Suberites*

Order HAPLOSCLERINA
Megascleres of one kind; spongin present.   *Haliclona, Spongilla*

Sub-class KERATOSA
Skeleton composed of spongin; spicules absent.   *Spongia*

2.3. Structure and Biology

The availability of sponges depends on several factors, most of which are out of the hands of the instructor running the practical class. In actual fact this matters very little, for the majority of practical exercises can be carried out with any sponge. All the suggestions which follow will occupy two three-hour periods. The fact that adequate living material may be difficult to obtain is taken into account As a general rule sponges are at their best in Britain in the summer months but travel best when the weather is cold.

Representative sponges should include the following:

(a) Skeletons of *Euplectella*, *Monoraphis* and *Hyalonema*, to show the complexity of the skeletons of the Hexactinellida. Unfortunately living sponges of this group are difficult to obtain.

(b) *Leucosolenia*, *Sycon* and *Leuconia* are suitable representatives of the Calcarea. It should be noted that the spicules are basically three-rayed and dissolve in acid.

(c) *Halichondria*, *Hymeniacidon*, *Myxilla* and *Cliona* represent the spicule bearing marine Demospongiae and *Ephydatia* and *Spongilla* freshwater forms. The marine forms can be obtained from the Marine Biological Association Laboratories at Plymouth and the freshwater forms, especially *Ephydatia*, are very common in clear freshwater, where they live encrusted on plants or stones. They are at their best in the summer.

(d) *Spongia*, the bath sponge, as a representative of the Keratosa.
In general, preserved sponges are hardly worth examining except as skeletons or for the preparation of spicules. Some of the structure of *Spongia* can be seen in thick slices which have been preserved.
Sectioned material, stained in haematoxylin and eosin or similar stains, should be available for the interpretation of sponge structure. This should be carried out in conjunction with work on living material and therefore the same sponges should be used. However, these are not always easy to cut and sections of *Spongia* (the bath sponge) are very useful. Sponges should be fixed in alcoholic fixatives.

2.3.1. EXAMINATION OF LIVING SPONGES

Attention should be paid to form and colour. The common British sponges, *Leucosolenia complicata*, *Sycon ciliatum*, *Grantia compressa*, *Hymeniacidon perleve* and *Halichondria panicea* are usually available. Water currents can be demonstrated using these sponges.

Place a sponge in a dish of seawater and allow it to settle so that the water becomes almost still. With the aid of a pipette direct a stream of fine carmine particles on to the sponge. Observe and note the stream of particles passing out of the osculum. Under a microscope it should be possible to see particles entering the ostia. When the sponge has been subjected to an environment of carmine for some time, it can be removed and slices cut. Examine these and note any accumulation of particles which should be seen in the flagellated chambers. Indian ink can be used instead of carmine. Suitable sponges can be obtained from the Marine Biological Association laboratories at Plymouth.

2.3.2. DISSECTION

Suitable sponges to show grades of organization are: *Leucosolenia complicata* (Ascon grade), *Sycon ciliatum* (Sycon grade) and *Leuconia fistulosa* (Leucon grade). These first two sponges can be dissected by cutting them open and spreading them out for microscopical examination. The flagellated chambers of *Leuconia* are best seen if they contain carmine particles.

2.3.3. SPONGE CELLS: DISSOCIATION AND AGGREGATION

It is possible to see and identify pinacocytes, amoebocytes, choanocytes and spicules by the examination of sponge cells obtained by teasing living sponges in sea water. *Grantia* is a particularly good sponge to examine as the choanocytes are very large and there are also large and active amoebocytes. *Leucosolenia* is also a rewarding sponge to examine. In general, the cells of siliceous sponges are much smaller than those of calcareous sponges.

It is not always desirable or convenient to separate sponges cell from cell, and far more can be seen of the sponge organization if specimens are gently teased with needles. Sponges such as *Leucosolenia* can be cut open and the walls examined under a microscope, *Grantia* can be sectioned with a razor and *Halichondria* can be sectioned tangentially to the surface. *Sycon* and *Leuconia* contain too many spicules to provide any really useful information if treated in this way. Amoebocytes, choanocytes and spicules can easily be demonstrated in this way. For further details see Jepps (1947).

Sponges provide useful material for a number of investigations, but probably the most interesting is that concerned with aggregation of dissociated sponge cells. The methods for dissociating cells and subsequently aggregating them are given on p. 47. Young sponges formed in this way can be used for the study of sponge morphology as well as the activity of the various cells. These young sponges also stain well with haematoxylin and provide useful demonstration slides.

If the cells of sponges are separated from one another, for example by squeezing the sponge through bolting silk they will, under suitable conditions, re-aggregate and may grow into functional sponges. If the cells of different species of sponges are mixed they will usually sort themselves out into their individual species, although some examples of hybridization are known. The phenomenon of aggregation in sponges is well illustrated in the paper by Moscona (1961).

The sponges *Halichondria panicea*, *Hymeniacidon perleve*, *Microciona atrasanguinea* and *Suberites domuncula* are suitable for this type of investigation. Cut up the sponges, make sure that they are living and undamaged. Immerse the fragments in 0.004M EDTA (ethylene-diamine tetra-acetate) in 0.55M sodium chloride buffered with 0.004M *tris*-hydrochloric acid buffer (2-amino-2-hydroxymethyl 1,3-propanediol-hydrochloric acid) for 24 minutes. Press the fragments gently to produce a dense suspension of cells. Suspend the cells in sterile filtered sea water to give between half a million and one million cells per mm³. Mix 5 ml of this suspension with 10 ml sterile seawater and allow the cells to settle on coverslips. Replace the seawater after one hour. Grow the sponges for about 20 hours at 16–17°C and then examine alive, fix and stain on coverslips or fix for sectioning. In this way sponges can be made available for a variety of purposes. *Microciona atrasanguinea* forms aggregates in about 3 hours, *Halichondria panicea* takes about 8 hours while *Suberites domuncula* and *Hymeniacidon* take 15 hours. For further details see Curtis (1962). Curtis obtained a number of mixed aggregates and these could provide the basis for several further studies.

2.3.4. SPICULES

Spicules of calcareous and siliceous sponges can be obtained by boiling the tissue in 10% potassium hydroxide. Permanent preparations can be made by washing the spicules in water, allowing them to settle, decanting off the water and then taking them up in 90% alcohol and allowing them to settle again before transferring them to a slide on which they can be dried over a flame. Mount in Canada Balsam. For a detailed account of the spicules of one easily obtainable species, *Leucosolenia complicata*, see Jones (1954a, b).

Spicules can also be obtained from teased sponge fragments. A variety of spicules may be found by examining the following sponges: *Leucosolenia*, *Sycon*, *Myxilla*, *Halichondria* and *Cliona*, all of which can be obtained from the Marine Biological Association laboratories at Plymouth. Reference should be made to the descriptions of spicules given in Hyman (1940) [pp. 296–301]. Reference slides, particularly of spicules from siliceous sponges, should also be available.

2.3.5. EMBRYOLOGY

The embryology of sponges is best known in the Calcarea, but stages in development are not easy to obtain in the laboratory. Slides of the amphiblastula of one of these sponges should be seen and the cells with the flagella on the outside recognized. The larvae of *Ephydatia* are often common in July in sponges brought into the laboratory. Gemmules, or asexual buds, occur in *Ephydatia* in the autumn.

In freshwater sponges, and in some marine forms, overwintering bodies or gemmules are formed. These gemmules are formed in the autumn and are able to resist freezing or desiccation. In the spring a mass of cells emerges from the gemmule and this forms a new sponge. Sexual reproduction occurs in July. Young sponges can be grown in the laboratory in this way and provide useful class and experimental material.

Collect gemmules of the freshwater sponges *Spongilla* and *Ephydatia* in the winter and keep in a refrigerator for some weeks. Transfer individual gemmules to coverslips in petri dishes of clean pond water and allow to hatch in the laboratory. The young sponges spread out over the coverslip and all the details of the organization of the sponge can be seen. For further information see Jepps (1946, 1947). Jepps collected gemmules in September and kept them in a refrigerator until December. On transference to Petri dishes they hatched in 2–3 weeks. A further batch took only 10 days to hatch in March. Most of the details of the organization of the young sponges were apparent within 10 days of hatching.

3. COELENTERATES

3.1. Introduction

Many coelenterates are transparent or semi-transparent and much of their morphology may be seen without special treatment. Large jellyfishes and actinians are best studied from thick slices or sagittally-cut specimens.

Many intertidal anemones may be kept successfully in aquaria providing the water is aerated and the anemones are fed from time to time on pieces of other marine animals, fragments of earthworm or raw meat. It is important to remove any fragments not eaten. Scyphistomas may be kept for years in small dishes in this way. Hydroids are more difficult to keep, but may be successfully maintained if their tank is kept clean and free from other animals, is well aerated and circulated, and the hydroids are fed. *Artemia* (brine-shrimp) nauplii which are easily cultured provide an excellent food.

3.2. Classification

The true coelenterates or Cnidaria comprise 3 classes: the Hydrozoa, the Scyphozoa (jellyfishes) and Anthozoa (sea anemones and corals). All are distinguished by possession of a gastrovascular cavity and by the possession of nematocysts. The following scheme is generally accepted. Demonstration of the range of form will depend largely on the species available.

Phylum COELENTERATA

Class 1. HYDROZOA

Polypoid or medusoid; polypoid phase prominent, often forming branched, sometimes polymorphic colonies; medusoid phase small, motile or reduced, bearing epidermal gonads. Mesoglea not cellular.

> **Order 1. TRACHYLINA***
>
> Predominantly medusoid. Polypoid phase diminutive. Medusoid often with submarginal tentacles.
>
> *Gonionemus, Geryonia, Cunina*
>
> **Order 2. HYDROIDA**
>
> Polypoid phase prominent. Medusoid may be absent.
>
> > **Sub-order 1. LIMNOMEDUSAE**
> >
> > Polypoid naked.
> >
> > *Hydra*
> >
> > **Sub-order 2. ANTHOMEDUSAE**
> >
> > Polypoid athecate. Medusoid usually bell-shaped.
> >
> > *Tubularia, Hydractinia*

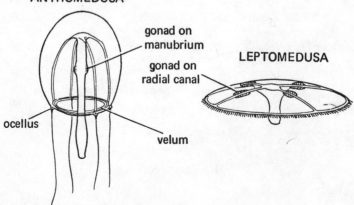

Figure 9   Hydroid medusae

> > **Sub-order 3. LEPTOMEDUSAE***
> >
> > Polypoid thecate. Medusoid usually saucer-shaped.
> >
> > *Obelia*
> >
> > **Sub-order 4. CHONDROPHORA**
> >
> > Pelagic polymorphic colonies of polypoids.
> >
> > *Velella, Porpita*

* trachymedusae and leptomedusae are really distinguishable only by the detailed structure of the marginal sense organs.

Order 3. ACTINULIDA

Interstitial micropolypoids. No medusoids.

Halammohydra

Order 4. SIPHONOPHORA

Pelagic polymorphic colonies of both polypoids and medusoids.

Physalia, Muggiaea

Order 5. MILLEPORINA

Tropical calcareous colonial (massive) corals liberating minute medusoids.

Millepora

Order 6. STYLASTERINA

Tropical calcareous colonial branching corals, essentially similar to *Millepora*, but without medusoids.

Stylaster

Class 2. SCYPHOZOA

Jellyfishes. Medusoid dominant, often large and long-lived. Polypoid phase relatively inconspicuous. Medusoid unlike Hydrozoa in never having a velum (acraspedote), in the manubrium frequently having 4 frilly lips (oral arms), in the gonads being endodermal, in the gastrovascular cavity being divided into a central gastric cavity and 4 pouches with gastric filaments, and in the marginal rhopalia (tentaculocysts) consisting of reduced and specialized marginal tentacles.

Order 1. STAUROMEDUSAE (LUCERNARIIDA)

Sessile; stalked. Recognized by virtue of the quadrangular manubrium and 8 (reduced) rhopalia.

Haliclystus, Lucernaria

Order 2. CUBOMEDUSAE (CARYBDEIDA)

Tropical and subtropical jellyfishes. Bell cuboidal with simple margins.

Carybdea

Order 3. CORONATAE

Jellyfishes with a deep corona or groove round the bell.

Atolla

Order 4. SEMAEOSTOMAE

Jellyfishes of variable shape with scalloped margins. Gastrovascular cavity with canals radiating to the margin. Margin usually with tentacles. The group includes most common northern species.

Aurelia

Order 5. RHIZOSTOMAE

Oral arms fused, enclosing canals opening by small pores. Bell margin without tentacles.

Rhizostoma

Class 3. ANTHOZOA

Solitary or colonial, often large, polypoids. No medusoid phase. Polypoid with mouth leading into a stomodeum invaginated into the gastrovascular cavity which is divided by radial septa. Gonads endodermal. Nematocysts without opercula.

Sub-class 1. ALCYONARIA (OCTOCORALLIA)

Mostly colonial. Polyp with 8 pinnate tentacles. Skeleton, when present, internal (spicular), occasionally with additional external skeleton. Arrangement of septa basically octamerous.

Order 1. STOLONIFERA

Polyps arising from basal mat. Skeleton, when present spicular, or forming platforms.

Clavularia, Tubipora

Order 2. TELESTACEA

Simple colonies with polyps arising laterally from a primary prostrate polyp. With calcareous spicules.

Telesto

Order 3. ALCYONACEA

(soft corals). Polyps embedded in gelatinous calcareous-spicule strewn matrix. Colonies often lobular.

Alcyonium (dead man's fingers)

Order 4. COENOTHECALIA

(blue coral). Indo-Pacific. Skeleton blue, but concealed by polyps. Massive, distinguished by typically octocorallian polyps.

Heliopora

Order 5. GORGONACEA

(gorgonians, sea fans, feathers, whips etc.). Horny axial skeleton. Colonies mainly upright branching, often in one plane. Rarely dimorphic (the red 'precious' coral, *Corallium*, which belongs here, is exceptional).

Gorgonia

Order 6. PENNATULACEA

(sea pens). Axial polyp with lateral secondary polyps forming a more or less fleshy colony anchored in sand by the stalk of the primary polyp. Dimorphic: autozooids and siphonozooids.

Pennatula, Renilla (sea pansy)

Sub-class 2. ZOANTHARIA (HEXACORALLIA)

Solitary or colonial. Polyp usually with many simple tentacles (more than 8). Skeleton, when present, morphologically external. The dominant actinians (anemones) and madreporarians (true corals) have a hexamerous arrangement of septa.

Order 1. ACTINIARIA

(sea anemones). Solitary. Without skeleton and usually with 2 siphonoglyphs.

Calliactis, Metridium, Anemonia, Taelia

Order 2. MADREPORARIA

(stony corals. 'true' corals). Mostly colonial with hard calcareous skeletons. Morphologically otherwise like actinians (but nematocysts different).   *Acropora, Porites*

Order 3. CORALLIMORPHARIA

Morphologically like madreporarians (nematocysts similar), but without skeleton.

Corynactis

Order 4. ZOANTHIDEA

Small, anemone-like anthozoans, without skeleton but with a single siphonoglyph and sometimes colonial. Often found attached to other invertebrates.

Epizoanthus

Order 5. ANTIPATHARIA

(black corals; thorny corals). Colonies branching, superficially gorgonid-like, with a thorny skeletal axis, but distinguished by typical zoantharian polyps.

Antipathes

Order 6. CERIANTHARIA

Sand-living solitary elongate actinian-like anthozoans with a single siphonoglyph, but with very many complete septa with weak muscles, quite unlike the hexamerous arrangement found in actinians.

Cerianthus

3.3. Structure

3.3.1. GENERAL ANATOMY

Most coelenterates are best preserved in 5% formalin, following narcotization with menthol or a 1:1 mixture of 7.5% $MgCl_2$: seawater. Small medusae or other coelenterates found in plankton may be fixed by gradual addition of formalin to the water while stirring until the concentration reaches about 3%. When the animals have settled, replace with 5% formalin made up in seawater.

For histology Bouin's fluid is adequate for general fixation, followed by Azan, Mallory or simply haematoxylin and eosin.

Coelenterates are easy to fix, embed and section in paraffin wax. *Hydra*, or the polyps of the alcyonarian *Alcyonium* are ideal for practice in routine histology. Fix in Bouin, dye with haematoxylin and eosin. Hydrozoa may be mounted and dyed with borax carmine or haematoxylin. Borax carmine works well, especially if well coloured and then well differentiated in 1% HCl in 70% alcohol. For whole mounts hydroids should be narcotized with menthol (overnight) or with 1:1, 7.5% $MgCl_2$: seawater. Transfer to 70% alcohol and colour for half an hour or more in borax carmine. Living specimens may also be placed directly in polyvinyl lactophenol containing lignin pink.

3.3.2. MACERATION

The diploblastic nature of the body may be appreciated from thin sections, but the cellular components can also be studied following maceration. Place in 1:1, 0.05% osmium tetroxide: 0.2% acetic acid in seawater for 3 minutes. Rinse in 0.1% seawater acetic, and leave overnight in a fresh portion of the same mixture. Small fragments of tissue should be placed on a slide and may be disintegrated with needles or simply by slight pressure on the coverglass. Examine uncoloured, or wash the acid off well and colour with 0.1% toluidine blue in seawater, preferably overnight. Rinse off excess dye and mount in seawater.

3.3.3. NEMATOCYSTS

Nematocysts may be collected from living specimens by presenting a coverslip with dried saliva to the animal. Allow the animal to respond and then gently pull the coverslip away. Some nematocysts will pull out. The cover may be mounted immediately and examined in seawater, preferably with phase contrast; or fixed and made into a permanent preparation as desired. Such covers may be quickly fixed in osmium tetroxide vapour. Nematocyst discharge may also be caused by addition of 1% acetic acid to the water near the animals. Squash individual tentacles or hydranths on a slide and examine immediately. A microscope of good quality is required to see details. Nematocysts of actinians are notoriously small. Those of *Hydra* are larger (10–20 μm) and readily obtained. Nematocysts of zoantharian corals may be large and, if available, those of the corallimorph *Corynactis* are particularly good for observation and experiment because of their large size (95 × 35 μm). Nematocysts from *Corynactis*, both discharged and undischarged can be collected by wiping a clean coverslip across the oral disc. Such covers may be kept dry in a desiccator over calcium chloride for years without further treatment. Undischarged nematocysts can be discharged under a microscope by addition of water. For details, see Robson (1953).

3.3.4. Nerve Net

Preparations of the nerve net are best made from actinian mesenteries. Batham, Pantin and Robson (1960) used those of *Metridium* and these are specially good. Their method is as follows.

1. Narcotize the animals in 1:1, 7.5% $MgCl_2$: seawater.

2. Cut out a mesentery and pin to paraffin wax half filling a Petri dish with non-metallic pins, stretching the mesentery as much as possible. Hedgehog, cactus or other plant spines can be used. Mount under the narcotizing mixture.

3. Dye by addition of a few drops of reduced methylene blue. The dye solution is prepared as follows. Dissolve 0.5 g methylene blue in 100 ml 1:1, 7.5% $MgCl_2$: seawater. Add 3 drops 24% HCl (24 ml conc. HCl+76 ml dist. water). Mix thoroughly. Filter. Dissolve 12 g 'rongalit' (formaldehyde-sodium sulphoxylate) in 1:1, 7.5% $MgCl_2$: seawater. Add 20 ml of this to 100 ml of the methylene blue mixture. Warm gently (do not boil) in a beaker, stirring constantly. When the blue colour has turned to a dirty green, remove from heat but continue to stir. After a few minutes the mixture clears and a yellowish precipitate forms. Set aside to cool. Filter. Stand 1–2 days before use. This solution will last 8–10 days.

4. Pour off the dye mixture. Rinse briefly in fresh $MgCl_2$: seawater. Fix in Susa, overnight.

5. Pour off fixative. Add 1% phosphomolybdic acid and leave until the tissue has become turquoise in colour.

6. Wash in distilled water, then in 95% alcohol with iodine (to remove Hg ions from the fixative). Dehydrate, Unpin. Clear in xylol and mount.

For silver staining and other details, see Batham, Pantin and Robson (1960).

3.3.5. Structure of jellyfishes

Much of the structure of a semaeostome such as *Aurelia* can be seen by virtue of the body's transparency (Fig. 10). Black paper under the specimen or a black dish helps. The sexes are separate. The female may be recognized by the convoluted edges of the oral lobes which retain the developing eggs. The relationship of the central gastric cavity to the pouches and radial canals can be investigated with a seeker. Ink may be injected through the mouth, but usually does not penetrate beyond the pouches. Jellyfishes well hardened in strong formalin, or better in dichromate (see p. 111) can be cut sagittally in various radial planes. Single tentaculocysts may be snipped out with scissors and examined under the high power of a dissecting microscope. Details are best seen if stained with haematoxylin and mounted (Fig. 11).

The gastrovascular system of *Rhizostoma* may be injected with ink by passing a large syringe through the exumbrellar surface into the central gastric cavity.

The stauromedusae, *Haliclystus* or *Lucernaria* occur on *Zostera* (eelgrass) or seaweeds around low water of spring tides on certain rocky shores. They provide excellent subjects for comparison with typical jellyfishes. *Haliclystus* may be embedded in gelatin. Thick horizontal slices (2 mm) are informative, especially if compared with a specimen halved

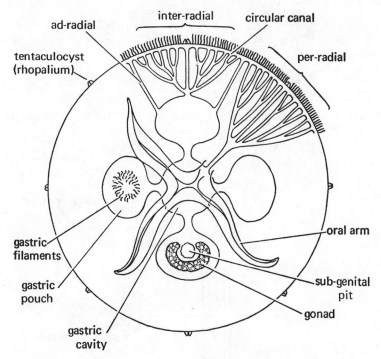

ad-radial

inter-radial   circular canal

tentaculocyst
(rhopalium)

per-radial

gastric
filaments

gastric
pouch

oral arm

sub-genital
pit

gonad

gastric
cavity

Figure 10 *Aurelia*, Sub-umbrellar view

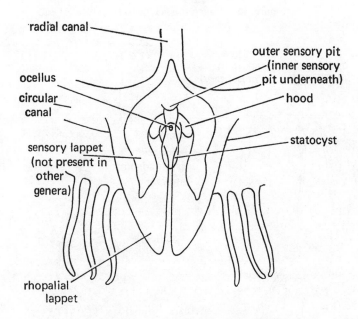

radial canal

outer sensory pit
(inner sensory
pit underneath)

ocellus

circular
canal

hood

statocyst

sensory lappet
(not present in
other
genera)

rhopalial
lappet

Figure 11 *Aurelia*, Tentaculocyst

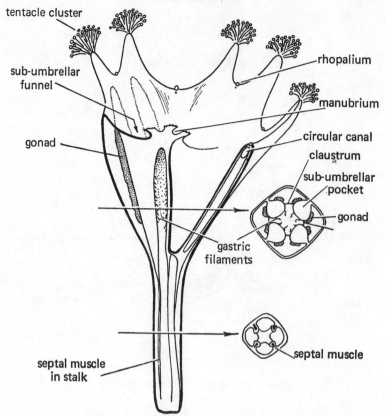

Figure 12 *Haliclystus*, General structure

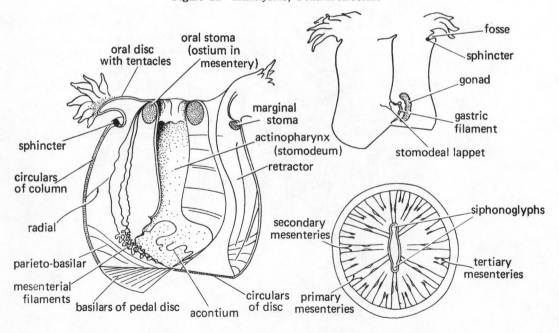

Figure 13 Anemone, General structure (partly after Batham and Pantin, 1951)

sagittally (Fig. 12). Narcotize with menthol or 1:1, 7.5% $MgCl_2$: seawater and fix in 5% seawater formalin. Wash off the formalin in running water. Embed in gelatin by transferring to 5% gelatin at 37°C for 24 hours. Then transfer to 10% gelatin for 12 hours, at the same temperature. Place in a fresh portion of 10% gelatin; then allow to cool. When set, put the block in 10% formalin to harden and to render the gelatin insoluble in water. The block may be kept indefinitely, and may be sliced with a single-sided razor blade under a dissecting microscope. Details may be seen from paraffin sections cut at 10 μm. Bouin fixes well. Sections stain well in Azan, Mallory or Masson's trichrome mixture.

The structure of the mesoglea may be studied with phase contrast or polarizing microscopes, or with thin slices of living tissue coloured with neutral red chloride or methylene blue.

Mesoglea of jellyfishes fixes quite satisfactorily for general purposes in Bouin, but better in Helly's mixture. Mallory and Azan are satisfactory staining mixtures. Thick sections (0.1 mm) can be cut by hand or better on a freezing microtome. Sections so cut from living tissue are quite informative if coloured with 0.1% toluidine blue for several hours. For other details and methods, see Chapman (1953).

The visco-elastic properties of the mesoglea can also be demonstrated, either with jellyfishes or, more manageably, with strips of column of *Metridium* or *Calliactis* in which the muscles are narcotized with 1:1, 7.5% $MgCl_2$: seawater. For details, see Alexander (1962).

3.3.6. STRUCTURE OF SEA ANEMONES

Anemones live well in aquaria and may form the subject of various experiments and demonstrations; see p. 49, above and p. 59, below. Structural details are very variable. Reference should be made to regional monographs such as those of Stephenson (1928, 1935).

The anatomy is best studied from cuts made in various planes. If possible narcotize the anemones with 1:1, 7.5% $MgCl_2$: seawater and fix in 5% seawater formalin. Hardened specimens are easier to slice. Cut sagittally (a) through the siphonoglyphs, (b) at right angles to this plane (Fig. 13), (c) horizontally through the middle of the column. Large specimens make the best demonstrations. Individual mesenteries may also be cut out, dyed with chlorazol black E, dehydrated, and cleared in benzyl alcohol. If small enough they may be mounted on large slides, but are usually too large for this and are best examined in benzyl alcohol in a Petri dish. Much of the musculature can be seen by oblique lighting from below. Rotation of the specimen between crossed polaroids is also helpful in distinguishing the muscles (see p. 111). Details can be obtained from histological sections: use smaller specimens of the same species. Fix in Bouin; embed in paraffin wax; stain with Azan. For nerve net see p. 54, above.

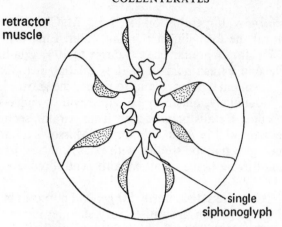

Figure 14 *Edwardsia*, Pattern of mesenteries

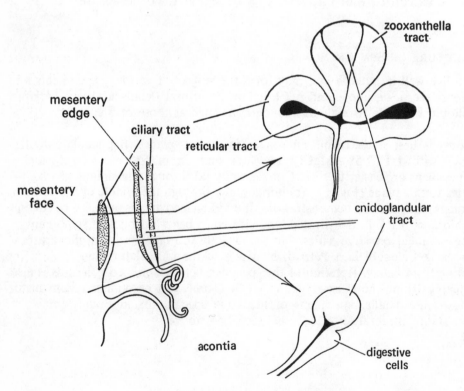

Figure 15 Anemone, mesenteries

The left-hand figure shows part of two mesenteries in side and edge-on views. The other figures indicate the distribution of cell types in mesentery (above) and acontium (below) in section.

3.4. Regeneration and Reconstitution

Simple experiments on regeneration and reconstitution are informative in relation to the tissue constitution of coelenterates; some are quite easily performed, and were first done by Trembley and described by him in his book of 1744. The fresh water *Hydra*, the marine *Tubularia* and the brackish water *Cordylophora* are ideal for this work.

3.4.1. *HYDRA*

Hydra may be cut up with sharpened needles into 30–50 pieces. If the pieces are heaped up they will reconstitute in 4–5 days. Conditions must be clean. Preferably sterilize needles by flaming before use. Water can be sterilized with a U.V. source. The fragments may be heaped into a small watchglass supported within a larger dish of natural water (crystallizing dishes are ideal), or into a slight depression made in a bed of agar in a Petri dish. The fragments must be kept together and easily located. A dimple in an agar bed can be made with a small warm glass rod. When cutting up the *Hydra* discard the hypostome and tentacles. Then add streptomycin, penicillin, or other antibiotic to the water.

Reaction of small fragments to each other can be observed on a slide. Endoderm cells alone fuse, but ectoderm and endoderm are necessary for reconstitution.

Hydras may also be turned inside out. To do this first make the animal contract by gentle poking. Then place the head down and press the middle of the pedal disc gently but firmly with a blunt needle so that the pedal disc is pushed through the mouth. Then ease the outer layer up. This was done by Trembley. Some hydras may not survive the treatment. Some will turn themselves right side out again. In others there will be an active migration of ectoderm and endoderm cells past each other to restore their natural relationship. This process is best seen in green hydras where the zoochlorellae are in the endodermal cells. All these processes may be seen within a few hours of turning the animal inside out. For details, see Roudabush (1933) and Papenfuss (1934).

3.4.2. *TUBULARIA* AND *CORDYLOPHORA*

Tubularia and *Cordylophora* have both been experimented with by various workers in recent years. Conditions must be clean and water preferably sterilized (see 3.4.1., above). For general references see Tardent (1963) and Tardent (1965; in Kiortsis and Trampusch). For design of various experiments using *Cordylophora* see Beadle and Booth (1938) and Moore (1952). Reconstitution masses (cut up, as for *Hydra*) require about 2 days for *Cordylophora*. Colonies of *Tubularia* can be grown by allowing the actinulae to settle on glass slides in a dish. Feed on *Artemia* nauplii. For details of colony growth, see Mackie (1966). *Cordylophora* experiments are best done in 50% seawater.

3.5. Experiments with Sea Anemones

3.5.1. REACTION TO MECHANICAL STIMULATION AND FEEDING REACTIONS

Anemones kept hungry for a few days or for a week or two will respond well to various stimuli. A water-colour paint brush may be used to stroke part of the disc, tentacles, column or pedal disc. Various species may be used to demonstrate the extent and type of reaction in different parts of the body. Response to food may be tested with a clean glass rod or paint brush dipped into glucose solution, or by using a wisp of cotton wool previously dipped into glucose solution. Bits of cotton wool similarly primed with different substances may be dropped onto the disc. For details, see Pantin and Pantin (1944).

3.5.2. ELECTRICAL STIMULATION AND PROPERTIES OF THE NERVE NET

The feeding responses described above can be amplified by electrical stimulation of isolated tentacles, using a 6 V source through a potentiometer. Clark (1966) gives a simple circuit and directions for such experiments. Muscular movements and internal pressures can be followed by hooking an anemone to a kymograph and recording pressure changes in a manometer connected to a canula passed into the gastrovascular cavity through the base of the column. For a simple apparatus see Batham and Pantin (1950).

Various anemones can be used to demonstrate the general properties of the nerve net.

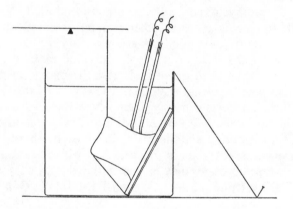

Figure 16 Set up for electrical stimulation of anemones (after Clark, 1966)

Calliactis parasitica found on whelk shells inhabited by hermit crabs is specially good for such demonstrations. The anemones must be coerced to attach themselves to pieces of slate, tile or similar base and allowed to accommodate for some days to these new conditions before use. If possible, set up in tanks in which the experiments are to be done. Stimuli are delivered from a 6 V source incorporating a potentiometer so that the strength of the stimulus can be varied. Electrodes made from glass tubing only slightly drawn out and filled with seawater agar, are best. Nylon thread, or (better) dental silk can be attached (hooked) into the edge of the disc sometime before the experiments, and related to an isotonic lever. Stimuli need to be delivered at rather long intervals (say 1/ s). If a stimulator able to do this is not available, then a metronome (rocking into mercury dash-pots) can be used. Pantin (1935a) gives the basic circuit (see also Clark, 1966). Kymograph drum speeds of 25–100 mm/min. are appropriate, according to test. With such an arrangement it is possible to demonstrate the basic properties of a nerve net: that a single shock does not produce a response unless the stimulus is very strong;

that 2 shocks given with different time intervals between them will produce a response (by facilitation) if the time interval is short enough; and by giving a train of shocks at 1 every 2–4 sec, the phenomena of facilitation and summation can be shown. Several individuals will be needed to demonstrate all these effects on a single occasion as the animals take time to expand again after each experiment. These experiments are all based on the classic papers of Pantin (1935a, b, c, d) to which reference should be made for further details. See also Robson (1961) and Kerkut (1968).

4. CTENOPHORES.

4.1. Introduction

The ctenophores are exclusively marine and, apart from the bottom-living Platyctenia, are planktonic.

Ctenophores may be kept for a time in glass tanks in the laboratory providing that they are fed with fresh plankton or *Artemia* nauplii.

4.2. General Methods

The anatomy of ctenophores is best seen in living animals or in specimens preserved in 5% formalin in seawater. When collected they may be killed by adding formalin very gradually to the seawater until a concentration of 1–2% formalin has been reached. The animals should be allowed to settle, the liquid poured off and replaced with 5% seawater formalin. The tentacles are almost invariably contracted.

Comb-jellies such as *Pleurobrachia* are easy to embed in paraffin wax and section. Bouin's fluid is quite adequate for general histology, followed by staining with haematoxylin and eosin. Azan, and Mallory's triple stain also work well. For details see Hyman, Vol. 1.

The characteristic colloblasts can be seen by pulling out a tentacle and squashing well on a slide. A microscope of good resolution is required and the squash examined under an oil immersion objective ($\times$ 100).

4.3. Classification

The Phylum is divided into two Classes. There seems to be general agreement that of these the Tentaculata may be further subdivided into four Orders. The ctenophores were at one time included with Cnidaria within the Phylum Coelenterata. They are distinguished from the cnidarians by the lack of nematocysts and possession of colloblasts. In practice they are easily recognized by the eight comb-rows.

Phylum CTENOPHORA

Class 1. TENTACULATA
With tentacles.

> **Order 1.** CYDIPPIDA
> Spherical or gooseberry-shaped. Tentacles rectractile.

Pleurobrachia (sea gooseberry)

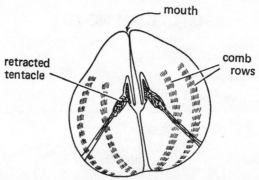

Figure 17 *Pleurobrachia*, in swimming posture, tentacles retracted

Order 2. LOBATA
Body somewhat flattened, 'lobed'—2 large oral lobes bearing 2 comb-rows, and 2 pairs of smaller 'auricles' with a tentacle between each pair.

Mnemiopsis

Order 3. CESTIDA
Body flattened and belt-like, with 2 main comb-rows along one edge, the 'tentacles' adnated along the opposite edge.

Cestus (Venus's girdle)

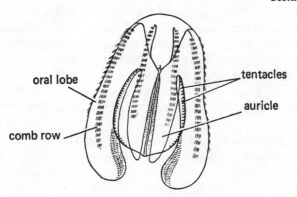

Figure 18 *Mnemiopsis*

Order 4. PLATYCTENIA
Body flattened. Comb-rows very reduced. Creeping animals recognized by their two tentacles with colloblasts.

Class 2. NUDA

Without tentacles. Thimble-shaped (flattening when preserved). Recognized by comb-rows.

Beroë

5. PLATYHELMINTHS

5.1. Introduction

Members of the phylum have a characteristic organization of the body, a muscular body wall enclosing parenchymatous mesenchyme tissue which surrounds the internal organs. The body wall and the parenchyma act together as a hydrostatic skeleton. The absence of a vascular system in these animals imposes limitations on the dimensions of the body and, almost without exception, the platyhelminths, as the name implies, are flattened. Specimens are thus amenable to examination as whole mounts and by suitable preparation their internal anatomy can be clearly seen. Sectioned material is a useful adjunct for the study of the body tissues and for building up a complete picture of the structure of these animals. Platyhelminths are not only flattened, but unsegmented and acoelomate. The intestine, when present, is without anal opening. They have a protonephridial excretory system, no vascular system, are hermaphrodite, and may be either free-living or parasitic.

5.2. General Methods

5.2.1. FIXATION

It is essential to fix these animals in an extended condition and this may be achieved in a variety of ways.

Turbellaria. Planaria and some smaller turbellarians may be fixed while swimming in an extended position by placing them in a small volume of water and dropping fixative directly onto them. Alternatively the animals can be relaxed by placing in a solution of menthol crystals (0.5 g/100 ml of water, or 1 drop per 100 ml of water of a solution of 24g/10 ml 95% alcohol) prior to fixation (Abdel-Malek, 1951).

Monogenea, Digenea and Cestoda. The parasitic platyhelminths should be washed in an appropriate saline after removal from the host. They can be extended by menthol relaxation or by allowing them to die in tap water.

However extension is achieved, the animals should always be compressed for some time during the process of fixation to maintain flattening. Place the fixative in a Petri dish and compress the specimen between two slides, held together by vaseline at each end, or use a piece of glass heavy enough to flatten the specimen onto the bottom of the dish (or onto a piece of filter paper in the dish) without causing the body to rupture. Leave for 15 to 30 minutes and then remove the compressing agent and allow fixation to continue for at least an hour, longer if the specimen is large.

Turbellaria, Monogenea and Digenea may be fixed in either Bouin or in A.F.A. (85% alcohol—85 parts, 40% formalin—10 parts, glacial acetic acid—5 parts). Cestodes may be fixed in 5% aqueous formalin. Specimens may be left for considerable periods in these solutions.

5.2.2. BLEACHING

Pigmented turbellaria may be bleached after fixation in Mayer's chlorine. Add a few crystals of potassium chlorate to a few drops of concentrated hydrochloric acid and then add 25–50 ml of 70% alcohol. Leave the specimens in the solution until colourless. Commercial solutions of sodium hypochlorite may also be used for this purpose.

5.2.3. STAINING

A variety of methods may be used to obtain satisfactory whole mount preparations.

(a) Borax carmine. Transfer from 70% alcohol or less and stain for 3–10 minutes. Differentiate in acid alcohol. This method can be used for all platyhelminths.

(b) Gower's carmine. Transfer from water to the acidified carmine and stain for about 24 hours. Destain in a bleaching solution. This method is good for digeneans.

(c) 'Rapid Haematoxylin'. Transfer from water to 3% iron alum solution and mordant for 10–30 minutes. Rinse and stain in Ehrlich's haematoxylin for a similar time. Differentiate in 1% iron alum solution. Can be used for all platyhelminths. Alternatively the full Heidenhain's haematoxylin method can be used, carrying out the staining procedure at 37°C, thereby reducing the time necessary.

(d) Dilute haematoxylin. Transfer from water to Ehrlich's haematoxylin diluted 1:50 or 1:100 with distilled water. Overstain and differentiate in acid alcohol. This method gives excellent results with most platyhelminths.

(e) Acetic haematoxylin (Chubb, 1962). After fixation in A.F.A. or formalin, stain in haematoxylin which has been made up with 10 ml of glacial acetic acid and diluted 1:3 with 45% acetic acid before use. Stain for 2–20 minutes according to size. Dehydrate in glacial acetic acid. Clear in methyl salicylate. This method is good for all parasitic platyhelminths.

Methyl salicylate is a very good clearing agent for these helminths.

5.3. Classification

The phylum has traditionally been divided into three classes Turbellaria, Trematoda and Cestoda, the Trematoda containing the Monogenea and Digenea. Many recent workers (see Llewellyn, 1965) consider that the Monogenea and Digenea are not closely related and that the former are closer to the Cestoda. Accordingly the phylum is here divided into four major classes, namely Turbellaria, Monogenea, Cestoda and Digenea. To these a fifth Class, the Cestodaria, may be added. Subdivisions of each Class are defined under the appropriate sections which follow.

5.4. Turbellaria

5.4.1. INTRODUCTION

The most useful turbellarians for general class purposes are the fresh water rhabdocoels and triclads. Both can be obtained easily from ponds and streams, rhabdocoels also commonly occur in aquaria. Polyclad turbellaria may be collected from sea shores, but are more difficult to maintain.

5.4.2. CLASSIFICATION

The classification of the Turbellaria is unsettled, but an arrangement into five orders (Hyman, Vol. 2) is convenient.

Class TURBELLARIA
Free living or commensal. Body covered by cellular epidermis, usually ciliated and containing rhabdites. No specialized adhesive organs in majority of species. Mouth ventral, pharynx well developed. Hermaphrodite, life cycles simple and usually direct.

    Order ACOELA
    Marine. Small forms of simple organization, no intestine or excretory system.
                                                Convoluta

    Order RHABDOCOELA
    Marine, fresh water and terrestrial. Small forms. Intestine sac-like.
                                    Dalyellia, Stenostomum,
                                    Temnocephala

Order ALLOEOCOELA

Marine and fresh water. Small forms. Intestine sac-like or diverticulated. A rather artificial group.

Prorhynchus

Order TRICLADIDA

Marine, fresh water and terrestrial. Large forms. Intestine with three branches.

Procerodes, Dendrocoelum,
Bipalium

Order POLYCLADIDA

Marine. Large forms. Intestine with many branches. Life cycle may involve a free-swimming larval stage.

Leptoplana

5.4.3. FRESH WATER TRICLADS:*Dendrocoelum*

Fresh water triclads ("Planaria") such as *Dendrocoelum*, *Dugesia* and *Polycelis* are suitable material both for live examination and whole mount and sectioned preparations. Living specimens may be kept in the laboratory and survive well if fed regularly with some form of fresh meat and if care is taken to prevent pollution of the water. Characteristic locomotory patterns, such as ciliary gliding along the bottom or on the surface film, muscular swimming and righting reflexes when disturbed, may be observed. The abundant secretion of mucus and its use in adhesion will also be evident. Starved animals show klinokinetic behaviour when food is placed in the water and move towards the food, testing the strength of the chemical stimuli on each side by alternate lateral movements of the head. By placing a mirror beneath a glass container it is possible to observe the protrusion of the pharynx and the position of the body while the animals are feeding. A useful summary of turbellarian behaviour is given in Hyman, Vol. 2.

Vital staining of triclads with methylene blue can be used to show sensory cells in the chemoreceptive organs on the head and to show the zone of adhesive gland cells around the margins of the body on the ventral surface.

Turbellaria show considerable powers of asexual reproduction and regeneration and the latter process may easily be studied in pigmented triclads such as *Dugesia* and *Polycelis*, in which the non-pigmented regenerated tissues are clearly visible. By means of suitable cuts across the body it is possible to demonstrate antero-posterior and lateral gradients of regeneration (see Hamburger, 1942 and Hyman, Vol. 2). It is essential to use sharp instruments for the incisions and to keep the regenerating animals in clean containers, changing the water regularly to prevent contamination.

Dendrocoelum lacteum

The living animal is elongate, flattened dorso-ventrally and white in colour. Anteriorly the well demarcated head bears a pair of pigment-cup ocelli on the dorsal surface and lateral sensory grooves which are chemoreceptive. The muscular pharynx is visible as a pale ovoid area in the middle of the body; the mouth opening can be seen on the ventral surface at the posterior border of the pharynx. It is sometimes possible to see the opening of the genital atrium behind the mouth. In specimens recently fed (on blood clots or liver) the outline of the intestine is visible through the body wall, one branch running anteriorly from the pharynx and two branches running posteriorly.

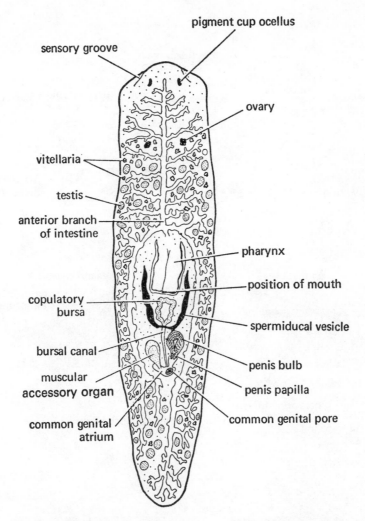

Figure 19 *Dendrocoelum lacteum.* Dorsal. Testes and vitellaria schematic

    The reproductive system is visible only in stained, whole mount preparations (Fig. 19).
There are two small ovaries anteriorly and numerous testes and yolk-cell producing
vitellaria in the lateral regions. Sperm are conveyed from the testes to the prominent
spermiducal vesicles which open into the base of the penis bulb. A muscular accessory
organ, the adenodactyl, adjoins the male organ. Ova and yolk-cells are conveyed along
paired ovovitelline ducts to the common genital atrium. During reciprocal cross fertilization the copulatory organ is protruded through the genital pore and enters the genital
atrium of the other individual. Sperm are passed into the atrium and enter the copulatory
bursa via the bursal canal. Fertilization and egg shell formation take place subsequently
in the atrium.

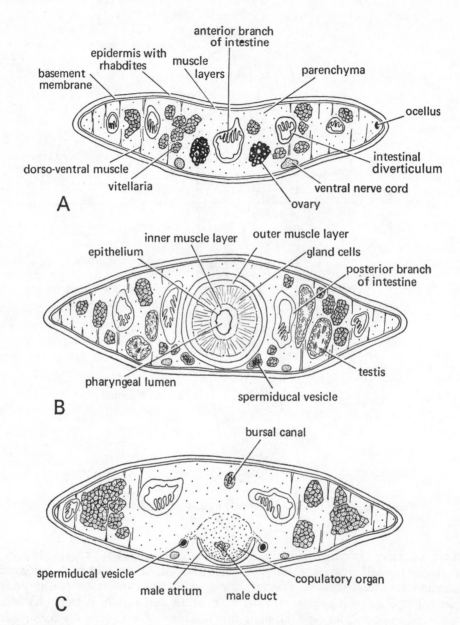

Figure 20 *Polycelis.* A, T.S. through ovaries; B, T.S. through pharynx; C, T.S. through copulatory organ

In transverse section (Fig. 20) the triclad epidermis appears as a single layer of ciliated cells, interspersed with mucous gland cells. Many epidermal cells contain rod-shaped rhabdites, formed in the underlying parenchyma and carried to the surface as cells are replaced. The function of rhabdites is possibly excretory, although a defensive role has also been suggested. The epidermis rests upon a distinct basement membrane, which plays an important part in limiting extension of the body during movement. Below the membrane lie the muscle layers, an outer circular and inner longitudinal layer with an oblique layer between. Dorso-ventral muscles run into the parenchyma and function in flattening the body. Above the muscle layers on the ventral surface lie the two major longitudinal nerve cords, which run posteriorly from the anterior cerebral ganglia.

The parenchyma contains numerous gland cells; at the lateral margins of the body eosinophilic subepidermal gland cells, which secrete an adhesive mucus, may be seen.

The cerebral ganglia, pharynx and components of the reproductive systems will be visible in sections taken from appropriate regions of the body.

5.4.4. Fresh Water Rhabdocoels: *Stenostomum*

Living rhabdocoels of the genus *Stenostomum* (common in ponds and aquaria) usually show the process of asexual reproduction by transverse fission, chains of individuals swimming together for some time before separating. Vital staining shows the anterior sensory ciliated pits on each individual and some stains, e.g. cotton blue, will stimulate the discharge of rhabdites. In unstained specimens it is possible to see the mouth, pharynx and simple intestine and also the prominent excretory tubule. *Stenostomum* may be cultured in the same way as ciliate protozoa.

5.5. Monogenea

5.5.1. Introduction

Monopisthocotyleans may be collected from the gills, skin and cloacal cavities of fish. *Gyrodactylus elegans*, a small viviparous form, is sometimes found on the gills of goldfish and sticklebacks.

Polyopisthocotyleans may be collected from the gills of marine teleost fish e.g. *Diclidophora merlangi* from the Whiting. Such forms show intimate adaptation of the opisthaptor to the gill structure of the host.

5.5.2. Classification

Class MONOGENEA
Mostly ectoparasites of aquatic vertebrates, some endoparasites. Ciliated epidermis lost in adult. Adhesive organs present anteriorly (prohaptors) and posteriorly (opisthaptors). Mouth anterior, intestine bifid. Excretory pore anterior. Hermaphrodite, life cycles simple and direct, involving one host only.

> **Sub-class MONOPISTHOCOTYLEA**
> Opisthaptors simple, either hooks and/or simple sucker-like organs. Genito-intestinal canal absent. Testes few.
>
> *Entobdella, Gyrodactylus*
>
> **Sub-class POLYOPISTHOCOTYLEA**
> Opisthaptors complex, hooks, suckers or clamps (modified suckers). Genito-intestinal canal present. Testes many.
>
> *Diclidophora, Polystoma*

5.5.3. *Polystoma*

Polystoma integerrinum occurs in the urinary bladder of anuran amphibia. When adult, *Polystoma* is endoparasitic (and therefore atypical), but the life cycle commences with larvae attached to the gills of an amphibian tadpole.

Living specimens are found attached by their opisthaptors to the wall of the bladder. The area of their feeding activity is often marked by haemorrhage where blood vessels have been damaged and blood may be visible in the intestinal caeca. The worms should be removed, washed in saline and fixed.

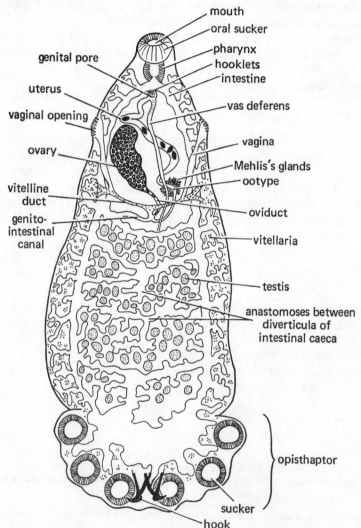

Figure 21 *Polystoma integerrinum.* Ventral. Testes and vitellaria schematic

The body of *P. integerrinum* (Fig. 21) broadens posteriorly where the opisthaptor, consisting of six suckers and two large hooks, is situated on a distinct muscular disc. Larval hooklets are also present on the anterior margin of the haptor, but usually are difficult to see. Much of the body is occupied by the diverticula from the intestinal caeca. Lateral prominences in the anterior region indicate the openings of the vaginae, which run back obliquely and join with ducts from the lateral vitellaria before opening, by a median duct, into the oviduct. Egg formation is carried out in the ootype, which is surrounded by Mehlis's glands or 'shell glands' (although they do not secrete the shell). Fine ducts convey the sperm from the testes to the vas deferens which leads into the male organ. Both male and female ducts open at the common genital pore, the position of which is marked by the circle of eight hooklets around the male organ. The genito-intestinal canal can usually be seen running between the oviduct and the right intestinal caecum, its function is uncertain.

Reproduction in *Polystoma* is initiated by hormonal changes associated with the breeding of the host amphibian. This ensures the availability of new hosts for the continuation of the parasite's cycle.

Jennings (1956) has given a method for the detection of *Polystoma* in living hosts, based upon a test for the presence of haemoglobin breakdown products in the urine.

5.6. Cestoda and Cestodaria

5.6.1. INTRODUCTION

The small class Cestodaria is restricted to phylogenetically ancient hosts such as sturgeons, chimaerid fish and chelonians. Its affinities are uncertain, although it is undoubtedly related to the Monogenea and the Cestoda. Some workers, however, consider the group to be artificial.

The class Cestoda is divided into several orders, some of which are well established. Authorities differ, however, on the validity of the smaller orders.

5.6.2 CLASSIFICATION

The classification given below is based on those in Hyman (1951), Wardle and McLeod (1952) and Grassé (Vol. 2) and includes only those orders which are generally accepted.

Class CESTODARIA

Endoparasites without mouth or intestine. Body not strobilate. Adhesive organs not on scolex. Hermaphrodite, male and female openings separate. Life cycle involving intermediate hosts, larvae have ten hooklets. In body cavity of host.

Amphiline, Gyrocotyle

Class CESTODA

Endoparasites without mouth or intestine. Body characteristically strobilate, divided into proglottids, each with one or more sets of reproductive organs. Body covered with thick, metabolically active tegument. Adhesive organs borne on anterior scolex and may be hooks, suckers, bothria (sucking grooves) or bothridia (muscular outgrowths). Parasitic as adults in vertebrate small intestine. Life cycles often complex, involving intermediate hosts.

Order TETRAPHYLLIDEA
Scolex with four bothridia. Vitellaria scattered. Uterus with pore or pores. In elasmobranch fish.

Acanthobothrium,
Phyllobothrium

Order LECANICEPHALA

Scolex without bothridia, may have suckers and/or retractile tentacles. Reproductive system tetraphyllid. In elasmobranch fish. Sometimes classified with Tetraphylliade.

Lecanicephalum

Order DIPHYLLIDEA

Scolex with two bothridia and large hooks. Scolex on spinose head-stalk. Vitellaria scattered. Uterus closed. In elasmobranch fish. Sometimes classified with Tetraphyllidea.

Echinobothrium

Order TETRARHYNCHIDEA

Scolex with bothridia and retractile spiny tentacles. Vitellaria scattered. Uterus closed. In elasmobranch fish.

Grillotia

Order PROTEOCEPHALA

Scolex with four suckers, sometimes a fifth apical sucker. Vitellaria scattered. Uterus with pores. In fresh water fish, amphibia and reptiles.

Proteocephalus

Order NIPPOTAENIIDEA

Scolex with single sucker. Strobila cylindrical, few proglottids. Vitellaria compact. Uterus closed. In fresh water fish.

Nippotaenia

Order PSEUDOPHYLLIDEA

Scolex with bothria. Vitellaria scattered. Uterus with pore. In teleosts and fish-eating vertebrates.

Diphyllobothrium,
Schistocephalus

Order CARYOPHYLLIDEA

Scolex poorly developed. Non-strobilate, one set of reproductive organs, pseudophyllid in pattern. In fresh water oligochaetes and fish.

Archigetes

Order CYCLOPHYLLIDEA

Scolex with four suckers and often a hook-bearing rostellum. Vitellaria compact. Uterus closed. In birds and mammals, occasionally in amphibia and reptiles.

Taenia, Hymenolepis

5.6.3. CESTODA

The first four orders enumerated above show a rigid host specificity and occur only in the spiral valve of elasmobranch fish. This specificity is reflected in physiological adaptations and, in the case of the Tetraphyllidea (Fig. 22) by morphological adaptations, the structure of the scolex being intimately related to the nature of the host's intestinal mucosa. Considerable variation in scolex structure is found in these orders, whereas the orders Pseudophyllidea and Cyclophyllidea show relatively little variation. The basic organization of the cestode body is best studied in pseudophyllid and cyclophyllid cestodes, which are the most easily obtainable forms.

(a) *Cyclophyllidea*

Taenia pisiformis occurs as an adult worm in the intestines of dogs, which acquire the infection by eating intermediate hosts (rabbits) infected with the larval stages. A description of this species may be taken as representative of the order.

The body of the adult worm consists of a small scolex, bearing four cup-shaped suckers and an armed rostellum (Fig. 23), a short undifferentiated neck region and a strobila of proglottids, the most anterior of which are small and immature. The proglottids show progressive maturation along the length of the strobila,

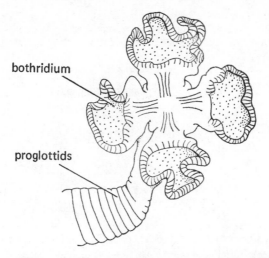

bothridium

proglottids

Figure 22 *Phyllobothrium (Anthobothrium)* scolex

the male reproductive system maturing before the female (protandry). After fertilization the female system atrophies, the uterus enlarges and becomes filled with eggs. These gravid proglottids break free from the strobila, pass out with the host's faeces and release the eggs by disintegration.

A stained preparation of a mature proglottid (Fig. 23) shows the lateral and transverse excretory canals and the lateral nerve cords, both systems running along the strobila from the scolex which contains the protonephridia and nerve ganglia. The male and female systems are also visible. The ovary is a prominent, bilobed organ, almost enclosing the ootype in which egg formation occurs. Below the ovary lies the compact vitelline gland. The uterus is small in mature proglottids and extends anteriorly from the ootype. Both male and female systems open, via a common genital atrium, at the lateral genital pore. A narrow vagina passes back from the pore and enters the oviduct, enlarging to form a seminal receptacle a little anterior to the junction. The numerous testes of the male system are situated laterally; sperm are conveyed by fine ducts to the vas deferens, which opens into the copulatory cirrus. Reproduction normally occurs by self fertilization.

The eggs of *T. pisiformis* (Fig. 23) can be seen in the posterior gravid proglottids. When ingested by the intermediate host the thick embryophore is broken down by the action of enzymes and the hexacanth escapes from the lipoidal membrane, penetrates the intestinal wall, enters the blood stream and is carried to muscle and connective tissues, in which it develops to the cysticercus stage. The cysticerci are frequently found in the mesenteries of laboratory rabbits, appearing as opaque white spheres. With strong transmitted light the region of the inverted scolex may be seen, but it is necessary to dissect away the wall of the cysticercus, or to flatten and stain the cysticercus, in order to see the scolex clearly. When ingested by the final host the scolex is evaginated and becomes attached to the intestinal wall. Proglottids then develop from the neck region.

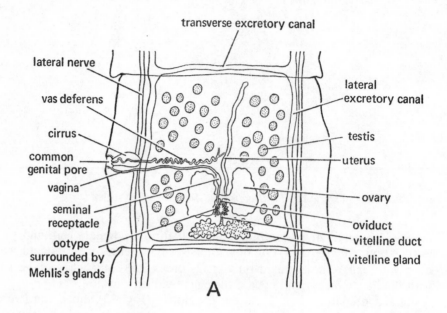

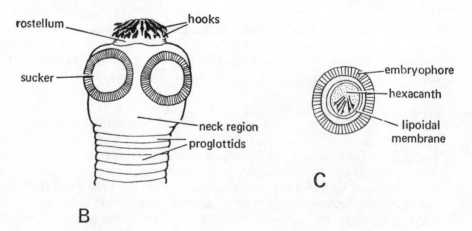

Figure 23 *Taenia pisiformis*, A, mature proglottid; B, scolex; C, egg

Living cyclophyllidean cestodes are easily obtained from the small intestines of laboratory rodents. The species commonly found in the rat and mouse is *Hymenolepis nana*, a cestode uniquely capable of completing its life cycle in the absence of an intermediate host, although arthropod hosts can be employed. Live worms should be studied in saline and the degree of motility of the scolex observed. The organization of the hymenolepid proglottid differs from that of the taeniid type in that the testes are compact and usually number three.

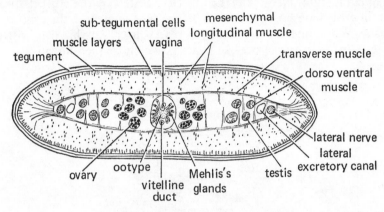

Figure 24   *Taenia* sp. T.S. proglottid

In transverse section (Fig. 24) cyclophyllidean cestodes show a basic platyhelminth organization, together with some characteristic specializations. The tegument is visible as a thick, apparently homogenous layer covering the outer surface, beneath lie the layers of circular and longitudinal muscle. Elongate cells below the longitudinal layer connect with the tegument. The loose parenchyma contains many inclusions and is divided into inner and outer regions by a layer of transverse muscle which surrounds the reproductive organs. The lateral excretory canals and nerve cords are situated just within this layer. In mature proglottids the lateral testes are visible in all sections, but the ovary and vitelline gland are found only in sections taken posteriorly. Sections through gravid proglottids show little other than the enlarged uterus and the enclosed eggs.

(b) *Pseudophyllidea*

Diphyllobothrium latum, the broad fish tapeworm, is the most familiar example of this order and is a common parasite of man in certain countries. Infection is acquired by eating fish which contain infective larval stages. The adult worm may reach a length of several metres.

The general organization of the body is similar to that of *T. pisiformis*, but the scolex is relatively undifferentiated and bears two weak bothria. The strobila contains a number of proglottids which have reached sexual maturity simultaneously, producing eggs which are liberated through the uterine pore into the host's intestine.

The structure of the reproductive system in *D. latum* is shown schematically in Fig. 25. The major differences from the taeniid pattern are the scattered vitellaria, the presence of the uterine pore and the mid-ventral genital pore. The eggs of pseudophyllid cestodes are exposed to the enzymes of the host's intestine before release to the external environment and therefore have a protective quinone-tanned shell. Unlike the eggs of *T. pisiformis* those of *D. latum* are operculate and hatch in an aquatic environment to release a ciliated free-swimming coracidium larva. This has to be ingested by the correct species of crustacean intermediate host (copepod) before development to the procercoid larva can occur. When infected copepods are eaten by a suitable fish host the procercoid develops into a plerocercoid larva, which is then infective to the final host.

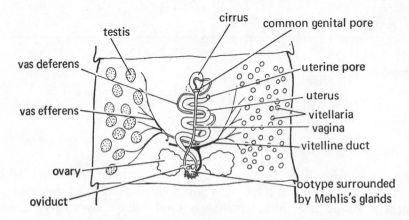

Figure 25   *Diphyllobothrium latum*. Mature proglottid. Scheme of reproductive organs (modified after Fuhrmann)

Plerocercoid larvae of *Schistocephalus solidus* may be obtained from naturally infected sticklebacks, which can be recognized by the swollen abdomen caused by the presence of the parasite in the body cavity. The final hosts of this species are fish-eating birds, but sexual maturity can be achieved relatively easily by *in vitro* culture (see Taylor and Baker, 1968).

5.7.   Digenea

5.7.1.   INTRODUCTION

Fasciola hepatica (the large liver fluke) is commonly described as an example of a digenean, but it is to some extent atypical in its large size and branched internal organs. The characteristic morphology of the group is more easily studied in smaller flukes and the detailed description (5.7.3., page 80) is based upon *Dicrocoelium dendriticum* (small liver fluke). On the other hand *F. hepatica* is suitable for the preparation of sectioned material and gives a good picture of the histological structure of digeneans (5.7.4., page 82).

Trematodes are readily available from animals commonly dissected in the laboratory and those from frogs (*Rana temporaria* and *R. esculenta*) are particularly suitable. Freshly caught frogs are best for this purpose. The frogs should be killed, the body cavity opened and the lungs, intestine and urinary bladder removed into saline. Each of these organs and the buccal cavity of the frog should be carefully examined for the presence of parasites, where possible checking under a microscope for small individuals. The worms should be observed *in situ*, removed carefully, washed, fixed and stained. Many species have been recovered from frog hosts; Dawes (1946) should be consulted for identification.

Larval stages of various digeneans can be obtained from common marine and fresh water gastropod snails by dissecting out and teasing apart the digestive gland. Sporocyst, redial and cercarial stages may be found and should be studied live, and after vital or permanent staining. Sporocysts and rediae show relatively little differentiation, apart from the oral sucker and short intestine in the latter, but contain balls of germinal cells which develop into the subsequent larval stage. Cercariae show many of the adult characteristics —oral and ventral suckers, pharynx and developing bifid intestine—as well as specifically larval characteristics. A key to cercariae from British freshwater molluscs is given by Nasir and Erasmus (1964).

Metacercarial larvae of *Diplostomum phoxini* commonly occur in the brain of the minnow and can be cultured *in vitro*, in glucose-saline mixtures at 40°C. Partial development of the adult body form can be achieved, complete development requires a more nutrient medium (see Taylor and Baker, 1968). The final hosts of this species are ducks.

5.7.2. CLASSIFICATION

The classification of digenean platyhelminths has been based largely upon characteristics of the adult worms and the major families are well established. There is no generally accepted scheme involving categories of higher taxonomic status, although some have been proposed on the basis of developmental characteristics. However, as students will generally be concerned with the adult organism, the established families provide a convenient classificatory system. Only those families which include species of particular interest from a zoological or economic standpoint, or species which are commonly encountered in laboratory animals are given below. The host range of each family is indicated by the letters F=fish, A=amphibia, R=reptiles, B=birds and M=mammals.

Class DIGENEA

Endoparasites. Ciliated epidermis lost in adult, replaced by well developed, metabolically active tegument. Adhesive organs usually an anterior oral and a mid-ventral sucker. Mouth anterior, intestine bifid. Excretory pore posterior. Hermaphrodite, with few exceptions, life cycle complex with several larval stages and involving at least two hosts, the first always a mollusc, the last a vertebrate.

Bucephalidae (F)
Mouth mid-ventrally placed, gut sac-like. In intestine.
Bucephalus

Strigeidae (RBM)
Body divided into anterior and posterior regions. Suckers feeble, large accessory adhesive organ. Life cycle often complex. In intestine.
Strigea

Diplostomatidae (BM)
Similar to above.
Diplostomum

Schistosomatidae (BM)
Unisexual. Elongate. Male with groove for holding female. Suckers feeble. In blood system.

Schistosoma

Echinostomatidae (BM)
Head bears stout spines. In intestine.

Echinostoma

Fascioloidae (M)
Large. Internal organs branched. In intestine.

Fasciola

Paramphistomatidae (FARBM)
Body thick and fleshy. Ventral sucker posterior. In intestine.

Paramphistomum,
Diplodiscus

Notocotylidae (BM)
Ventral sucker absent. Groups of gland cells on ventral surface. In intestine.

Notocotylus

Plagiorchidae (FARBM)
In lungs and intestine.

Dolichosaccus, Haplometra,
Macrodera, Plagiorchis

Dicrocoelidae (ARBM)
Small. Body translucent. Ovary behind testes. In liver and intestine.

Dicrocoelium, Brachycoelium

Troglotrematidae (BM)
Body fleshy. Suckers feeble. In cysts in sinuses, lungs and skin.

Paragonimus, Troglotrema

Gorgoderidae (FAR)
Small. Suckers powerful, ventral sucker well developed. In urinary system.

Gorgodera, Gorgoderina

Opisthorchidae (RBM)
Small. Suckers feeble. In liver.

Opisthorchis

Heterophyidae (BM)
Small. Suckers feeble, accessory genital sucker. In intestine.

Heterophyes, Metagonimus,
Cryptocotyle

Note

The Aspidogastrea, a small group parasitic in fish, chelonians, molluscs and crustaceans have not been included in the classificatory scheme. Although usually and variously classified with the Digenea they differ in several respects, notably the possession of a very large, divided adhesive organ which occupies the ventral surface, and may merely show convergence.

5.7.3. MORPHOLOGY: *DICROCOELIUM*

The body of *D. dendriticum* (Fig. 26) is elongate and tapers markedly to the anterior end. By careful focussing on the margins of the body it should be possible to distinguish the tegument and underlying muscle layers.

The large, slightly lobed testes lie posterior to the ventral sucker. It is normally not possible to see the vas deferens which conveys sperm from the testes to the seminal vesicle, the latter opening into the copulatory cirrus. Posterior to the testes lies the ovary and on each lateral margin of the body lie the vitellaria. Right and left vitelline ducts unite to form a median duct which joins the oviduct immediately before the latter opens into the ootype. During cross fertilization sperm are passed along Laurer's canal and stored in the seminal receptacle. Eggs are formed in the ootype, shell material being provided by the vitelline cells, and are stored in the uterus, which runs posteriorly and then anteriorly to open at the genital pore. The eggs show the typical digenean operculate condition (Fig. 26B).

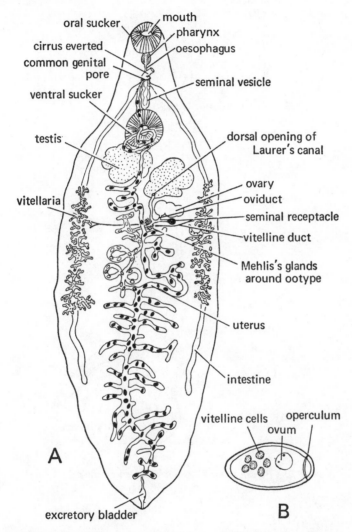

Figure 26 *Dicrocoelium dendriticum.* A, Ventral view; B, egg

In the life cycle of *D. dendriticum* the eggs are ingested by terrestrial snails, in which hatching takes place. The cercariae are released in slime balls and develop into infective metacercariae in ants. The final host is infected by ingestion of these intermediate hosts.

Variation in the morphology of adult digeneans is relatively limited and most species are referable to the pattern described in *Dicrocoelium*. The most important exceptions are the schistosomes, elongate and unisexual blood flukes, which live in the mesenteric veins of the final host and in which the adult male and female live in permanent copula (Fig. 27). Schistosomiasis is one of the most important parasitic diseases of man.

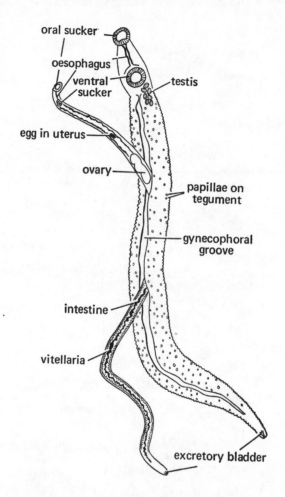

oral sucker
oesophagus
ventral sucker
egg in uterus
ovary
testis
papillae on tegument
gynecophoral groove
intestine
vitellaria
excretory bladder

Figure 27 *Schistosoma mansoni*. Female and male *in copula*

5.7.4. HISTOLOGY: *FASCIOLA*

The basic platyhelminth organization of the Digenea, together with the specializations typical of the class, is well shown in transverse sections of *Fasciola hepatica* (Fig. 28). The tegument is visible as a thick covering to the body and contains prominent spines. Below the tegument are the outer circular, middle oblique and inner longitudinal muscle layers; muscles also run dorsoventrally, particularly in the lateral regions of the body. As in triclad turbellaria there are prominent longitudinal nerve cords above the ventral muscle layer.

As a result of the extensive branching of the intestine and testes in *Fasciola*, portions of these organs will be found in many sections through the body. The intestinal wall consists of a thin epithelial layer of irregular cells, surrounded by fine muscle layers. The branches of the testes appear in section as small clusters of sperm-forming cells. Sections through the ovary will be found in the anterior third of the body. In mature specimens the folds of the uterus are extensive and occupy many of the internal spaces of the body.

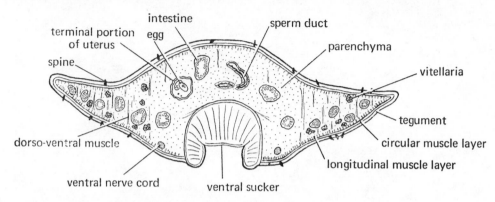

Figure 28 *Fasciola hepatica.* T.S. through ventral sucker

If living *Fasciola hepatica* are available the excretory system can be demonstrated by injecting a suitable dye, such as Prussian Blue, into the posterior excretory pore by means of a fine hypodermic needle or glass capillary. The animal should then be fixed and mounted.

Johri and Smyth (1956) have given a useful account of histochemical methods which may be used in the study of helminth morphology.

6. NEMERTINES

6.1. Introduction

The nemertines have elongate, somewhat flattened and highly extensile bodies. They are commonly recognizable by the proboscis which lies in a cavity dorsal to the gut. The anus is terminal. Nemertines are usually dioecious and have a protonephridial excretory system. Most of them are free-living and are found mainly in the sea.

Living specimens can be maintained for considerable periods in the laboratory and if possible living animals should be examined and the general movement and extensible nature of the body observed in addition to external features such as the anterior eyes, sensory cephalic grooves, mouth and proboscis openings. The cerebral ganglia and cerebral organs may also be visible in unpigmented specimens.

6.2. General Methods

Methods for examination of nemertine worms are essentially similar to those used for Platyhelminthes (p. 66). Animals may be narcotized by adding a solution of 14% $MgCl_2$ to the water. The proboscis may be everted by narcotizing the animal and subjecting it to gentle pressure between slides. Fixation can then be carried out and stained whole mounts made. The smaller, less pigmented species are the most suitable for study, although darker forms such as *Lineus* can be used if the specimen is bleached prior to staining.

6.3. Classification

The phylum is relatively small, containing less than 700 species, and is divided into two classes.

Phylum NEMERTINA

Class ANOPLA

Mouth posterior to brain. Nerve cords subepidermal or within muscle layers. Proboscis unarmed.

Order PALAEONEMERTINI

Body wall muscles in 2 or 3 layers, if 3 then inner layer is circular. Littoral.

Tubulanus

Order HETERONEMERTINI

Body wall muscles in 3 layers, inner layer longitudinal. Littoral.

Lineus

Class ENOPLA

Mouth anterior to brain. Nerve cords internal to muscle layers.

Order HOPLONEMERTINI

Proboscis armed. Marine forms littoral and pelagic.

Tetrastemma, Prostoma (fresh water), *Geonemertes* (terrestrial).

Order BDELLONEMERTINI

Proboscis unarmed. One genus of commensal forms.

Malocobdella

6.4. Structure

In stained whole mount preparations most of the major body organs are visible (Fig. 29). The proboscis is long and extends well back into the body when retracted. It is attached anteriorly, in front of the cerebral ganglia, and can be everted through the narrow rhynchodaeum and proboscis pore by the contraction of muscles in the proboscis sheath, which increases pressure within the rhynchocoel. Retractor muscles attached to the posterior end of the proboscis help to bring about inversion. The proboscis in Hoplonemertini has three regions, the anterior two being separated by a muscular diaphragm which bears a large stylet. Secretions from the posterior region are passed through the narrow interconnecting canal. The proboscis is used both for the capture of prey and for defensive purposes.

In many Hoplonemertini the proboscis and digestive tract open at a common anterior pore. A short oesophagus leads into a wide stomach; the intestine gives rise to an anterior caecum. Numerous lateral diverticula from the intestine fill much of the interior of the body.

An anterior ring of nervous tissue, consisting of dorsal and ventral ganglia connected by commissures, surrounds the proboscis. Large lateral nerves pass posteriorly from the ventral ganglia.

Characteristic of nemertines are the chemosensory (and possibly neurosecretory) cerebral organs which are closely applied to, and innervated from, the dorsal ganglia.

The gonads in mature worms can be seen as rounded sac-like bodies lying between the intestinal diverticula and opening to the exterior by simple ducts. Development of the fertilized eggs is usually direct but in heteronemertines there may be a free swimming pilidium larva.

In marine nemertines the excretory system is generally little developed. Paired excretory tubules lie in contact with the lateral blood vessels and open at pores in the anterior region.

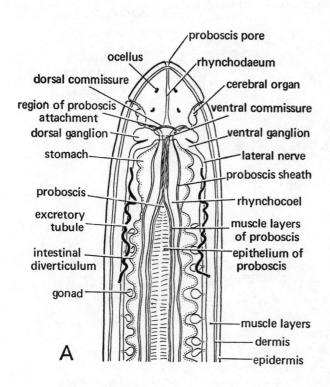

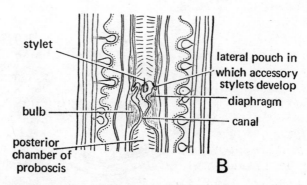

Figure 29 *Tetrastemma* sp. A, anterior end; B, stylet region

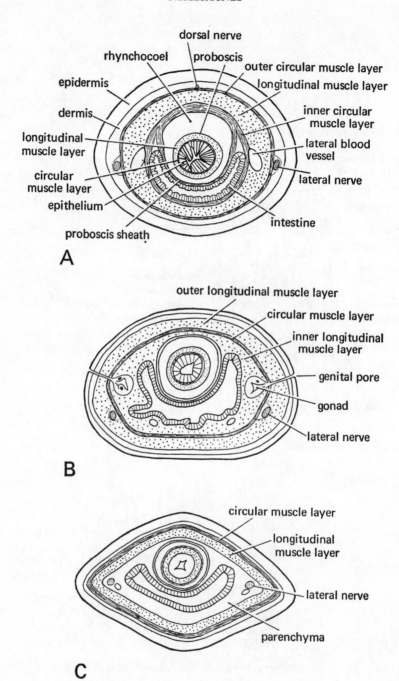

Figure 30 A, a palaeonemertine (*Tubulanus*). T.S.; B, a heteronemertine (*Lineus*) T.S.; C, a hoplonemertine (*Amphiporus*) T.S.

Transverse and longitudinal sections through various regions of nemertine worms should be studied in order to gain a complete picture of their anatomy. The orders show a fundamental similarity of structure, but each has a characteristic organization of dermal, muscular and nervous tissues (Fig. 30). In palaeonemertines the thick glandular epidermis rests on a well developed dermis, within which lie the nerve cords. Beneath the dermis are the circular and longitudinal muscle layers, an inner layer of circular muscle investing the proboscis and intestine. Heteronemertines show an inner and outer longitudinal muscle layer and a middle circular layer, the nerve cords lying in the outer muscle layer. In hoplonemertines there is an outer circular and an inner longitudinal muscle layer, the nerve cords lying internal to the latter. In many nemertines there is a layer of spiral muscle between the circular and longitudinal layers and this plays an important part in body movement. In each order the proboscis, being essentially an invagination of the body wall, has a similar, though reversed, arrangement of tissues.

The large lateral blood vessels which are seen in transverse section connect anteriorly and posteriorly, *via* large lacunae, and contain contractile elements in their walls.

7. PSEUDOCOELOMATES

7.1. Rotifers

7.1.1. INTRODUCTION

Rotifers are characterized by an anterior ciliated corona, used in locomotion and feeding. The pharynx is modified to form a muscular mastax bearing internal jaws.

A phytoplankton net (180 meshes per inch) towed through almost any body of freshwater will yield a variety of rotifers. The Freshwater Biological Association supplies preserved samples of Windermere plankton which contains a variety of rotifers, some of which are illustrated in Fig. 31. Bdelloid rotifers of the type illustrated in Fig. 31,A are often common among mosses. For further details see Voigt (1957) and de Beauchamp (1965).

A few strands of cotton wool and a cover slip will often serve to restrain a rotifer sufficiently for examination under a monocular microscope. Alternatively the cover slip may be supported by minute fragments of Plasticine and then pressed down just enough to trap the rotifer.

7.1.2. CLASSIFICATION

In modern classifications the rotifers are regarded as a class of the Phylum Aschelminthes. Within the class three orders can be distinguished.

Class ROTIFERA

    Order SEISONACEA

        Very elongated, with long neck region between corona and trunk. Males similar in size to females. Gonads paired, the ovaries without vitellaria. Epizoic on the gills of *Nebalia*.

                                        Seison

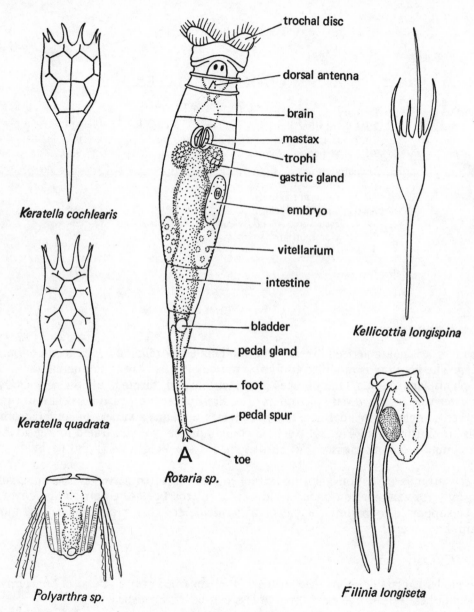

trochal disc

dorsal antenna

brain

mastax

trophi

gastric gland

embryo

vitellarium

intestine

bladder

pedal gland

foot

pedal spur

toe

Keratella cochlearis

Keratella quadrata

Polyarthra sp.

A

Rotaria sp.

Kellicottia longispina

Filinia longiseta

Figure 31 Rotifers

Order BDELLOIDEA

Corona usually with two trochal discs. Males unknown, females with two ovaries including vitellaria.

Rotaria, Philodina, Mniobia

Order MONOGONONTA

Females with a single ovary. Males usually much smaller than the females.

Keratella, Filinia, Polyarthra, Asplanchna

7.1.3. STRUCTURE: *Rotaria*

(a) *External anatomy*. At the anterior end there is a corona with two trochal discs. The stout cilia on the trochus serve both as a locomotory and as a feeding organ. Just behind the trochal disc are two red eyes, and in the mid-dorsal line behind the eyes a small projection forms the dorsal antenna. The trunk is cylindrical in form and tapers down to a narrow telescopic foot which has three toes at the end and two spurs a short distance from the end.

(b) *Internal anatomy*. When a living specimen of *Rotaria* is viewed by transmitted light the most conspicuous internal organ is the modified pharynx or mastax. This muscular organ contains the trophi which act as jaws. The mastax is linked to the food collecting apparatus of the corona by a ciliated tube. Immediately behind the mastax the gut dilates to form the gastric glands which open into the combined stomach-intestine. Some rotifers have the stomach and intestine visible as separate entities. The intestine opens via the anus in the mid-dorsal line near the junction of the trunk and foot. Before opening to the exterior the intestine receives the oviducts and the protonephridial tubules. The latter open into a bladder, and in a living specimen the bladder can be seen contracting at regular intervals of a minute or two.

The only part of the nervous system that is easy to see is the brain lying just in front of the mastax.

Rotaria is a member of the Bdelloidea, so that it has two germovitellaria. It is also viviparous, and it is sometimes possible to see a young rotifer inside its mother alongside the intestine.

The foot contains pedal glands which open on the toes, and secrets an adhesive substance which enables the rotifer to attach itself temporarily to a substratum.

7.2. Nematodes

7.2.1. INTRODUCTION

The nematodes are among the most abundant of animals and occur in every habitat. The majority of species are free-living, but many are parasitic in plants and animals. The relatively uniform morphology of the class is a consequence of the possession of a fluid-muscle hydrostatic skeleton based on the antagonism of the cuticle and longitudinal musculature.

The nematodes are commonly associated with the rotifers to form the Phylum Aschelminthes.

7.2.2. GENERAL METHODS

Permanent preparations of nematodes are not satisfactory and are not necessary for the study of the structure of small, transparent species. Larger specimens, fixed in 10% formalin, may be cleared and mounted in lactophenol plus cotton blue.

The characteristic morphology of nematodes is most easily studied in small transparent species. Rhabditid nematodes, which can be obtained without difficulty, are particularly suitable. Earthworms are commonly infected with larval stages of *Rhabditis maupasi*, the adults of which are free-living saprobionts in soil. The larvae lie free in the nephridia or encysted in the coelom or coelomic 'brown bodies'. Development of these larvae to the adult is inhibited in the earthworm, presumably by inadequate nutrition. After the death of the host the larvae feed on bacteria in the decaying tissues and mature rapidly. In the presence of adequate nutrition the cycle can continue outside the host. Reinfection of the host earthworm is thought to take place possibly through the nephridiopores, coelomic pores and genital openings. Male *R. maupasi* occur less frequently than female and it has been suggested that functional male and female adults co-exist with protandric hermaphrodites.

Cultures of the nematode can be set up by allowing pieces of earthworm body wall, containing infected nephridia or seeded with brown bodies, to decay on agar or moist filter paper in Petri dishes. After a week or so all stages in the life cycle of the parasite should be present.

Living worms show the dorso-ventral flexures of the body, characteristic of almost all nematodes, and on suitable surfaces the translation of these movements into forward progression can be observed. Worms should be killed by heat before further examination.

7.2.3. CLASSIFICATION

The classification of the nematodes is still largely one of convenience; only relatively recently have attempts been made to provide a phylogenetic basis for classification. That given below is not complete, including only the more familiar groups, and is based upon the schemes given in Chitwood (1950) and Grassé [Vol. 4 (3)].*

Two subclasses are recognized, although the diagnostic characters of each are not easily identifiable, being based primarily on the nature of small organs anteriorly (amphids) and posteriorly (phasmids).

Class NEMATODA

Elongate, cylindrical, unsegmented, pseudocoelomate. Cilia absent. Body covered by cuticle, often of complex nature. Body wall musculature longitudinal only. Excretory system glandular. Sexes usually separate.

Sub-class ADENOPHOREA

Amphids well developed. Phasmids absent. Caudal and hypodermal glands present. Excretory organ without lateral canals. Free-living in all habitats and parasitic.

Infra-class CHROMADORIA

Free-living forms.

Infra-class ENOPLIA

Order DORYLAIMIDA

Free-living and some plant parasitic forms.

*Useful keys to nematodes parasitic in vertebrates are published by the Commonwealth Institute of Helminthology, St. Albans, Herts, U.K.

Order ENOPLIDA

Super-family MERMITHOIDEA
Pharynx thin, embedded in cells. Larvae parasitic in invertebrates, adults free-living.
Mermis

Super-family TRICHUROIDEA
Pharynx as above. Eggs lemon-shaped with polar plugs. Single ovary in female. Parasites of vertebrates.
Trichuris, Trichinella,
Capillaria

Super-family DIOCTOPHYMATOIDEA
Large forms parasitic in kidneys of vertebrates.
Dioctophyma

Sub-class SERCENENTEA
Amphids poorly developed. Phasmids present. Caudal and hypodermal glands absent. Excretory organ with lateral canal or canals. Saprobiotic or parasitic.

Order RHABDITIDA
Small transparent forms. Pharynx with posterior bulb and often a pseudobulb. Free-living, saprobiotic, parasitic.
Rhabditis, Rhabdias

Order TYLENCHIDA
Important plant parasites. Small forms. Mouth armed with stylet. Super-families Tylenchoidea, *Ditylenchus, Heterodera,* and Aphelenchoidea, *Aphelenchoides.*

Order STRONGYLIDA
Animal parasites. Male with copulatory bursa supported by rays.

Super-family ANCYLOSTOMATOIDEA
Large buccal capsule with teeth or cutting plates.
Ancylostoma, Necator

Super-family STRONGYLOIDEA
Large buccal capsule with or without teeth.
Strongylus, Syngamus

Super-family TRICHOSTRONGYLOIDEA
Small buccal capsule. Intestinal parasites.
Trichostrongylus, Haemonchus

Super-family METASTRONGYLOIDEA
Small buccal capsule. Lung parasites.
Metastrongylus, Dictyocaulus

Order ASCARIDIDA
Mouth characteristically with three lips. Intestinal parasities of arthropods and vertebrates.

Super-family ASCARIDOIDEA
Often large forms. Lips large. Intestine often with caecum.
Ascaris, Ascaridia

Super-family HETERAKOIDEA
Lips small. Precloacal sucker present in male.
Heterakis

Also included in the order are the super-families Cosmocercoidea, Seuratoidea and Subuluroidea.

Order OXYURIDA
Pharynx with posterior bulb. Parasites of large intestine.
Super-family OXYUROIDEA
Oxyuris, Enterobius,
Syphacia

Order SPIRURIDA

Pharynx of two parts, anterior muscular and posterior glandular. **Parasites** of arthropods in larval stages and of vertebrates when adult.

> **Sub-order** CAMALLANINA
> Intermediate hosts copepods.
>
> **Super-family** DRACUNCULOIDEA
> Long slender tissue parasites, female viviparous.
> *Dracunculus*
>
> *Also* super-family Camallanoidea
>
> **Sub-order** SPIRURINA
> Intermediate hosts usually insects.
>
> **Super-family** FILARIOIDEA
> Long slender tissue parasites, female releases microfilariae.
> *Wuchereria, Loa*
>
> *Also* super-families Spiruroidea and Physalopteroidea.

7.2.4. STRUCTURE: *Rhabditis* and *Ascaris*

(a) *Rhabditis*

The cuticle of the worms is quite transparent and allows the internal organs to be studied without further preparation. (Fig. 32). The female may be distinguished by its larger size, vulval opening and presence of eggs; the male by the cuticular expansion around the cloaca. The digestive and reproductive systems are the most obvious body organs, but it is sometimes possible to see the excretory pore and terminal part of the excretory duct. Little can be seen of the nervous system other than the circumpharyngeal ring.

In both sexes the digestive system consists of an anterior pharynx, long intestine and terminal rectum. The pharynx is muscular and functions as a suctorial organ, a valvular apparatus in the end bulb regulating the passage of food into the intestine. In the male the terminal part of the system joins with the vas deferens and opens at a common cloaca.

The male reproductive system consists of a single testis, reflexed upon itself and continuous with the vas deferens. Ventral to the cloaca is a small pouch which contains two sclerotized spicules, used during copulation. The female system contains two ovaries, continuous with the uteri, the parts being arranged symmetrically on either side of the vulva.

(b) *Dissection of Ascaris*

Further study of the internal anatomy of nematodes should be carried out using the large roundworm of the pig *Ascaris suum*. After examination of external features such as the mouth and lips, ventral excretory pore immediately posterior to the mouth, lateral chords, female ventral vulva, male cloaca and spicules, open the body cavity by a dorsal incision (dorso-lateral in the male) and pin out the body wall. Identify the component parts of the digestive and

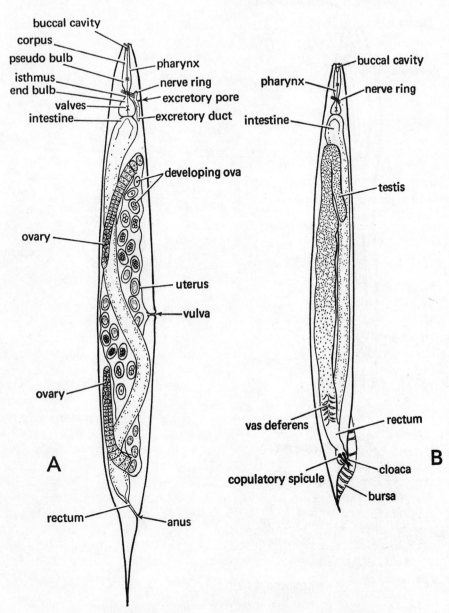

Figure 32 *Rhabditis maupasi*. A, female; B, male

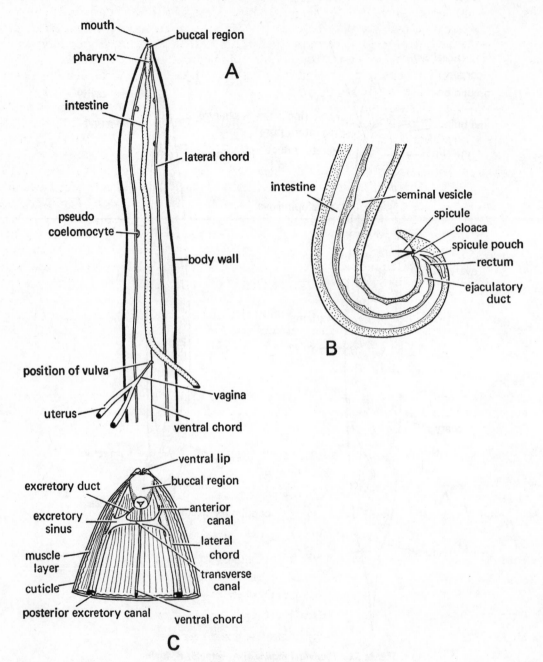

Figure 33 *Ascaris suum*

A, female: dissection of anterior region; B, male: dissection of posterior region; C, female, excretory organ

reproductive systems. The gonads in each sex are very long and convuluted and are continuous with the terminal ducts. As in *Rhabditis* the female gonads are paired and the male single (Fig. 33). Remove the digestive tract posterior to the buccal region and trace the excretory organ and duct, if necessary using a binocular microscope. The pseudocoelomocytes, large cells attached to the lateral chords, may be excretory in nature, although other metabolic functions have been attributed to them.

When the dissection has been completed, remove a portion of the body wall and strip off a layer of cuticle. Under the microscope it is possible to see the three fibre layers, which form a "lazy tong" system and limit radial expansion of the body.

In transverse sections of *A. suum* (Fig. 34) it is possible to study the relationships of the cuticle, hypodermis and longitudinal muscle cells. Several layers of the cuticle should be visible. The underlying hypodermis forms a thin layer beneath the cuticle, except at the dorsal, ventral and lateral chords, in which are concentrated the hypodermal nuclei, nerve cords and the lateral excretory canals. The peculiar structure of the muscle cells, with a protoplasmic and a fibrillated region, can be clearly seen. Anterior sections show the triradiate lumen and muscular wall of the pharynx; more posteriorly the thin-walled intestine, with the characteristic "brush border" of microvilli, is visible. The tubular gonads can be followed along their lengths and all stages in the formation and maturation of gametes seen.

7.2.5. COMMON HOST SOURCES

Parasitic nematodes of other nematode groups are easily obtainable from laboratory animals and from animals commonly dissected. The majority of those listed below are small and can be examined directly.

Host animal	Nematodes
Cockroach	Several Oxyuroidea in intestine
Dogfish	*Proleptus* (Physalopteroidea) in spiral valve
Frog	*Oswaldocruzia* (Trichostrongyloidea) in intestine
	Rhabdias (Rhabditida) in lungs
Fowl	*Ascaridia* (Ascaridoidea) in intestine
	Heterakis (Heterakoidea) in caeca
	Capillaria (Trichuroidea) in intestine
Pigeon	*Capillaria* (Trichuroidea) in intestine
Mouse	*Aspiculuris* (Oxyuroidea) in colon
	Syphacia (Oxyuroidea) in caecum
Rat	*Trichosomoides* (Trichuroidea) in urinary bladder
Rabbit	*Graphidium* (Trichostrongyloidea) in intestine
	Passalurus (Oxyuroidea) in caecum

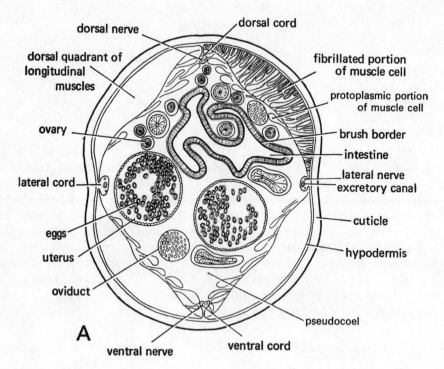

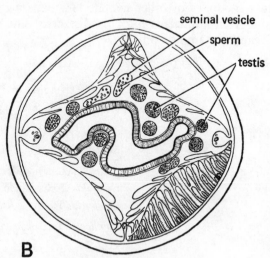

Figure 34 *Ascaris suum*, A, female, T.S.; B, male, T.S.

7.3. Acanthocephala

Acanthocephala may be obtained from the intestines of many vertebrate animals, fresh-water fish, amphibia and birds are good sources of material. Larval stages can frequently be found in freshwater gammarids. Owing to the opacity of the body wall, little internal structure can be seen in untreated specimens. Care must be taken to evert the proboscis before fixation. This can be achieved by menthol narcotization followed by gentle pressure. Acanthocephala can be fixed in A.F.A. or glacial acetic acid followed by 70% alcohol and stained by the acetic haematoxylin method. Unstained fixed specimens cleared in methyl salicylate show internal organs quite clearly. The process of clearing can be expedited by making a small hole in the body wall to allow rapid penetration.

The following description is based upon specimens of *Prosthorhynchus* a palaeoacan-thacephalan from the intestine of blackbirds. (Fig. 35).

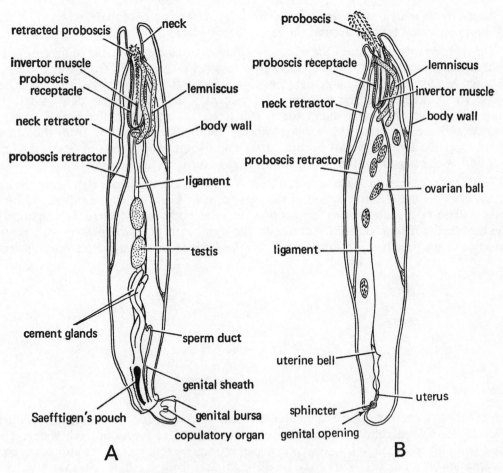

Figure 35 *Prosthorhynchus* sp. A, male; B, female

The body is divided into an anterior presoma, consisting of the retractile proboscis and neck region, and the posterior metasoma or trunk. Covering the body is a thick body wall made up of an outer cuticle (with epicuticle), three underlying layers and a basement membrane. The lowest of the three layers contains nuclei and an inter-connecting system of lacunae and is considered by some workers to be the epidermis. Below the basement membrane are circular and longitudinal muscle layers. The body wall is intimately involved in the uptake of nutrients from the host's intestine.

The proboscis acts as the organ of fixation and in life is embedded in the intestinal mucosa of the host. Eversion is produced by contraction of muscles in the proboscis receptacle increasing the internal fluid pressure, while inversion and retraction are brought about by the combined action of the invertor and retractor muscles. The lemnisci may act as reservoirs for the fluid contained in the proboscis apparatus, although a function in nutrition has also been suggested. The ganglia of the nervous system are situated in a ganglionic mass on the ventral wall of the proboscis receptacle.

An excretory system is present only in the order Archiacanthocephala, where it consists of protonephridial tubules around the reproductive system.

The reproductive systems of the male and female worms are contained in ligament sacs. In the female the ovarian tissue breaks down to form ovarian balls from which the eggs develop. Mature females contain many eggs lying free in the body cavity. The male system consists of paired testes and a complex of cement glands. Sperm are conveyed to the muscular copulatory organ along the sperm duct. In copulation the genital bursa is everted and used to grasp the posterior end of the female. Secretions from the cement glands seal the female genital opening after copulation. The pouch of Saefftigen may function as a reservoir for the fluid involved in the eversion of the bursa.

Fertilized eggs mature to the acanthor stage before release, immature eggs being retained in the body by the selective action of the uterine bell. When the eggs are ingested by an intermediate host the acanthor larva bores into the body cavity, there developing into the acanthella. The acanthella matures to the cysticanth or juvenile form and is then infective to the final host. In *Prosthorhynchus* the intermediate host is a terrestrial isopod.

7.4. Endoprocts

(*Kamptozoa, Calyssozoa*)

The endoprocts are often encountered when searching for small animals amongst hydroids and weed collected from the seashore. A few are found in fresh water. They are commonly attached to hydroids and algae, or may be found on other animals such as sponges and crustaceans. They may be distinguished from the lophophorate bryozoans by their habit of rolling the tentacles inwards (as in hydroids) when they retract (Fig. 36).

Endoprocts are best examined alive. They may be narcotized with MS 222 or with

menthol. Bouin, Helly and Zenker all fix well. Paraffin wax sections colour well with
Azan. Vertical sections are most informative. For details see Grassé, Vol. 5(1) and
Hyman, Vol. 3. For feeding, see Atkins (1932).

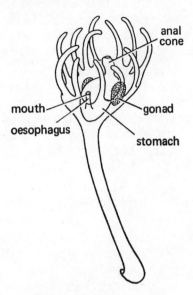

Figure 36 *Loxosoma* sp.

8. ANNELIDS

8.1. Introduction

Species of ragworm (*Nereis*) and *Nephthys* (=*Nephtys*) are of world-wide distribution and are generally obtainable from sand and muddy sand or under stones, according to species. *Nephthys* is characteristic of 'clean' (unstable) sand; *Arenicola* (or *Abarenicola*), the lugworm or lobworm, is characteristic of muddy sand and is usually abundant where it occurs. *Sabella* or related genera may be locally abundant in muddy sand or attached to pier-pilings. *Pomatoceros* is also widely distributed and is very common on stones or rocks between tidemarks on British shores. Various species of *Amphitrite*, *Lanice* or *Terebella* are locally common; *Myxicola* and *Chaetopterus* are of more variable availability. The oligochaete *Tubifex* is abundant in foul ponds and rivers, and is commonly available from aquarists ('bloodworms'). Various leeches may be found in freshwater, but the larger *Haemopis* and *Hirudo* (the medicinal leech) suitable for dissection, are best got from professional suppliers. These are the annelids perhaps most suited for laboratory study.

Earthworms are commonly studied in elementary classes and their basic morphology is not described here. *Eisenia foetida* may be mentioned as an example of the group obtainable in large numbers from rotting garden refuse. It is readily cultured and has easily recognizable cocoons. For ideas for physiological work, see Laverack (1963).

8.2. General Methods

Large annelids are best dissected fresh. They may be killed with chloroform or by addition of alcohol to the water so as to bring the concentration to 30%. If killed in this latter way the worms should be removed as soon as movements cease. Dissections are best done under water, seawater or invertebrate saline.

Annelids are best narcotized with MS 222. 0.75% w/v in seawater is excellent for marine species. Magnesium chloride (1:1, 7.5% $MgCl_2$: sea water) and menthol also work well. Worms may be kept in 5% formalin for a short time, but for longer periods they are better kept in a mixture of 1:1:1, formalin: glycerol: 95% alcohol, but this mixture is expensive. Failing this, specimens should be transfered to 70% alcohol for storage. expensive. Failing this, specimens should be transfered to 70% alcohol for storage.

For histological work large specimens or parts are best fixed in Dubosq-Brasil rather than Bouin, though quite good results may be obtained with Bouin, and for small specimens or parts with Helly or Zenker. Tissues stain well with Azan, or with 0.1% aqueous toluidine blue.

8.3. Classification

The annelids are divided into four Classes: Polychaeta, Oligochaeta, Hirudinea and Myzostomaria. The last is sometimes included within the Polychaeta. For details, see Dales (1962), Day (1967), Fauchald (1977) for polychaetes; Stephenson (1930), Edwards and Lofty (1977) for oligochaetes; Mann (1962) for Hirudinea) Grassé, Vol. 5(1) for Myzostomaria.

Phylum ANNELIDA

Class 1. POLYCHAETA

Chaetae borne in sheaves, commonly in two groups (dorsal notochaetae; ventral neurochaetae). Parapodia prominent, biramous, uniramous or sometimes reduced. Prostomium commonly well developed, bearing eyes, tentacles, but highly modified in sedentary forms. Commonly dioecious. Usually marine; some estuarine, a very few in fresh water or even terrestrial (tropical).

Class 2. OLIGOCHAETA

Chaetae borne singly (various arrangements). No parapodia; rarely (some aquatic forms) with gills. Prostomium small, conical, without eyes or tentacles. Hermaphrodite (cross-fertilizing); arrangement of gonads and ducts specialized, segmentally restricted. Some segments forming a clitellum (seasonally prominent) to secrete cocoon. Usually fresh water or terrestrial; a few estuarine.

Class 3. HIRUDINEA

Without chaetae (one exception). Segments secondarily annulated. Anterior sucker around mouth, variably developed; posterior sucker prominent. Prostomium insignificant. Hermaphrodite (cross-fertilizing). Arrangement of gonads and ducts specialized. Some segments forming a clitellum (seasonally recognizable). Usually fresh water; a few terrestrial; some marine.

Class POLYCHAETA

The polychaetes are commonly divided into two groups, the *Errantia* and *Sedentaria*. This is somewhat arbitrary, and from a phylogentic point of view the many families are better considered as forming a series of more or less distinct Orders. For discussion of this see Dales (1962). The Errantia include the more actively motile worms; the Sedentaria the microphagous tube-dwellers and many permanent burrowers. For well-illustrated Keys to the families see Day (1967) and Fauchald (1977).

The main families are listed below. Those with common members are shown in heavy type.

1. **Phyllodocidae**
2. Alciopidae
3. **Tomopteridae**
4. Typhloscolecidae
5. **Aphroditidae**
   (Often split into several families)
6. Chrysopetalidae (Palmyridae)
7. **Glyceridae**
8. Goniadidae
9. Sphaerodoridae
10. Pisionidae
11. **Nephtydidae**
12. **Syllidae**
13. Hesionidae
14. Pilargiidae

15. **Nereidae**
16. **Eunicidae**
    (Often split into several families)
17. Amphinomidae
18. **Capitellidae**
19. **Arenicolidae**
20. Scalibregmidae
21. Maldanidae
22. Opheliidae
23. Sternaspidae
24. **Spionidae**
25. Disomidae
26. Poecilochaetidae
27. Longosomidae
28. Paraonidae
29. **Chaetopteridae**
30. **Sabellariidae**
31. Magelonidae
32. **Cirratulidae**
33. **Ariciidae**
34. Oweniidae
35. Pectinariidae
36. Ampharetidae
37. **Terebellidae**
38. Flabelligeridae (Chloraemidae)
39. Psammodrilidae
40. **Sabellidae**
41. **Serpulidae**

The most important are:

Errantia

Phyllodocidae
Slender crawlers; prostomium with 2 eyes; without palps; proboscis without jaws; 2–4 pairs of tentacular cirri; parapodia uniramous, cirri leaf-like.

Aphroditidae
Body with notopodial scales (concealed by hair-like chaetae in the sea-mice, revealed in scaleworms), scales alternating with notopodial cirri; palps elongate.

Tomopteridae
Pelagic, transparent; parapodia biramous paddles without chaetae. Prostomium fused to following segment which has long cirri supported by acicula.

Glyceridae
Body circular in cross-section, unpigmented apart from blood pigment. Prostomium slender, conical, with 4 very short antennae at the tip. Proboscis balloon-like, with 4 jaws.

Nephthyidae (Nephtydidae)
Body somewhat rectangular in cross-section. Prostomium small, pentagonal, with 4 very short antennae. Proboscis papillate, with 2 jaws. Parapodia biramous with a sickle-shaped recurved gill between.

Syllidae
Small; prostomium with simple palps which are sometimes fused, 3 antennae. 2 pairs of tentacular cirri. Parapodia uniramous.

Nereidae (ragworms)

Prostomium with 2 short antennae and commonly 4 small eyes, palps biarticulate. Peristomium with 2 pairs tentacular cirri. Proboscis with 2 toothed jaws and horny denticles (paragnaths).

Eunicidae (rockworms, palolos)

Proboscis ventral, with 4 or more jaws. Segments often with dorso-lateral gills. Prostomium with 0–7 (often 5) antennae; palps simple.

Sedentaria

Capitellidae

Small, very slender worms without appendages, easily breaking up. Hooded dentate hooks amongst the chaetae.

Arenicolidae (lugworms)

Segments annulated, middle region of body with gills; 'tail' often present. Proboscis papillate. Prostomium simple, reduced.

Maldanidae (bamboo worms)

Body with relatively few segments which are commonly much longer than broad. Head without appendages, obliquely truncate. Pygidium often with a corona of papillae, entire collar or obliquely truncate.

Spionidae

Single pair of anterior feeding tentacles ("palps"). Segments often with dorso-lateral gills. Both dorsal and ventral parapodial cirri lamelliform.

Chaetopteridae

Anterior tentacles present but short in *Chaetopterus*. Anterior segments with uniramous parapodia; those in the middle region biramous. Very heavy chaetae on the 4th segment.

Sabellariidae

Anterior segments forming an operculum with a crown of large terminal chaetae and a mass of feeding tentacles. Simple dorsal parapodial cirri forming gills. Tail reflexed.

Cirratulidae

Body circular in cross-section, tapering at both ends. Commonly many elongate tentacular gills arising dorso-laterally from many segments.

Ariciidae

Prostomium without appendages. Parapodia with short, simple, dorsal gills. Body more or less divisible into 2 regions, the anterior somewhat larger and flattened, the posterior narrower and cylindrical.

Amphictenidae (Pectinariidae)

Body delicate, short. Worm living in a conical tube built from sand and shell fragments. With an anterior comb of large chaetae, and a short leaf-like tail.

Ampharetidae

Body with 3–4 pairs of simple cirrus-like anterior gills. Feeding tentacles retractable within the mouth.

Terebellidae

Body with a large number of extensible feeding tentacles arising from prostomium. 1–3 pairs of (usually arborescent) gills immediately behind the tentacles.

Sabellidae (fan worms, feather-duster worms)

Prostomium expanded into a feathery crown. Tube built from sand and shell fragments cemented with mucus.

Serpulidae

Prostomium expanded into a feathery crown. one filament of which is commonly (not always) modified into an operculum. Tube calcareous.

Class OLIGOCHAETA

It is not proposed to describe in detail here the separate families of oligochaetes. Reference should be made to Stephenson (1930), Grassé Vol. 5(1) and identification guides such as those of Brinkhurst (1962) and Edmondson (1959). Families are separated by differences in arrangement of the reproductive organs and their ducts, and by their chaetae.

The Class may be divided into four Orders.

Order 1. PLESIOPORA PLESIOTHECATA

Openings of vasa deferentia on the segment following that containing the testes. Spermathecae in testicular segments or proximal to testicular segments.

Aelosoma, Tubifex, Nais

Order 2. PLESIOPORA PROSOTHECATA

Openings of vasa deferentia on the segment following that containing the testes. Spermathecae anterior to the testicular segments.

Enchytraeus

Order 3. PROSOPORA

Vasa deferentia opening on the testicular segments. If there are two adjacent testicular segments, then the openings occur on the segment containing the posterior pair of testes.

Lumbriculus, Branchiobdella

Order 4. OPISTHOPORA

Vasa deferentia opening several segments posterior to the testicular segments. Earthworms.

Lumbricus, Allolobophora, Eisenia

Class HIRUDINEA

The leeches are divided into four Orders. It is not proposed to enumerate here the different families. Reference should be made to Mann (1962). Keys to the families and to genera of marine leeches will be found in this book. For British freshwater leeches see Mann (1954b); for North American leeches see Edmondson (1959).

Order 1. ACANTHOBDELLIDA

With chaetae. A single genus found on salmonids in Finland and Russia.

Acanthobdella

Order 2. RHYNCHOBDELLIDA

Leeches provided with an eversible proboscis. Mouth small, in middle of oral sucker. Glossiphoniids (fresh water); piscicolids (fish leeches).

Glossiphonia

Order 3. GNATHOBDELLIDA

Leeches with 3 cutting teeth (sometimes reduced); Mouth large, almost confluent with rim of oral sucker. Commonly 5 pairs of eyes. *Hirudo* (medicinal leech). *Haemopis* (horse leech).

Order 4. PHARYNGOBDELLIDA

Similar to gnathobdellidans but pharynx jawless. 6–8 pairs of eyes.
Mainly freshwater leeches feeding on insect larvae worms and molluscs.

Erpobdella

8.4. Basic Structure

The body of all annelids consists of a series of segments bounded by a presegmental prostomium and a postsegmental pygidium. Some idea of the gross variations may be obtained by comparison of thick slices of hardened worms. Comparison of *Nereis, Nephthys, Arenicola*, earthworm and *Sabella* are instructive (Figs. 37, 38, 39), if it is desired to study these variations in greater detail some of these species at least are well documented. For muscles of *Nereis* see Mettam (1967); *Nephthys*, Mettam (1967), Clark and Clark (1960); *Arenicola*, Wells (1944); earthworm, Newell (1950), Arthur (1965); *Sabella*, Nicol (1930). Also instructive is a comparison between the parapodial development of bottom-living and epitokous "heteronereis". See Clark (1961) and Defretin (1949) for details. For general purposes it is best to compare the main structures of the segments; parapodia have complicated musculatures. Details are given in the papers cited above. For these purposes thick slices of whole segments are more informative than thin sections which, seen in isolation are often difficult to interpret.

8.4.1. METHOD FOR HARDENING AND SECTIONING

Harden specimens with 3% potassium dichromate, preferably for several weeks. Cut off single segments with a scalpel, or use a single-sided safety-razor blade. Cut several at different levels (both through, and between segments), and take from different parts of the body. Dehydrate, clear in benzyl alcohol (relatively cheap, but muscles become brittle) or cedar wood oil (expensive, but muscles remain somewhat more pliant). Mount in deep cells or support coverglass with 'Plasticine' columns or with rings of aluminium or hard plastic tube O.D. 7/8 inch (22 mm) to fit circular covers for permanent preparations; or, transfer to 1:1 alcohol: glycerol to preserve for demonstration in solid watch glasses, if preferred.

Segments cleared in benzyl alcohol may be dissected under polaroid (one piece supported between objective and specimen, the other beneath specimen which is rotated on stage). Segments may also be dissected while being stained with chlorazol black or methylene blue. Add dye from a fine pipette to stain smaller muscles and rinse off during dissection in alcohol. Details may be supplemented from thin sections. Embed in paraffin and cut at 10 μm, stain Azan or haematoxylin and eosin. Sections cut at 25 μm and coloured with haematoxylin alone and well differentiated, are also useful.

Segments may be separated by septa. Septa serve as complete or partial bulkheads and restrain increase in girth. Septa are lost in certain regions. In polychaetes, muscles connecting gut to body wall are derived from septa or from dorsal or ventral mesenteries; other muscles, not relating gut to body wall are oblique muscles or extrinsic parapodial muscles. The basic segmental arrangement is one of circular muscles outside and longitudinal muscle blocks inside. The septa when well developed subdivide the coelom, but not the body wall musculature. A comparison of thick transverse sections or whole segments of various polychaetes alongside whole preserved specimens can make an instructive demonstration of many morphological variations within the group.

In Polychaeta, the parapodia may be modified in various ways, especially as gills. The prostomium, being also asegmental, is modified in different worms, particularly for feeding. Demonstration of these variations will depend on specimens available.

8.4.2. SEGMENT OF *Nereis diversicolor*

The muscles, especially the oblique muscles , vary in detail from one species to another. Fig. 37 refers to *Nereis diversicolor*. The muscles are interpreted according to Mettam (1967). For details see that paper. For vascular system see Nicoll (1954). The longitudinal muscle blocks are paired, the ventral pair folded. This arrangement is characteristic of nereids. Circular muscles are not obvious in a transverse section of the body and are not easily recognized in thin sections (10 μm). They are best seen by (1) dissecting dorsally, removing the superficial vessels and following round and behind the parapodium on each side, when the circular muscles may be seen passing dorso-ventrally posterior to the extrinsic parapodial muscles; and by (2) longitudinal vertical section (preferably using a well hardened specimen) followed by dissection.

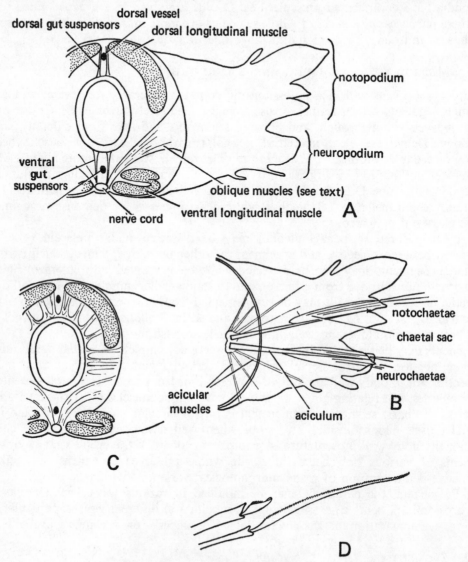

Figure 37 Segment of *Nereis diversicolor*
A, general aspect; B, acicula and muscles; C, septum (anterior face); D, chaetae: heterogomph falciger (above), homogomph spiniger (below)

In a transverse section the circulars ('dorso-ventrals') lie behind (*i.e.* posterior to) the oblique muscles and acicula. Mettam distinguishes 8 pairs of oblique muscles arising from the body wall mid-ventrally on each side of the ventral nerve cord and inserted around the base of the parapodia or between the parapodia; one on each side passes through the septum to the base of the parapodium behind. These muscles should not be confused with the diagonal muscles crossing the aperture to the parapodial coelom. The thin septa have mainly radial fibres. The dorsal and ventral gut suspensors are presumably derived from the dorsal and ventral mesenteries.

Chaetae are compound, with the distal parts short (falcigers) or long (spinigers), held in asymmetrical (heterogomph) shaft-heads or symmetrical (homogomph) shaft-heads.

If 'heteronereids' are available the form of the parapodia and chaetae of a metamorphosed segment may be compared with this basic pattern.

8.4.3. SEGMENT OF *NEPHTHYS HOMBERGI*

The longitudinal muscles of a segment from the mid-body region form a pair of closely apposed dorsal blocks and a pair of ventral blocks with the nerve cord between (Fig. 38A). The circular muscles are represented by the dorso-ventrals running round the sides of the dorsal longitudinals and behind (*i.e.* posterior to) the pseudoseptum. This forms a partial bulkhead and is derived from the largest oblique muscles. It is absent in the most anterior segments. In fresh animals glistening white striped ligaments will be seen arising from the connective tissue sheath, arising from two nodes of attachment in each segment and inserted in the intersegmental groove at the top of the neuropodium. They presumably brace the body against deformation. The ligaments may be confused with muscles in preserved specimens. For details see Clark and Clark (1960), and Mettam (1967).

8.4.4. SEGMENT OF *ARENICOLA MARINA*

The segments in the mid-body region are annulated, one annulus including the noto- and neuropodia; the others are simpler (Fig. 38B, C, 40).

In a simple annulus the circular and longitudinal muscles form almost complete cylinders; the longitudinals while divided into blocks are separated by narrow lines: ventral, dorsal, at the level of the notopodia, and along the line of the nephridiopores. Some longitudinal vessels follow these lines. Oblique muscles arise on each side of the nerve cord and are inserted in the circular muscles along each notopodial line. It is these muscles which are removed in a dorsal dissection to reveal the nephridia.

In a chaetigerous annulus the extent of the neuropodium and size of the notopodium differs according to position. In the posterior branchiate segments, the neuropodium forms a ridge in a ventro-lateral position enclosing a narrow groove from which the neurochaetae emerge. Anterior to the branchial region part of the neuropodium of each side extends dorsally and ventrally to form a complete annular ridge. These ridges are specially prominent in the 1st 3 chaetigerous segments (see later). The neurochaetae are crotchets and arise ventrally and are paid off dorsally, the oldest regressing at the dorsal end of the series. In a fresh animal these regressing chaetae appear as green masses. Neuropodial muscles may be seen in horizontal sections (10–15 μm). See Wells (1944) for details. Notopodia increase in size posteriorly. As in the neuropodium, chaetae arise singly from individual follicles, the follicles being paid off dorsally. The notochaetae project from a cleft in the distal part of the notopodium. For details of the notopodial muscles, see Wells (1944).

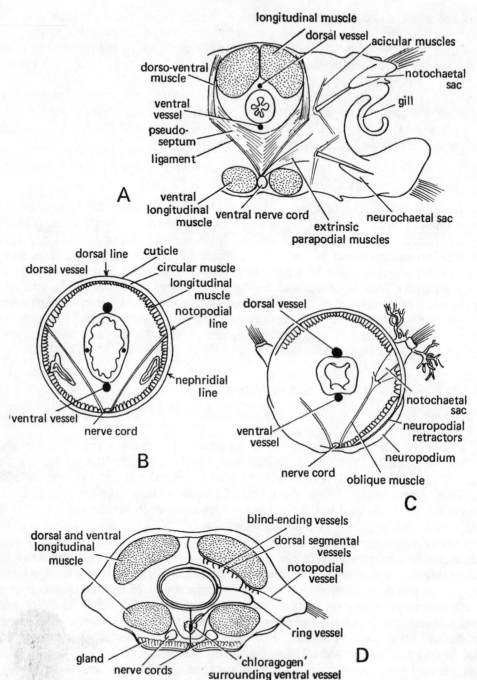

Figure 38 **Segmental structure**

A, *Nephthys hombergi*. Posterior face mid-body segment. Not all the muscles are shown (details of extrinsic and intrinsic parapodial muscles omitted); B, C, *Arenicola marina:* B, simple ammulus; C, parapodial annulus. Intrinsic and extrinsic parapodial muscles not shown; D, *Sabella penicillus*

8.4.5. SEGMENT OF *Sabella penicillus*

In the mid-body region, the longitudinal muscles form 4 discrete blocks: a dorsal pair and a ventral pair overlying the glandular ventral shields (Fig. 38D). The nerve cord is double with large giant fibres. The ventral vessel has brownish 'chloragogen' tissue around it. The neuropodia form ridges with the hook-like neurochaetae emerging from a cleft. The notopodia project has lobes with the longer notochaetae emerging from a cleft in the distal part. The relative positions of neuro- and notopodia are reversed in the 1st 8 chaetigerous segments, the neuropodia being apparently dorsal. In a whole segment numerous blind-ending finger-like vessels projecting into the coelom will be seen coming off the segmental dorsal and notopodial vessels. These may be dark brown ('chloragogen') in colour in older worms.

8.4.6. SEGMENT OF EARTHWORM

The general pattern will be familiar from previous studies and is well illustrated in elementary textbooks to which reference should be made for details, or to Grassé, Vol. 5(1).

8.4.7. SEGMENT OF LEECH (*Hirudo* or *Haemopis*)

Sections through the mid-body region of a leech provide a marked contrast with those of both oligochaetes and polychaetes in the reduction of the coelom to sinuses, and by the complete loss of septa (Fig. 39). The circular and longitudinal muscles form complete cylinders, the longitudinal fibres being bundled into groups by connective tissue and separated laterally by dorso-ventral muscles.

The main difficulties in interpretation of thin (10 μm) sections in isolation is in the reproductive segments. Sections rarely show the gut with a caecum symmetrically cut through on each side. Reference should be made to Figs. 50, 51 showing the complete anatomy of these leeches to interpret isolated sections. Thick sections of whole leeches are very helpful in supplementing dissection (see below). Leeches should be well narcotized and fixed straight; they should be hardened for as long as possible before slicing.

8.5. Larval Development: *Pomatoceros*

Ripe oocytes and sperm are free in the coelom. Artificial fertilizations are readily made by slitting open worms and mixing the gametes. Sperm should be motile. In Britain, March-April is best, but fertilizations may be achieved at any time of year. Females are violet-brown when they contain ripe oocytes; males are yellowish-white from the contained sperm, when mature.

Wash the oocytes into clean seawater in a dish and add a pipette-full of sperm; stir. After half an hour decant the water and surplus sperm and add fresh seawater. Clear plastic boxes with lids (obtainable from kitchen departments of stores) or crystallizing dishes are suitable. Larvae are very sensitive to some reagents and cleaning agents. Glassware should be absolutely clean. Plastic boxes reserved entirely for the purpose and rinsed only with seawater are best.

When making fertilizations add very little sperm. Larvae will settle and metamorphose after three weeks at 20°C if fed on algal cultures (such as *Isochrysis*). Oocytes are about

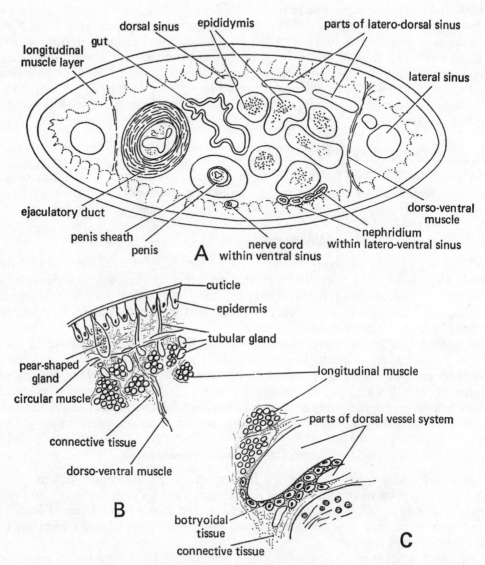

Figure 39 Segment of *Haemopis*

T.S. through clitellar region. A, general aspect; B, detail of skin and muscle layers; C, botryoidal tissue

75 μm across when mature. Cleavage follows the spiral pattern typical for the group; ciliated larvae are formed in 16 h, and typical trochophores in 2–3 days. For details see Segrove (1941). For general features of Polychaete development, see Anderson (1966).

8.6. Regeneration

Many annelids regenerate well, others do not. Among polychaetes, sabellids such as *Sabella* or *Myxicola*, serpulids such as *Hydroides*, and the cirratulid *Dodecaceria* all have very well developed powers of regeneration. With *Hydroides* or *Sabella* crowns may be removed, and new ones will be regenerated in a few weeks. With *Hydroides* the operculum alone may be removed. *Myxicola* will regenerate missing segments anteriorly and posteriorly from a short piece taken within the first 14 segments of the body. In sabellids generally, if an incision is made so as to cut the ventral nerve cord and this is hooked out, a new head may be induced to form around the wound. Some freshwater oligochaetes also regenerate well. Experiments should be performed with clean instruments on healthy worms. Pass the instruments through a flame before operating. Regenerating animals should be placed in dishes of clean seawater containing antibiotic such as streptomycin. Sabellids from which the crown has been removed should be left in their own tubes and may remain in the aquarium. *Myxicola* must be removed from the tube since without crowns they asphyxiate.

The effect of removal of the prostomium or brain, and of implantation of ripe gametes into an immature worm can also be studied in nereids. Clark (1966) describes these experiments. For other details, see Herlant-Meewis (1961).

8.7. Polychaetes

Some of the variations in structure have been outlined on p. 110 above. Other adaptations to different modes of life within the Polychaeta can be best appreciated by detailed dissections of *Arenicola* (burrowing), a terebellid such as *Terebella*, *Amphitrite* (or *Neoamphitrite*), or *Lanice* (tentacle feeders), and a microphagous filter feeder such as *Sabella* or a serpulid such as *Pomatoceros*. *Arenicola* is the best single dissection: it is large and usually easily obtained. Further, much of its physiology and behaviour is known, so that the structures observed may be related to function. On the Pacific coast of North America, *Abarenicola* differs only in detail. Large species of *Neoamphitrite* are also relatively easy to dissect, but are local in distribution. The same general methods outlined below may be used with other genera as available.

8.7.1. BURROWERS

(a) *Dissection of Arenicola*

> This account refers to *A.marina*. The same method may be used with other species or with *Abarenicola*, but differences in detail will be found.

Fresh specimens should be killed immediately before dissection by placing in 30% alcohol and removing as soon as dead. Dissect under water. If specimens are to be preserved for dissection, Wells's methods are best. These are:

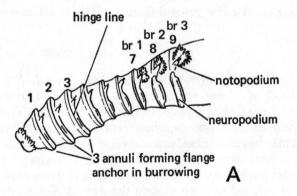

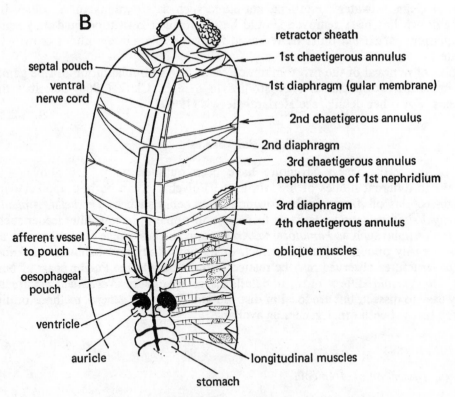

Figure 40 *Arenicola marina*

A, anterior segments: external, showing parapodia (after Wells, 1944); B, anterior segments: internal

Method 1. 'Relaxed' worms.

Narcotize in 7.5% $MgCl_2$. This may take 5–6 hours if worms are large. Place in 4–5% formalin made up in 75 g/l. $MgCl_2$. Do this in a large dish (such as a pie dish with straight sides, straightening the worm against the side as much as possible for a few minutes. Leave in this fluid.

Method 2. Distended worms.

Narcotize as under Method 1. Connect a separating funnel with tubing to a glass cannula. Fill the funnel with 4–5% formalin in 7.5% $MgCl_2$ and support 100–150 mm above the bench. When the worm is fully narcotized pass the cannula into the tail coelom (cannula pointing forward). Place the worm into formalin/$MgCl_2$ and at once open the tap to distend the worm. Ligature the tail in front of the cannula which may then be removed. Leave in this fluid.

Dissection

Before dissecting locate the external features (Fig. 40).

1. Open by a mid-dorsal incision, remembering that the gills are dorso-lateral. Start the incision just anterior to the tail and cut forward to about 10 mm behind the prostomium. This is best done with scissors, the worm being held in the hand. Coelomic fluid will spurt out. If a fresh worm is being dissected collect a drop of the fluid on a slide and examine under the high power of a microscope for coelomocytes. Oocytes or sperm may be present also, according to sex and season.

2. Pin out under water. Much of the general anatomy can be made out by inspection (Figs. 41, 42). Behind the proboscis the *gut* consists of oesophagus leading to stomach, intestine and rectum. The stomach occupies most of the main body cavity and merges imperceptibly into the intestine at about the 14th segment. The intestine passes into the rectum at the 19th segment. The rectum lies in the posterior septate region (tail). Further subdivisions of the gut may be recognized from histology (see Kermack, 1955). Note the three anterior 'diaphragms'.

3. There are 6 pairs of *nephridia* in *A. marina*. These open by pores on the 4th–9th chaetigerous annuli just dorsal and posterior to the neuro-podia. The oblique muscles (see 8.4. above) overlie them and must be removed. Slit the muscles near the nerve cord and base of the notopodia. The nephridial funnels are frilly and well vascularized. The main part of the nephridium becomes darker with age from contained 'chloragogen'. The gonads are attached to the posterior tips of the nephrostomes (except 1st pair). They are small (0.5 mm) and pink in a fresh worm.

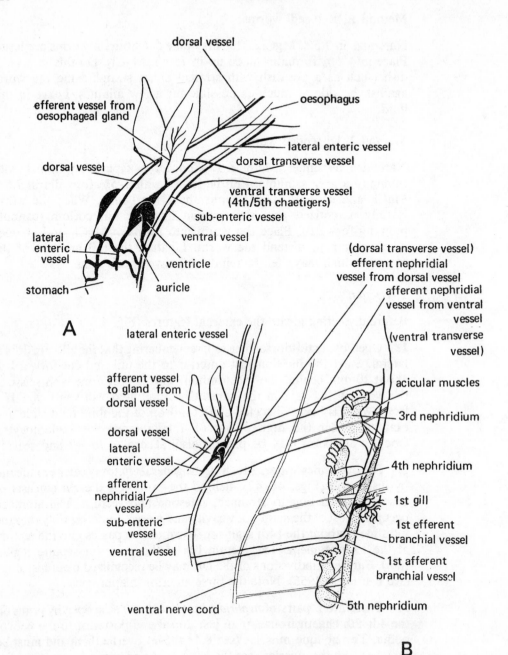

Figure 41 *Arenicola marina*

Anterior vascular system: A, showing relationship of hearts to the ventral and enteric vessels (the transverse vessels around the heart itself are not shown); B, afferent and efferent vessels to the nephridia and first branchia

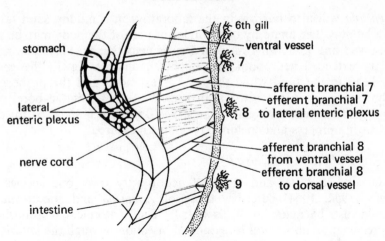

Figure 42 *Arenicola marina*

Relationship of branchial vessels to the ventral vessel and lateral enteric plexus in the mid-body region

4.   *Nervous system.* Leaving the dissection pinned out under water extend the dorsal incision forward into the prostomium. Most of the *nerve cord* can be seen by deflecting the gut. Better, open the head end of another specimen mid-ventrally. To display the *brain* and connectives, cut through the oesophagus and carefully remove the diaphragm, retractor and gut. The brain (supra-oesophageal ganglia) lies within the prostomium and is related to the ventral nerve cord by circum-oesophageal commissures in front of the 1st diaphragm. Otocysts are innervated by short nerves arising from these commissures (connectives). The otocysts are about 1 mm across in a large worm and are usually yellowish in a fresh specimen. The brain and associated structures may also be seen by making median sagittal sections of well hardened specimens (for method see 8.2. above). Dissection of the nervous system is aided by pouring off the water and flooding with 95% alcohol. Leave for 20 minutes or longer if possible.

(b) *The burrowing apparatus*

Polychaetes burrow by a combination of body wall and proboscis activity. The proboscis is withdrawn by retractor muscles and its protrusion by coelomic fluid pressure normally prevented by keeping the mouth closed or by a similar sphincter-like action of the anterior body wall muscles. When these are relaxed, the proboscis may be protruded (Fig. 43).
Nephthys has a large balloon-like proboscis which aids in burrowing. This may readily be observed in the laboratory or in the field. Large specimens may be dissected from the dorsal side when the large retractor muscles can easily be seen. Horizontal sagittal sections of hardened or simply preserved specimens are also instructive.

Arenicola will also burrow in the laboratory into muddy sand taken from its own habitat. The anatomy of the anterior part of the body may be studied from preserved and preferably hardened specimens (see p. 111). The ridges formed by the extended neuropodia are important in forming the 'flanges' anchoring the head in the sand while the head is extended and the proboscis extruded. "Burrowing" may be seen within a glass tube of suitable bore, or in muddy sand held between two glass plates. See Trueman (1966a, b) for details of the burrowing process and coelomic pressures measured.

(c) *The proboscis of Arenicola*

Details of structure and mode of action vary from one species to another. Wells (1952, 1954) describes those of *A.marina* and *A.ecaudata*. Reference should also be made to Wells (1950). The anatomical relationships are best appreciated by horizontal longitudinal bisection through the anterior end of the body, supplemented by sagittal section. Worms should be hardened (see 8.4., above) for at least 2–3 weeks. Slice the specimen with a single-sided razor blade (horizontally or vertically, as desired) and then cut off the rest of the body behind the 3rd chaetigerous segment. The whole apparatus consists of buccal mass, pharynx, post-pharynx and oesophagus. The first 'septum' arises from the body wall just anterior to the 1st chaetigerous annulus. It is split into two sheets: an anterior retractor sheath inserted at the junction of the buccal mass and pharnyx, and the gular membrane inserted on the oesophagus. The gular membrane is extended into septal pouches. The 2nd and 3rd diaphragms are perforated, so the 'head coelom' and 'body coelom' are separated by the gular membrane.

Observations

Observe the movements of healthy worms. Proboscis movements are more easily watched by ligaturing the body at the 6th–7th chaetigerous segments and pinning down in a dish just behind the ligature. Burrowing movements may also be seen by placing the worm in a glass funnel filled with seawater and with a stem wide enough to just accommodate the worm. Timing of the various movements can be recorded with an electrical key actuating a marker on a kymograph drum revolving at about 1.25 mm/s.

Coelomic pressure

Make a longitudinal dorsal cut in segments 7–8. Cut the gut and push the front end back into the anterior part of the body. Insert a piece of tubing (cannula) into cut end and tie. Connect cannula to a wider tube (30 mm outside diameter) and suspend in a bath of seawater. The coelomic pressure can be adjusted by raising or lowering the wide tube. Full extensions of the proboscis only occur with 20–100 mm pressure (*A.marina*). See Wells (1954) and Trueman (1966a, b) for an account of the pressures during burrowing. The cycles of activity observed may be related to the sequence of contractions shown by isolated extrovert preparations.

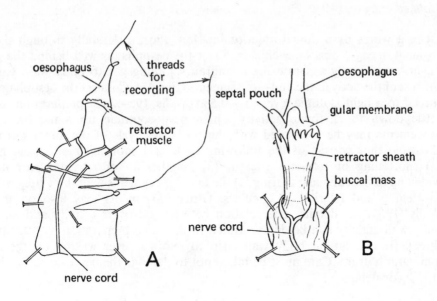

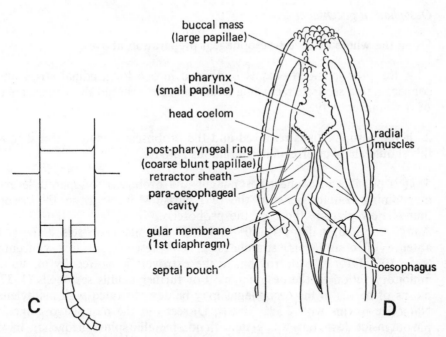

Figure 43 *Arenicola*

The proboscis apparatus. A, Preparation for simultaneous recording from oesophagus and body wall; B, 'isolated extrovert'; C, simple apparatus for demonstrating the effect of small increases in pressure on the proboscis apparatus; D, horizontal L.S. through the anterior end (*A. marina*) Interpretation follows that of Wells. A, after Wells (1937): B, C, Wells (1954).

Isolated extrovert

Open a worm by a dorsal anterior incision. Pin out laterally through the 1st segment, using 2 pins on each side. Sever the oesophagus well behind the gular membrane. Gently separate the membranes joining gut to body wall. Cut off the rest of the body at the 2nd–3rd segments (Fig. 43). Connect the oesophagus by thread to a light isotonic lever (0.3–0.5 g) (Wells, 1954). The preparation should settle down to rhythmic activity which may continue for some time. The movements may be correlated with the proboscis cycle of the intact worm. In *A.marina* they consist of the following cycle: (1) shortening of buccal mass; (2) shortening of retractor sheath; (3) lengthening of retractor sheath; (4) widening of buccal mass. During (1) the first 2–3 segments may show longitudinal shortening and relation of circulars. Other movements may also occur. See Wells (1954) for details. The relation between extrovert activity and anterior body wall activity can be demonstrated in the same preparation by attaching a thread from a lateral body wall strip to another lever writing on the same kymograph drum. Care must be taken not to damage the nerve cord. See Wells (1937) for details.

Oesophageal pacemaker

Using the whole extrovert-oesophageal preparation above:

1. If the proboscis-oesophagus is divided into 4 longitudinal strips and each connected to a separate recording lever, each strip should show the same pattern of rhythmic activity.

2. If the oesophagus is removed and the proboscis alone connected to a lever, this should show an irregular continuous activity.

3. It is possible to connect oesophagus and proboscis to separate levers while maintaining contact between the two parts. Movements in the oesophagus immediately precede those of the proboscis.
Addition of adrenalin $1:1 \times 10^6$—$1:1 \times 10^5$ inhibits proboscis activity in the absence of oesophagus, stimulates in its presence. In low concentrations $(1:5 \times 10^7$—$1:1 \times 10^8)$ the rhythm in the extrovert is slowed by an apparently inhibitory action of the oesophagus. For further details see Wells (1937). The plexus of nerves in the oesophagus may be demonstrated with methylene blue. Narcotize worms with 7.5% $MgCl_2$. Dissect out the oesophagus, open it and pin out inside down on a wax plate with non-metallic spines (hedgehog, hawthorn, cactus or rose spines will do). Wax poured into a Petri dish so as to half fill it will serve. Immerse in seawater and add 1 ml of reduced methylene blue (see p. 54). Leave 3 hours or more, preferably overnight. A permanent preparation can be made by transferring to 8% ammonium molybdate for 3 hours. With care the muscles can be dissected away. Dehydrate, clear and mount. For further details see Whitear (1953).

8.7.2. Tentacle Feeders: Terebellids

Various terebellids such as the larger species of *Amphitrite*, *Neoamphitrite* or *Thelepus* provide interesting general dissections and demonstrations of some aspects of the structure of tentacle feeders. Distribution is often local but some species may be numerous where they occur. The larger terebellids suitable for dissection are commonly not tube-building but live in permanent burrows in mud. Some tube builders such as the European *Lanice conchilega*, on the other hand, may be maintained in aquaria and may be dissected under a binocular microscope. Other terebellids will live if kept cool and if they are

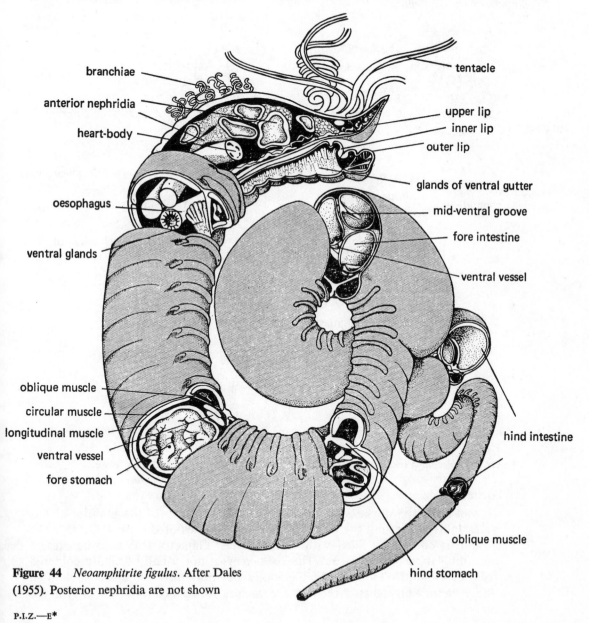

Figure 44 *Neoamphitrite figulus*. After Dales (1955). Posterior nephridia are not shown

provided with mud with which to build a prostrate tube on the bottom of the tank. It is essential to keep them in the dark until this is done.

The prostomium is reduced save for the feeding tentacles (Fig. 44). These are ciliated and may be extended by ciliary creeping. For details see Dales (1955). For anatomy of *Neoamphitrite figulus* (*Amphitrite johnstoni*) see Thomas (1940).

Terebellids irrigate their burrows or tubes by peristalsis of the body wall. They will do this readily in glass tubes of correct bore. Straight pieces, long enough to enclose the body when extended, placed in seawater in a dish are quickly adopted by healthy *Neoamphitrite*, or *Thelepus*.

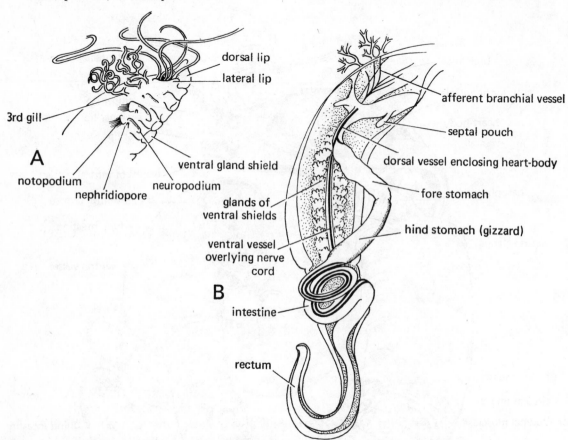

Figure 45 *Neoamphitrite figulus*. A, anterior region; B, plan of gut and main vessels

Dissection

Dissect from the dorsal side. Open slightly to one side of the mid-dorsal line so as to avoid cutting the dorsal suspensory muscles of the gut (Fig. 45). The coelom is commonly filled with coelomocytes. Gametes may also be present according to sex and season. The *coelomocytes* are usually brightly coloured from carotenoid dissolved in fat globules but may also contain haemoglobin and excretory pigments. Note that the coelom forms a continuous cavity in the main part of the body within which the gut may be thrown into coils. There is an *anterior diaphragm* separating a head coelom. Septa are well developed between rectum and body wall towards the hind end of the body. In *Neoamphitrite* the anterior diaphragm has 4 septal pouches. The anterior nephridia are commonly very large, decreasing in size posteriorly. They have large funnels. The oblique muscles overlie the nephridia, and these muscles have to be slit to display them.

The *gut* is well differentiated. The oesophagus leads into an anterior thin walled, usually yellowish stomach leading to a muscular hind stomach or gizzard itself leading to an intestine (often coiled) passing to the rectum.

The *dorsal vessel* enlarges above the oesophagus and contains a 'heart-body', large and dark in colour in old worms, but insignificant in young worms. Vessels are given off anteriorly to the gills, and efferent branchials from the gills return blood to the ventral vessel.

The *nerve cord* is double and unganglionated. It may be seen in a general display of the internal organs under the ventral vessel, but the cerebral ganglia and commissures are small and not easily demonstrated.

Dissection may be supplemented by study of thick sections of hardened specimens. Sagittal sections of the anterior region are informative and may be compared with transverse slices of anterior, mid and posterior segments.

Histological sections may be coloured with Azan, or simply with 0.1% aqeuous toluidine blue. The ventral gland shields colour vividly with mucus stains such as mucicarmine. Tissues fix well in Duboscq-Brasil, Helly and Zenker.

8.7.3. FILTER FEEDERS: SABELLIDS AND SERPULIDS

Sabellids and serpulids are widespread and many species are common. The European *Sabella penicillus* (*S.pavonina*) and the serpulid, *Pomatoceros triqueter* though small, are the most readily obtainable for study. Some other sabellids and serpulids available elsewhere are larger (*Sabella spallanzanii* ('*Spirographis*'), or *Eudistylia* and easier to dissect.

Worms live well in large tanks and will survive for years without circulation if fed with unicellular algae (*Phaeodactylum, Isochrysis* etc.) or with powdered nettle leaves (available from some "health food" shops). Sabellids may be collected from mud banks in the outer reaches of estuaries or attached to pier pilings or between rocks according to species.

Worms will adopt glass or transparent plastic tubes of suitable bore, within which the peristaltic movements of the body wall used to irrigate their tubes may be seen.

For histology use Duboscq-Brasil as fixative. Azan and mucicarmine work well.

(a) *Sabella penicillus*

Sabella penicillus has a relatively simple crown of slender filaments with pairs of pinnules (Fig. 46). The crown is in two halves separated mid-ventrally by the ventral sac which participates in tube building. The two halves of the crown

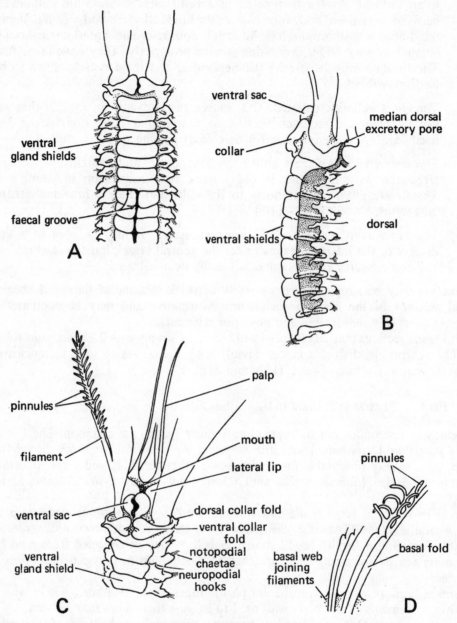

Figure 46 *Sabella penicillus*

A, anterior ventral view, showing reversed parapodia and faecal groove; B, ventro-lateral view, showing extent of anterior nephridia; C, crown structure; D, crown filaments

are joined dorsally at the base to form a membranous collar, and is extended distally into numerous filaments with alternating pinnules which, when spread, together form a wide funnel. The 'palps' which stand up slightly within the circle of filaments when the crown is fully open are simple, not pinnate, and thus easily distinguished. The mid-ventral faecal groove turns dorsally between the 8th and 9th chaetigerous segments, the groove ending at the base of the palps so that the faecal pellets are caught in the ejection stream from the crown. Posterior to the 9th chaetigerous segment the rows of uncini are dorsal to the ventral sheaf of chaetae and similar to those which are notopodial in chaetigers 1–8.

Dissection

Open the *body cavity* by a mid-dorsal incision. The longitudinal body wall muscles and straight gut (undifferentiated externally) are easily recognized. Open the *gut*, and note that the ventral ciliated groove commences at the anterior end of the intestine. The folds on each side of the groove have longer cilia than the other epithelial cells of the gut. The *gut* appears to be a simple ciliated tube but may be subdivided into different functional regions by histology and by enzyme activity (see Nicol, 1930). Worms may be fed on saccharated ferric sulphate and the absorptive regions located by Prussian blue reaction (Perls's test) on representative sections.

The gut has a vascular sinus related in each segment to the ventral vessel by paired S-shaped vessels (Fig. 47). These give off lateral connectives to the segmental dorsal vessels from which blind-ending vessels arise. The blind-ending vessels, a characteristic feature of many annelids, hang down freely into the coelom. 'Chloragogenous' tissue lies on the segmental vessels near the ventral vessel. This tissue which contains haematins derived from chlorocruorin-synthesis accumulates throughout life and may render these vessels very dark in old worms.

The single pair of *nephridia* lie in the most anterior segments and their ducts join to open mid-dorsally. Fluid escaping from the common pore is no doubt carried away in the rejection stream from the crown. The nephridia are very delicate and tend to break up easily. They are commonly dark brown in colour from accumulated haematins.

The ventral *nerve cords* lie close together and may be seen by removing the gut. The segmental nerves are best seen by pinning out a few segments in a stretched condition.

The *vascular system* may also be studied from thick slices of hardened specimens (see p. 111). For details, see Ewer (1941). Advantage can also be taken of the benzidine reaction for haemoglobin. Worms should be fixed in 4% formalin in seawater for 2 days, washed in running water to remove formaldehyde and then stored in seawater saturated with thymol. Thick slices may be cut or 100–250 μm sections cut with a freezing microtome. For details see p. 342.

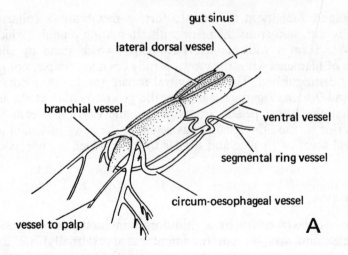

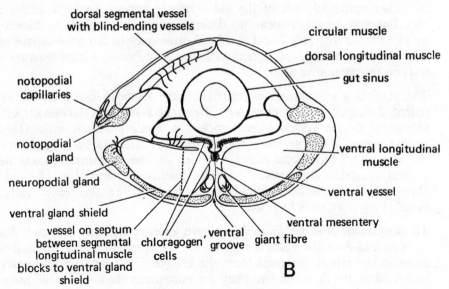

Figure 47 *Sabella penicillus*

A, arrangement of the main anterior blood vessels; B, segmental vessels in an abdominal segment

Feeding currents

Feeding can readily be demonstrated with powdered carmine or some other finely divided non-toxic insoluble material. Filter-feeding rates may be demonstrated with colloidal carbon suspensions and measuring transmittance at intervals using a densitometer (see Jørgensen, 1966). Direction of ciliary beat on the lips and crown can be seen by using carmine and watching under a binocular microscope. Currents pass from the outside up the centre of the spread crown. For details see Nicol (1930).

Skeleton of the crown

The crown skeleton is best seen by soaking the whole crown overnight in 0.4 M MgCl$_2$, and then brushing off the epithelium and other tissues with a fine water-colour brush.

(b) *Pomatoceros triqueter*

Pomatoceros triqueter is the largest of the common serpulids on British coasts. Other species and genera may be collected elsewhere. *Pomatoceros* secretes its limy tubes on large pebbles, shells and rocks and can be collected intertidally. For details of the tube see Hedley (1958). It lives well in the laboratory and may be maintained for years in small aquaria if fed on unialgal cultures or dried foods (see p. 127).

Feeding currents can be observed under a binocular microscope as in *S.penicillus* Reaction to touch, vibration, light etc. can also be demonstrated. If a worm be removed from its tube without injury it will adopt a glass tube of suitable bore. The tube must be open at both ends and preferably tapered. Glass tubing is easily drawn out into a tube like the worms' own tube. A current which irrigates the tube is drawn over the ventral side of the body from the posterior opening, turns to the dorsal side and flows out of the anterior opening into the rejection stream of the crown. The direction of these currents can be demonstrated with powdered carmine.

The *prostomium* is extended into a branchial crown as in sabellids, but unlike sabellids there is a peduncular stalk with operculum which closes the tube when the worm retracts. *Pomatoceros* is not easy to dissect and the internal anatomy is best made out from thick slices, preferably from hardened specimens (see 8.4., above), in both transverse and sagittal planes. Tissues are soft and these worms break up easily when dissected. For histology fix in Duboscq-Brasil. Azan works very well with sections fixed in this way.

The circular muscle layer is poorly developed; the longitudinal muscles form large dorsal bands and small ventral and lateral bands; the obliques are poorly developed. Septa are present and the most anterior septum has a pair of pouches. Dorsal and ventral mesenteries are well developed, supporting the gut and further subdividing the coelom. The vascular system is not unlike that of *Sabella*. For details see Hanson (1950). The single pair of nephridia lie anteriorly and open mid-dorsally by means of a common duct. These nephridia may be very dark in old worms.

8.8. Oligochaetes: *Tubifex*

It is not proposed to describe in detail here the structure of common earthworms since these are almost invariably studied at school and will already be familiar. Some general features of structure have already been dealt with under 8.4., above.

Tubificids are common in mud of ponds and streams, especially where natural organic waste is concentrated. These worms are commonly available from aquarists' shops where they are supplied as food for fish ('bloodworms'). Various species of *Tubifex* and *Limnodrilus* occur together and may be confused. In England, *Tubifex tubifex* occurs with *Limnodrilus hoffmeisteri*. In *Tubifex* the penes have chitinous sheaths and there are hair-like chaetae in the dorsal bundles. In *Limnodrilus* there are no hair chaetae, and the penis sheaths are thin and more elongate (Fig. 48).

For general anatomy examine in water, or (better) invertebrate saline. Worms may be narcotized with MS 222. Dissect with plain or borradaile needles in a solid watchglass or large cavity slide. Reproductive organs can be teased out; nephridia may also be disentagled, but nephrostomes may only be seen through the body wall of more transparent specimens. Addition of a little methylene blue can assist recognition of the nephridia. Reproductive organs are best developed in late autumn.

Fix in Duboscq-Brasil, holding the worm straight against the side of a slide. Worms may be narcotized first. The benzidine method may be used for small specimens or sections (see p. 129). To examine chaetae of *Tubifex* or other small aquatic oligochaetes examine in lactophenol. Permanent mounts can be made in polyvinyl lactophenol.

For details of *Tubifex*, see Dixon (1915).

8.9. Leeches: *Hirudo* and *Haemopis*

Leeches are found under stones or amongst plants in freshwater; occasionally on fish or other hosts. For British freshwater leeches see Mann (1954), and in North America, Moore (1959; in Edmondson). Mann (1962) gives a Key to the families, and Knight-Jones a key to all the marine genera in the same book. Medicinal leeches (*Hirudo*) can be obtained from various suppliers.

Medicinal leeches can be fed on one's own arm. If you do this, allow it to drop off when fed (this may take 20 minutes), do not try to pull it off. If you want to remove the leech before it is fully gorged hold a cigarette or match near it and it will let go. Be warned that the wound will bleed slowly and steadily for several hours owing to the salivary anticoagulin preventing normal clotting. Absorbent pads attached with plaster will have to be replaced repeatedly but in fact little blood will be lost. A gorged leech must be handled carefully. It may be fixed if a ligature is tied behind the mouth and in front of the anus. Fix in Duboscq-Brasil.

Leeches may be fixed in a relaxed state by addition of alcohol to the water so as to bring the concentration to about 30% over a period of half an hour. MS 222 also works well. Then flatten with the fingers into a straight-sided dish and keep flat with a small glass plate or microscope slide while fixative is added. Duboscq-Brasil certainly fixes best, but Bouin, Helly or Zenker are good and may be followed by Azan, Masson's trichrome or Mallory's triple stain.

The medicinal leech, because of its large size, is most suitable for dissection. *Haemopis*, the "horse leech" is also readily available in Britain and differences are noted here. Details of the Indian cattle leech, *Hirudinaria* are given by Bhatia (1941), and *Haemopis* by Mann (1954). Segment numbers are commonly in Roman figures.

For dissection, leeches may be killed by addition of alcohol to the water, or by chloroform. They are best dissected under invertebrate saline.

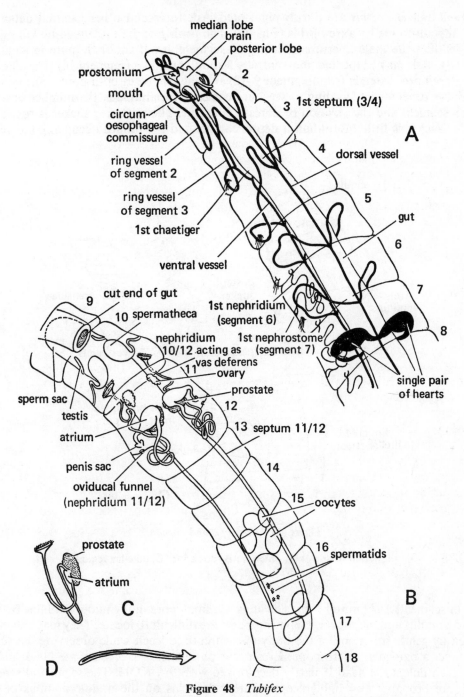

median lobe
brain
posterior lobe
prostomium
mouth
circum-
oesophageal
commissure
ring vessel
of segment 2
ring vessel
of segment 3
1st chaetiger
ventral vessel
1st septum (3/4)
dorsal vessel
gut
1st nephridium
(segment 6)
1st nephrostome
(segment 7)
single pair
of hearts

A

cut end of gut
spermatheca
nephridium
10/12 acting as
vas deferens
ovary
prostate
septum 11/12
sperm sac
testis
atrium
penis sac
oviducal funnel
(nephridium 11/12)
oocytes
spermatids

B

prostate
atrium

C

D

Figure 48   *Tubifex*
A, anterior; B, posterior; C, *Limnodrilus*, atrium and prostate, for comparison; D, *Tubifex*, spermatophore
(1-2 mm)

Hirudo and *Haemopis* are darkly pigmented. Apertures and other external details are often difficult to see because of this (Fig. 49). The penis may be protruded by killing, and this identifies the male aperture at least, on segment 10. If the male pore is identified, then the female may be located mid-ventrally 5 annuli behind it (segment 11). The clitellum, when developed, extends from segment 9–12, and thus includes both pores. The clitellum is variably developed according to season. Segments are annulate, the number of annuli in each segment and the position of pores is constant. The posterior sucker is round; the anterior sucker is little more than a depression around the mouth occupying the ventral

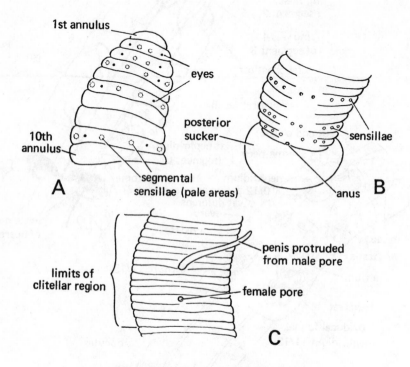

Figure 49 *Haemopis* and *Hirudo*.

External features: A, anterior; B, posterior; C, clitellar region

parts of segments 1–4 (annuli 1–6). The anus is a small pore in the mid-dorsal line between the last annulus and the sucker. Nephridiopores are difficult to locate. They may sometimes be seen by gentle squeezing of the body of a narcotized leech while observing the ventral side under a binocular microscope. Eyes occur on segments 1–5. They are black and not easily seen unless the head is partly decolorized with 5% KOH. The segmental sensillae which are probably also light receptors are pale areas on the middle annulus of each segment. The papillae, which probably function as tactile organs, may be raised if the specimen is contracted. They are often best seen by throwing a living leech into alcohol so that it contracts strongly.

8.9.1. DISSECTION

1. The body is best opened by means of 2 lateral incisions on the dorsal side (Fig. 50). The incisions are best made with fine scissors with a limp leech drooping over the fingers. Start cutting just in front of the posterior sucker and continue forward as far as possible. Join the 2 incisions with a transverse cut just in front of the posterior sucker.

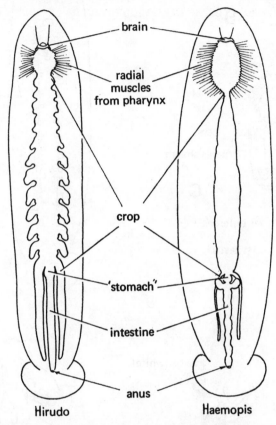

Figure 50 *Haemopis* and *Hirudo*. Comparison of guts

2. Pin down in a wax dish under invertebrate saline. Pin through the anterior sucker. Stretch the leech slightly and pin through the posterior sucker. Pick up the mid-dorsal strip with forceps and carefully peel the skin forward using a borradaile needle or fine scalpel to separate the skin from the underlying tissues. It is important to do this with great care, and is best done under a binocular microscope. Pin out carefully along the sides.

3. The course of the *gut* can be made out with a little further dissection consisting mainly of carefully picking away connective and botryoidal tissue. Two pairs of forceps and gentle irrigation with a fine pipette are best. Note the dorsal sinus on top of the gut and the lateral contractile vessels on each side giving

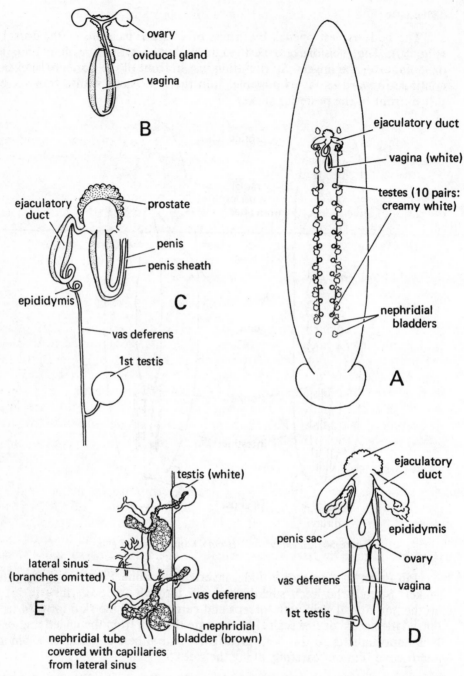

Figure 51 *Haemopis* and *Hirudo*.

Reproductive and excretory systems. A, general plan; B, diagram of female system; C, diagram of male system; D, appearance *in situ* on removal of gut and connective tissue (*Haemopis*); E, relationship of nephridia, lateral sinuses and male system (*Hirudo*)

off transverse vessels which break up into minor branches within the botryoidal tissue.

4. Cut the radial muscles from the pharynx to the body wall. Cut through the pharnyx behind the brain and dissect away from the surrounding tissues to reveal the reproductive system and ventral nerve cord surrounded by the ventral sinus.

5. To display the *excretory system* pour off the saline and leave in 5% formalin for 15 minutes (Fig. 51). Individual nephridia may be teased out in ammoniacal 70% alcohol on a slide, and cleared in 5% KOH. Wash well afterwards to make a permanent preparation.

6. The *nervous system* may be differentiated by pouring 95% alcohol on to the dissection. Leave at least 5 minutes. The ventral nerve cord lies within the ventral sinus. The anterior segmental ganglia are concentrated into a sub-oesophageal ganglion (ganglia of segments 2–5 inclusive); the ganglion of segment 1 is more or less dorsal and fused to the prostomial ganglion or brain. In *Haemopis* the the brain lies in segment 7, in *Hirudo* in segment 5.

8.9.2. SECTIONS

Thick slices of well hardened specimens (see p. 111) are also instructive, especially through the reproductive segments. Isolated thin sections are often difficult to interpret (Fig. 39). Many text-book figures are derived from an original illustration in which the gut and a pair of caecae were cut through. In practice, sections do not often look like this. Correct interpretation of gross structure can best be done by reference to a series of sections and by deduction from the known disposition of the structures, amplified if possible, by a study of thick slices. Fig. 39 shows a section through the clitellar region of *Haemopis*. In *Haemopis*, the penis and penis-sac are folded.

8.9.3. BEHAVIOUR AND LOCOMOTION

Some aspects of locomotion in leeches are readily demonstrated in the laboratory. Lightly narcotize with 0.1% MS 222. Pass loops of thread (dental silk, recommended by Clark, 1966, is best) round the body, at each end, so that the leech may be suspended horizontally by both loops, or by either thread. A running noose will do if not pulled tight. Suspend horizontally in fresh water. When the effect of the narcotic has worn off the leech will swim. Stimulate the underside of the body with a water-colour brush, or touch one or other sucker. Present one sucker with a coverslip. When the leech has gripped it, offer another coverslip to the other sucker. The first should be released if the second is gripped. For further details see Gray, Lissman and Pumphrey (1938).

8.10. Additional Experiments

8.10.1. BEHAVIOUR OF *ARENICOLA MARINA*

Some aspects of the biology of *Arenicola marina* have already been referred to (see p. 122). A number of other demonstrations can easily be made if healthy worms are available. Worms live best in cool (12°C) water, and in the dark, with some (but not much) aeration, and with sufficient muddy sand to enable the worms to burrow. Worms quickly die if aerated in a bowl of seawater in the light. Worms may be kept for long periods in a seawater flow system with the water kept level with that of a bed of muddy sand 50 mm or so deep.

For laboratory demonstrations, irrigation cycles may be demonstrated by worms placed in glass U-tubes. The tube must fit the worm and must be constricted at each end to prevent the worm from crawling out. For apparatus see Wells (1949a). A slow speed drum is required. Independence of the cycle from O_2 or CO_2 is easily demonstrated. Pin a worm to a cork sheet at the bottom of a dish and hook threads into each end of the main part of the body. The cycles will continue whether the water is aerated or not, or if a slow circulation if maintained through the dish. The cycles may be recorded kymographically, as before (Wells, 1949a). Recording of activities in sand requires good conditions. (see Wells, 1949b).

8.10.2. REGULATION OF WEIGHT AND IONIC REGULATION IN *NEREIS* AND *ARENICOLA*

Weight regulation by species of *Nereis* may be demonstrated in the laboratory with very simple equipment. An experiment to demonstrate weight regulation in a good weight-regulator such as *N. diversicolor* will need to be continued over two days. Sensitive, single pan balances or a torsion balance are best because weighing should be done as rapidly as possible. Worms must be carefully handled and it is best to have several worms in each experimental dish to allow for individual differences. Worms may be scooped up with waxed paper for weighing, and this will be found easy if the dishes used have straight sides. Worms must clearly not be picked up with forceps or drooped over a seeker! Surplus water can be shaken off the waxed paper scoop and the worm weighed on a slip of weighing paper. It is instructive to acclimate different lots of *Nereis diversicolor* to 100% seawater and 30% seawater for at least 3 days before the experiment and then follow the weight changes in various dilutions and concentrations (say, 125% seawater). Concentrate seawater by evaporation. It is also instructive to compare different species in their ability to regulate their weights, in relation to the conditions under which they occur in nature.

Dilutions of seawater may be done with distilled water. If done with natural water rich in Ca^{2+} relatively poor regulators such as *Perinereis cultrifera* will behave more like *Nereis diversicolor*. Similarly, *N. diversicolor* can be poisoned with CN^- and the effect partly corrected with Ca^{2+}. For details, see Beadle (1937).

Permeabilities of different species can be compared by placing them in conductivity water and measuring the increase in conductance as the salts leak from the body.

Estimations of Cl^- can be made on the coelomic fluid and on the medium at the beginning and end of the weight regulation experiments (using control worms for the initial Cl^- values). It is instructive to do this also with *Arenicola marina*. This worm has a large volume of coelomic fluid and can be acclimated to 50% seawater, preferably for 3 days before the experiment. *Arenicola* blood is also obtainable in sufficient quantity to enable freezing point determinations to be made, and these values may be compared with that of the medium. Slit open the body cavity and allow the coelomic fluid to drain into a 50 ml beaker for separate estimations. Fold the body between the fingers so that the hearts are over the edge of a clean 50 ml beaker; puncture the hearts, slit the large vessels around the gut (taking care not to puncture the gut itself) and allow the blood to drain. Transfer to a small centrifuge tube and spin at 3,000 rpm for a few minutes.

Determination of Cl^- under class conditions is probably best done by silver nitrate titration. The method depends on the precipitation of the halides in the seawater by a standard solution of $AgNO_3$, using potassium chromate ($K_2Cr_2O_4$) as indicator (Mohr's method). The method can be scaled down to the volume of body fluid that can be obtained from the animal. Microburettes or micrometer syringes can be used. The solution must be stirred effectively to avoid the precipitated silver chloride forming a mass which leads to considerable error. A magnetic stirer should be used if available. On a micro- or semi-micro scale a stream of air delivered by a fine glass jet from an air-line or aerating pump is quite adequate. The end-point is not sharp but the method is perfectly accurate if the worker is consistent in what is judged the end-point. Proteins interfere, and if blood or fluids known to contain appreciable protein are to be determined this must be precipitated first by addition of 2 × sample volume with 5% barium hydroxide (Ba $(OH)_2 8H_2O$) plus 2 × sample volume of 0.5% anhydrous zinc sulphate. Filter and wash through repeatedly until the filtrate is Cl^- free.

$AgNO_3$ analytical grade should be used and a standard solution prepared by direct weighing. 0.1 N $AgNO_3$ contains 17.0 g/l (Equiv.=molec. wt.=170), and since $Ag^+ + Cl^- \rightarrow$ AgCl, and "normal" seawater (34.33‰ at 20°) contains 19.440 g Cl^-/l, it is a relatively simple matter to adjust the molarity of the $AgNO_3$ used by dilution to the volume of the sample and the burette. The potassium chromate should be about 5.0% (w/v). Add a few drops to the sample at the beginning of the titration. As the end-point is approached the precipitated AgCl tends to flocculate leaving the supernatant liquid practically clear. The greenish-yellow colour changes to yellow and then a definite pale red (end-point just passed). A blank may be run omitting the sample. For more precise determinations see Strickland and Parsons (1968).

8.10.3. GIANT FIBRE RESPONSE AND BEHAVIOUR

Giant fibre reflexes wane rapidly. That this is due to habituation may be shown by tactile or photic stimulation using *Megalomma vesiculosum* or other sabellids (see Krasne, 1965). Stimulate repeatedly; after a few responses no response will be obtained. For light reactions of *Megalomma* (*Branchiomma*) see Nicol (1950). Such experiments may be done also with *Pomatoceros*. Arrange in a tank and first observe to which stimuli the worms give a good startle response: touch (use a needle or watercolour brush); light increase; light decrease (switch on/off a bright light placed immediately above the worms); vibration (drop a weight from a measured height onto a pad on the bench near the tank). Then when the worms have re-emerged, repeat the stimulus. If a number of *Pomatoceros* are present (as is usually the case) in each experiment, the number of worms reacting should be noted. Such an exercise lends itself to simple statistical treatment. If the failure to react, after repeated stimulation using the same stimulus, is due to behavioural habituation rather than physiological fatigue, then the same reaction should be obtained with a different stimulus.

Nereis, and various worms showing startle responses, may be subjected to the same kind of experiment. A class demonstration of habituation in *Nereis* is given by Clark (1966).

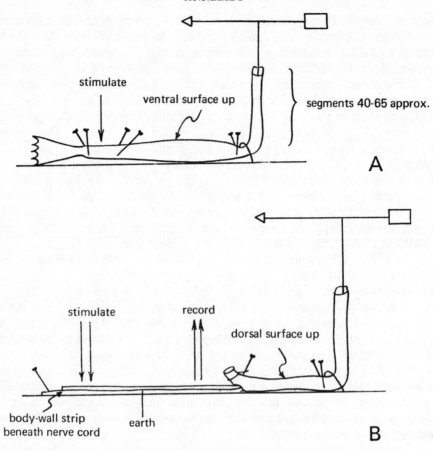

Figure 52 *Myxicola.*

A, preparation for recording longitudinal muscle contractions in response to stimulation of the anterior
dorsal body wall; B, nerve-muscle preparation. After Roberts (1962a)

For details of *Myxicola* preparations see Roberts (1962a). A simple whole animal
preparation or a nerve-muscle preparation can be set up as in Fig. 52. *Myxicola* having
the largest (single, compound) giant nerve fibre is particularly good for demonstrating
the physiological properties of such fibres. Because of the large size of the nerve cord in
Myxicola it is not difficult to dissect the body wall away leaving a strip of muscle underneath the cord itself. This giant fibre gives very large action potentials. A fairly stiff lever
(1 g tension=1 mm excursion) may be used. For electrical stimulation of controlled
intensity a signal-marker contact-maker can be used (up to 3 stimuli/sec) (see Hall
and Pantin, 1937) or for higher frequencies a neon lamp stimulator (see Clark, 1965).
Roberts used a turntable fitted with a variable number of contacts for giving multiple
volleys. *Myxicola* is local in distribution; other large sabellids such as *Eudistylia* may be
used, or even *Nereis* or earthworm. In most of these there are several giant fibres which
may conduct in different directions and give action potentials in relation to their diameters.
For earthworm see Roberts (1962b, c). The potentials can usually be distinguished by

their size. Fine, blunt electrodes can be used for recording with a condenser-coupled pre-amplifier and a double-beam cathode ray oscilloscope. The lower beam can give a 50 Hz hum to provide a time trace. For details of circuits and preparations see the papers cited above.

8.10.4. RESPIRATION IN *TUBIFEX*

Tubifex clump together if placed in a dish without mud, but the tails perform the writhing corkscrew-like movements observed in nature. The rate of these movements is inversely proportional to the O_2 content of the water, and this can be simply demonstrated. The effect of high CO_2 as distinct from low O_2 can be demonstrated by bubbling washed cylinder CO_2 gas through the water while following the CO_2 content with a pH meter. The O_2 uptake/CO_2 output may be followed in a Warburg apparatus, or in closed vessels in which the O_2 content can be measured from time to time by Winkler's method. Fox and Wingfield's (1938) micromethod is best, but requires some practice. Various other types of respirometer can, of course, be used, or the pH and O_2 contents followed with a meter.

Tubifex can withstand 2–3 days without oxygen. O_2-uptake rate is increased after such a period to pay off the debt. For details of Warburg's apparatus see Dixon (1943), also described by Clark (1966). Similar details may be found in other textbooks of Comparative Physiology.

8.10.5. RESPIRATORY PIGMENTS OF *SABELLA* AND *ARENICOLA*

While haemoglobins do occur within cells in certain worms, annelids characteristically have haemoglobin or chlorocruorin dissolved in the plasma. Chlorocruorin is confined to a few families including the Sabellidae. The absorption spectra of these pigments may be examined with a hand spectroscope or, if available, a Hartridge reversion spectroscope. *Arenicola* is a good source of haemoglobin, as it is large; *Sabella* for chlorocruorin. *Sabella* chlorocruorin can be obtained by snapping off the crown at the base and allowing the blood to drain from the body. The characteristic pyridine haemochromogens can be demonstrated by adding 20–30% sample volume of pyridine and a knife-point of fresh sodium dithionite. If a Hartridge is available the maxima will be clearly seen to be different in the two pigments. The absorption spectrum of the pigments may be plotted if a spectrophotometer is available. *Arenicola* blood can be obtained in sufficient quantity to determine the oxygen-dissociation curve. A simple tonometer can be made with a cuvette fitting a spectrophotometer (Fig. 53). Only 3 large *A. marina* are needed to provide enough plasma for a 10 mm cuvette. The haemoglobin will have to be diluted to enable readings to be taken, and if this is done with a buffer, the O_2-dissociation curve may be determined at 2 pH values: say, 7.0 and 6.8, to determine the Bohr effect. For details see Manwell (1963).

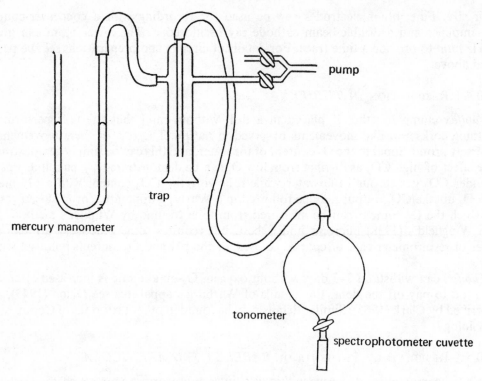

Figure 53 Apparatus for determining O_2-dissociation characteristics of blood pigments

8.11. Myzostomaria

The *Myzostomaria* are very easily recognized. They are confined to crinoids and some brittle stars. They are best fixed in Duboscq-Brasil, or in Bouin for preparation of whole mounts, colouring with borax carmine and differentiating in 1% HCl: 70% alcohol, clearing in benzyl alcohol or cedarwood oil. For details see Grassé, Vol. 5(1).

9. ECHIUROIDS

Echiuroids are found in mud or sand, and in mud-filled crevices or old shells on rocky shores. Though widespread they rarely occur in large numbers. *Echiurus*, *Thalassema* or, in Central California the large *Urechis* may be used to demonstrate the internal anatomy or activity. *Urechis* and *Echiurus* are found in muddy sand-flats where they live in U-shaped burrows which they irrigate by peristalsis of the body wall. For details of biology of *Urechis* see McGinitie and McGinitie (1949).

To display the internal anatomy, simply open the body cavity by a mid-dorsal incision. The concave side of the proboscis is ventral. Turn the animal with the convex or un-grooved side of the proboscis uppermost and cut forwards in the mid-line almost to, but not quite as far as, the base of the proboscis. A loop from the ventral nerve cord and extensions from the ventral and dorsal vessels pass forward to the anterior end of the proboscis. These are best displayed by careful dissection after the main body cavity has been explored and the organs displayed. For details see Grassé, Vol. 5(1).

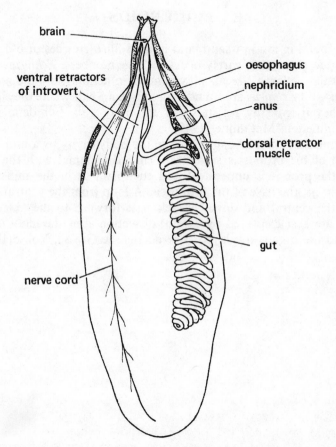

brain

oesophagus

ventral retractors
of introvert

nephridium

anus

dorsal retractor
muscle

gut

nerve cord

Figure 54 *Golfingia.*

General anatomy as displayed by opening the body cavity from the right dorso-lateral side

10. SIPUNCULIDS

Sipunculids are exclusively marine, and most are found in mud or muddy sand. Though widespread they seldom occur in large numbers. *Golfingia* spp. or *Dendrostomum* are sufficiently numerous in some localities to be used for dissection or experiment. The large Mediterranean *Sipunculus* are very suitable for demonstration of internal anatomy, but they are not numerous. The internal anatomy of the smaller *Golfingia* spp. may be best displayed by horizontal longitudinal section. For details see Grassé, Vol. 5(1), Hyman, Vol. 5.

If living specimens are available the coelom should be opened and the fluid drained into a centrifuge tube. The corpuscles may then be centrifuged out and examined under the oil immersion objective of a microscope. Coloured cells containing haemerythrin are 20–30 μm across. If a large quantity is available it is possible to hemolyse them with distilled water to extract the haemerythrin and to compare its absorption spectrum with that of haemoglobin. Amoebocytes and ciliated urns also occur in some (but not all) genera. They occur in *Golfingia* and *Sipunculus*.

To kill specimens with the introvert extended either for subsequent dissection or simply preservation, it is essential to narcotize first. MS 222 is perhaps best, but menthol or magnesium chloride may also be used.

Fixation for histological examination is best done in Duboscq-Brasil. Azan works well for general histology.

Some sipunculids can easily be maintained in aquaria. They withstand poor aeration and *Dendrostoma* and *Golfingia* are excellent subjects for experiments on respiration. Burrowing may be observed in either (see Zuckerhandl, 1950). Pressures may be measured by a water manometer connected to a hypodermic needle passed through the body wall into the coelom. The introvert can be extended by a quite small increase in pressure.

Sipunculids should be opened by a dorsal incision slightly to one side of the mid-dorsal line. If living animals are available for dissection narcotize first and distend the introvert. Locate the anus, which is mid-dorsal in position, at the base of the introvert. Cut to one side with scissors, being careful not to cut too deep so as to avoid damaging the gut, and extend the cut forwards to the tip of the introvert and backwards as far as possible. The body wall may then be pinned out. The appearance of *Golfingia minuta* opened in this way is shown in Fig. 54. There is little variation in layout in the different genera.

11. MOLLUSCS

11.1. Introduction

"Molluscs are easily caught, patient of study and wonderfully diverse in their adaptations". Not only, by their size and complexity, do they make good dissections. They first call for study in life. Placed in an aerated aquarium, most of the types described here readily display their movements and activity. Gastropods and chitons on glass sheets exhibit the locomotor wave of the foot. The actions of tentacles and siphons, and of the proboscis and radula should also be studied. Bivalves should be allowed to burrow in their natural substrata, and can often be so observed from the side wall of a glass tank. The rhythmic extensions and contractions of the foot should be studied, as also the length of siphons, and the nature of their orifices and currents.

11.2. General Methods

11.2.1. CILIA

Most molluscs have important ciliary and mucous equipment, in the mantle cavity and gut, that may easily be studied in life. Explore the currents of ciliated surfaces by adding from a fine pipette suspensions of carmine-stained starch grains, or a fine grade of carborundum powder. The animal should be displayed in plenty of water and time allowed when necessary to recover from the shock of opening up. The suspension should never be so dense as to adhere in clots, or prove too heavy to move, or obscure or smother the surface being examined. Follow the results where necessary with the naked eye, supplemented by observation with the binocular microscope.

11.2.2. PRESERVING

For general dissection, well-expanded fixed material is much preferable to fresh, where muscular contraction and distortion, and lavish production of mucus greatly impede operations. Narcotize an animal before fixing: this is generally a matter of trial and error with a particular species. Magnesium sulphate or menthol crystals may be effective; so alternatively may gradual tincturing of the water with alcohol. Fix when no longer strongly responsive, taking great care to catch before internal disintegration of tissues has begun, sometimes preceding complete abolition of muscular response. It is a good rule not to remove from the shell until after fixation. Wedge open bivalve shells, and carefully crack gastropods, (visceral spire as well as body whorl), to allow penetration. For removal of fresh gastropod material, sufficient pressure can be put on heavy shells with a bench vice, to pick away the pieces without injury to the animal beneath. Leave till last the removal of the columellar muscle from its shell attachment.

Formalcohol is a suitable general fixative and preservative. But one of the best for molluscs is Bouin's picro-formol-acetic, in which material should be left from 24–48 hours according to size, then briefly rinsed with water, and stored in 70%–80% alcohol. A hazard of Bouin's is its staining of the hands and clothes if spilled. Its virtue is in the firm, well-preserved and often differentially stained material, equally available for dissecting, thick-slicing or microtomy. Phenoxetol is finding increasing use as a preservative, keeping tissues well-expanded, flexible for dissection, and avoiding shrinkage or hardening.

11.2.3. THICK-SLICING

Molluscs removed from the shell are soft and compact, without hard skeleta, and able to be sectioned in any direction. The viscera are frequently close-packed by connective tissue and show to advantage *in situ* by the method of thick-slicing, without preliminary unravelling. Wherever possible, this method has been recommended as a preliminary or an alternative to classical dissection. Its advantage, as well as in economy of time, lies in the appreciation made possible of spatial relations and mutual topography. With the gut and genitalia it enables simultaneous study to be made of the interiors of hollow organs.

Make slices not with a wedge-shaped scalpel blade but in a single clean cut with a new safety razor blade.

11.2.4. INJECTING

An important auxiliary to dissection is the injection of the vascular system, large and complex enough in molluscs to repay detailed study. Use only fresh material, well-relaxed after narcotizing. The arteries may be reached by injection through the ventricle. The gastropod venous system should be injected through the foot, and for lamellibranch veins, see instruction, p. 194.

Suitable injection media include (i) soluble Prussian Blue, solidified after injection by plunging into 70% alcohol; (ii) gelatin coloured by carmine or water soluble pigments, introduced warm with a cannula, under a head of gravity; (iii) various coloured latex media, ammonia-soluble and injected cold. The disadvantage of all these is their frequent difficulty in flow, and tendency to solidify early. A better flowing medium is artist's oil paint, mixed with turpentine and injected with a hypodermic needle.

11.2.5. HISTOLOGY

Special organs should be examined in stained slides as directed. Whole sections of small animals can with advantage be studied with a binocular microscope. After Bouin's fixation, stain with Weigert's iron haematoxylin counter-stained with van Giesen, or with Masson's trichrome or Mallory or Azan triple stain. (See p. 339).

11.2.6. RADULAE

Remove the buccal mass, or (for a small animal) the whole head, and gently raise to boiling in an evaporating dish with a 10% solution of caustic potash. After digestion of the soft tissues, isolate with a needle the radula intact on its membrane. Rinse in alcohol, stain with picric acid and mount in glycerine jelly; or with small radulae clearing, staining and mounting can be done as one operation in polyvinyl alcohol coloured with lignin pink.

11.3. Classification

Phylum MOLLUSCA

Class GASTROPODA

Asymmetrical molluscs with a well-developed head and—at least primitively—a broad flattened foot. The shell is in one piece, spirally coiled, at least in the young stages. The mantle cavity and visceral mass shift through 180° in relation to the head-foot by torsion. Pallipericardial complex generally reduced and one-sided.

Sub-class PROSOBRANCHIA

Generally aquatic gastropods with sexes separate. Visceral mass with pronounced torsion and visceral nerve loop in a figure of 8. Shell closed by operculum. Mantle cavity primitively with two ctenidia, but generally reduced to one (post-torsional left); heart posterior to this. Single gonad opening on right, either through right kidney, where left suppressed, or through renal genital duct where left retained and functional. In latter case genital ducts elaborate. Frequently a free-swimming veliger larva.

Division DIOTOCARDIA (= ASPIDOBRANCHIA)

Order ARCHAEOGASTROPODA

Prosobranchs with some indication of their original bilateral symmetry. Ctenidia primitively paired (ZEUGOBRANCHIA) and bipectinate. Heart with two auricles (diotocardian), and right renal organ always a functional kidney, but—as well—conveying genital products. Nervous system little concentrated, with a long pedal ladder.

Super-family ZEUGOBRANCHIA

Archaeogastropoda with primitively paired ctenidia and generally with depressed or limpet-like shell.

Haliotis, Fissurella
Emarginula.

Super-family PATELLACEA

More specialized Archaeogastropoda, with full loss of spirality. Pallial cavity extending as a narrow space around the whole circumference with simple secondary gills. In addition there may be a single true ctenidium (Acmeidae) though in the Patellidae this is lost.

Patella, Acmea

Super-family TROCHACEA

Intermediate between Zeugobranchia and Monotocardia. Left ctenidium alone remains.

Trochus, Turbo

Division MONOTOCARDIA (= PECTINIBRANCHIA)

Prosobranchs retaining only the left palliopericardial organs. (Left) ctenidium monopectinate. Left kidney alone survives, right kidney incorporated into genital duct, which has added to it a complex pallial extension, producing (in female) spawn jelly or egg capsules, and serving as prostate in male. A penis and internal fertilization. Habits and shell form diverse. Nervous system variously concentrated.

Order MESOGASTROPODA

A large group of varying habits and diet. Some are herbivores, some carnivorous resembling whelks, a few ciliary feeding. Generally with spawn mass or egg capsules and a free-swimming larva.

Littorina, Crepidula

Littorina is described in full detail by Fretter and Graham (1962). *Crepidula* is chosen here as a supplementary type, placed after the Neogastropoda (p. 168).

Order NEOGASTROPODA

The most advanced prosobranchs, all carnivorous in habit, with an evaginable proboscis, feeding on dead or living animals. Shell with spout-like anterior canal and mantle cavity with an inhalant siphon. Osphradium large and bipectinate. Eggs in horny capsules, with free-swimming stage usually suppressed, intracapsular embryos sometimes cannibalistic on others.

Buccinum Thais

Sub-class OPISTHOBRANCHIA

Marine and hermaphrodite Gastropoda, with the shell reduced, becoming internal and finally lost. Detorsion of visceral mass proceeding to completion, with mantle cavity moving back along right side and widely opening before final loss. Gill never a ctenidium. With loss or torsion and asymmetrical shell, an eventual return to virtual bilateral symmetry externally. Great adaptive range of form, colour, feeding and locomotion. Usually with a (reduced) swimming veliger stage.

Order CEPHALASPIDEA (BULLOMORPHA)

Opisthobranchs with shell still moderately large. Pallial cavity well-developed with a single plicate gill. Head frequently forming a large shield for burrowing. Parapodia prominent and foot broad.

Bulla, Philine

Order ANASPIDEA (APLYSIOMORPHA)

Opisthobranchs with the shell retained, though already reduced and internal. Mantle cavity a small recess on the right side, with a plicate gill, covered by overlapping or fused parapodia. Head with paired cephalic tentacles and a pair of rhinophores.

Aplysia, Barsatella,
Dolabrifera

Order THEOCOSOMATA

Shelled planktonic pteropods; mantle cavity retained. Large parapodia.

Limacina

Order GYMNOSOMATA

Naked planktonic pteropods with neither shell nor mantle cavity.

Clione

Order NOTASPIDEA (PLEUROBRANCHOMORPHA)

Slug-like opisthobranchs with shell (in most forms) reduced and internal. No mantle cavity, but a naked plume-like gill overhung by the notum on the right side. Otherwise almost symmetrical externally. Rhinophores form (scrolled) head tentacles.

Pleurobranchus

Order NUDIBRANCHIA (ACOELA)

The largest and most slug-like of the opisthobranchs. Shell entirely lost. No mantle cavity but a naked upper surface or 'notum'. Highly adapted for various types of "grazing carnivorous" diet. With or without secondary gills. Anus median dorsal, or lateral, genital and renal openings on right side.

Doris, Aeolidia

Sub-class PULMONATA

Snail-like hermaphrodite gastropods, with no ctenidium, but with mantle cavity largely closed and vascularized as a lung, with a small aperture, the pneumatostome. Detorsion seldom complete, but nervous system so concentrated as to lose all trace of chiastoneury. Primitively spiral in shell and visceral mass, but may assume slug form. Genitalia with complex accessory glands.

Order BASOMMATOPHORA

Mainly marine and freshwater pulmonates, with a pair of single, non-invaginable head tentacles, carrying eyes at the base. Secondary branchiae may be acquired, within or outside the mantle cavity.

Lymnea

Order STYLOMMATOPHORA

Thc higher and terrestrial branch of the pulmonates. Pallial cavity always with a lung, and head with two pairs of invaginable tentacles, the eyes at the tips of the hinder pair.

Helix, Achatina

Class AMPHINEURA

Elongated bilaterally symmetrical molluscs with mouth and anus terminal. Mantle very extensive, covering the whole dorsal surface and sides. Heart dorsal and posterior with ventricle and lateral auricles. Nervous system with longitudinal pedal and pallial ganglionic cords, with cross anastomoses.

Sub-class POLYPLACOPHORA (LORICATA)

Flattened littoral or sublittoral Amphineura, with a broad ventral foot. The mantle bearing eight articulated transverse shell valves, its border forming a firm spiculose or scaly girdle. Ctenidia multiplied into numerous pairs, forming a functional curtain dividing mantle cavity into outer inhalant and inner exhalant regions. Renal organs complexly branched tubes; genital ducts paired and separate, from median gonad.

Order CHITONIDA (characters of the sub-class).

Chiton

Class BIVALVIA (=PELECYPODA)

Bilaterally symmetrical molluscs with rudimentary head, without buccal mass or radula. Generally ciliary feeders with greatly enlarged ctenidia and labial palps. Two mantle lobes completely enclosing the body, and secreting a shell of two calcified valves, united by a dorsal ligament. Foot compressed and generally without plantar surface, either adapted for burrowing, or reduced in favour of the byssus. Fertilization external, usually a long larval life.

Sub-class PROTOBRANCHIA

Primitive bivalves having ctenidia with simple leaf-like, filaments, and hypobranchial glands retained. Feeding primarily by extensile tentacles from large labial palps (palp proboscides). Foot with an expansible ventral surface.

Nucula

Sub-class LAMELLIBRANCHIA

Bivalves in which the gill filaments are folded and the adjacent filaments joined, either by tissue or simply by cilia.

Mytilus, Pecten, Cardium, Ostrea

Sub-class SEPTIBRANCHIA

Bivalves with gills modified into a septal pump.

Poromya

Class SCAPHOPODA

Bilaterally symmetrical Mollusca, mantle and shell elongated, uniting ventrally to form a tapered tube open at either end. Foot cylindrical and pointed. No ctenidium. Head without eyes but carrying paired clusters of food-catching *captacula*.

Dentalium

Class CEPHALOPODA

Bilaterally symmetrical Mollusca with a circlet of tentacles round the head, sucker-bearing in modern forms. Epipodium modified to form a pallial funnel through which passes the concentrated exhalant current that serves for jet propulsion. Nervous system greatly concentrated and highly organized. Efficient visual and tactile senses, and of higher metabolism and activity than other molluscs.

Sub-class NAUTILOIDEA

A single surviving genus, *Nautilus*. An external many chambered siphunculate shell, coiled or (extinct examples) straight. Head with numerous tentaculate appendages, retractile and lacking suckers. Funnel of two separate folds. Ctenidia, renal and pericardial organs in *Nautilus* increase to two pairs. Eyes open without cornea or lens.

Sub-class COLEOIDEA

Cephalopods with the mantle naked (in living forms), forming a sac covering the viscera and sometimes containing a more or less rudimentary shell. The head has always eight sucker-bearing arms and there may be in addition a pair of longer rectractile tentaculate arms, between the third and fourth short pairs. Funnel always a closed tube. Ctenidia, renal and pericardial organs always a single pair. Eye with a crystalline lens and closed or open cornea. Ink-sac present.

Order DECAPODA

Squids and cuttlefish, with tentacular arms, and with the eight normal arms shorter than the body. Suckers pedunculate with horny rings.

Internal shell rather well developed, a trabeculate shield with narrow camerae in Sepiidae, and a light chitinous pen in Loliginidae.

Sepia, Loligo

Order OCTOPODA

Octopus and their allies with no retractile arms and the eight normal arms longer than the body. Suckers non-pedunculate and without horny rings. The mantle encloses the viscera in a rounded muscular sac. Internal shell lacking, though the female *Argonauta* has a fragile external "shell", secreted by the dorsal arms.

11.4. The Basic Molluscan Plan:

Fissurella and *Emarginula*

Slit limpets or keyhole limpets (Fig. 55) give an admirable introduction to the general design of the molluscs. A well-fixed animal should be removed by detaching the U-shaped shell muscle. Then make a single sagittal razor cut, dividing the specimen into two halves.

The three elements of molluscan organization clearly appear. The muscular *head-foot* has an extensive sole, and a snout carried with mouth against the ground. Within the buccal mass is an *odontophore* carrying the *radula*. A band of *shell muscle*, radially disposed, rises from the foot to the overlying shell. Dissect one half laterally sufficiently to show these muscles enclosing the viscera in a spacious basin.

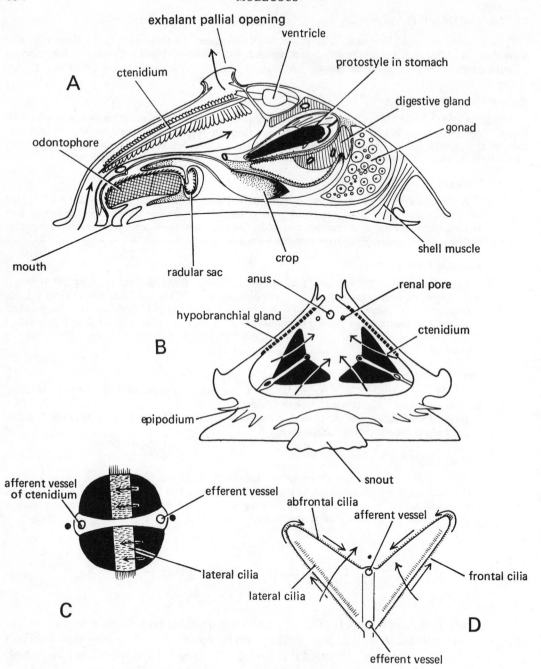

exhalant pallial opening

ventricle

ctenidium

protostyle in stomach

A

digestive gland

gonad

odontophore

mouth

radular sac

crop

shell muscle

anus

renal pore

hypobranchial gland

ctenidium

B

epipodium

snout

afferent vessel of ctenidium

efferent vessel

abfrontal cilia

afferent vessel

lateral cilia

frontal cilia

lateral cilia

C

efferent vessel

D

Figure 55 Early Molluscan Organization

A, *Fissurella* with internal structure revealed by a simple longitudinal razor slice; B, Transverse section of the mantle cavity and ctenidia; C, Gill filaments of a chiton (see p. 185); D, Gill filaments of a primitive prosobranch, (*Haliotis*) showing arrangement of ciliary currents

The *visceral mass* consists of the tube of the gut, the digestive gland and the unpaired posterior gonad. Observe the *oesophagus* with *crop*, *stomach* and *style sac* and *intestine*, tracing out the configuration of the whole gut.

Note the all-inclusive skirt of the *pallium* or *mantle*. The anterior pallial cavity has two *ctenidia*, paired and equal, and two *hypobranchial glands* and *osphradia*. (This primitive symmetry is evidently secondarily restored.) Note the two rows of filaments in each gill, and the relation of gills, anus, and renal openings, to the pallial exhalant aperture. This is an apical hole in *Fissurella*, an open slit in *Emarginula*, or in some fissurellids merely a shallow anterior sinus.

Locate within the *pericardium* the *heart*, with *ventricle* traversed by the rectum, two *auricles* with efferent branchial vessels, *renal organs* and *reno-pericardial openings*.

In living material, the ciliation of the pallial cavity may be studied as for *Haliotis*, p. 158.

11.5. Diotocardian Gastropods

11.5.1. An Early Prosobranch: *Haliotis*

Though not found alive on British shores, ormers are ideal archaeogastropods for dissection, being large and with a primitive organization straightforwardly shown (Fig. 56).

(a) *The living animal*

1. Attach an ormer to a large sheet of glass. Record from below the nature and direction of the locomotor wave. *Haliotis* combines a tenacious attaching power, with ability of rapid movement and agility of turning. It has the mobility of a top-shell and the tenacity of a limpet. (See Lissmann (1944) for analysis of locomotion).

2. Examine a well-expanded animal from above. Note the action of the tentacles; two *cephalic tentacles* with eyes at the base; the profuse fringe of tentacles from the *epipodium*; and the *pallial tentacles* emerging from the shell foramina.

3. With a suspension of carmine, locate the inhalant and exhalant sites of the pallial cavity.

4. Observe from below the snout pressed close to the glass, and the working of the *radula*, rolling forward on the expanded *odontophore*, sweeping the surface and retracting.

(b) *Pallio-pericardial Organs*

Dissect first the pallial cavity and viscera in a well-fixed specimen; and later return to a living specimen for the pallial cavity.

1. Dislodge the animal from the shell with pressure on the right side of the foot, or with a scalpel blade cut the large (right) shell muscle at its insertion on the shell. Pin the animal dorsal surface up at several places through the edge of the foot.

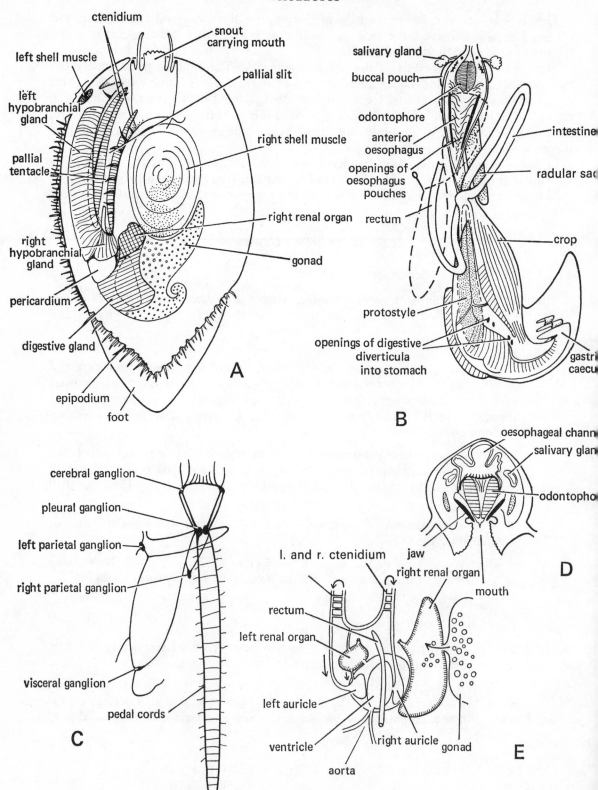

A

ctenidium
snout carrying mouth
left shell muscle
pallial slit
left hypobranchial gland
right shell muscle
pallial tentacle
right renal organ
right hypobranchial gland
gonad
pericardium
digestive gland
epipodium
foot

B

salivary gland
buccal pouch
odontophore
intestine
anterior oesophagus
radular sac
openings of oesophagus pouches
rectum
crop
protostyle
openings of digestive diverticula into stomach
gastric caecum

C

cerebral ganglion
pleural ganglion
left parietal ganglion
right parietal ganglion
visceral ganglion
pedal cords

D

oesophageal channel
salivary gland
odontophore
jaw
mouth

E

l. and r. ctenidium
right renal organ
rectum
left renal organ
left auricle
ventricle
right auricle
gonad
aorta

2. In the intact animal, note the extent of the pallial cavity, pericardium, visceral mass and shell muscles. With enlarged body whorl and tiny spire, *Haliotis* is an early experiment in limpet design. The pallial organs are paired and sub-symmetrical. An archaic feature is the pallial slit, lying beneath the line of shell foramina. In *Pleurotomaria* the same early organization co-exists with a coiled spire like a top-shell and an open shell slit, found also in young ormers.

3. Open the pallial cavity with a longitudinal cut between the ctenidia. Reflect and pin the flaps to either side. Note the form of the *ctenidia*, with paired rows of triangular filaments; the right and left *hypobranchial glands*; the *anus* and the right and left *renal organs* with their openings. Look for the simple *osphradia*.

4. In a thick section of a small entire animal, appreciate the natural disposition of the pallial organs.

5. Injection of the blood system should be performed—if desired—on a fresh intact specimen. The anterior and visceral aortae can be reached from the ventricle. The efferent branchial vessels may be injected from either auricle, or the venous sinuses, with afferent branchial vessels reached from deep injection of the foot. (See Fig. 58 for plan of circulation; and for details refer to Crofts, 1929).

(c) *Internal Structure*

1. Remove the mantle flaps and ctenidia to expose the floor of the mantle cavity and the head. Open the haemocoele by a careful crescentic cut, at some distance to the left of the large muscle, and running from snout to visceral mass.

2. Carefully conserve the nerve ring and visceral loop during dissection. The perioesophageal ring encircles the gut behind the buccal mass. A pleuroparietal connective runs obliquely above the crop from the right pleural ganglion to the left parietal ganglion near the insertion of the left gill. Conversely a connective beneath the gut runs to the right parietal ganglion.

3. Locate each of the parts of the alimentary canal, which is capacious and primitively complete. Note the *buccal mass* and the small *salivary glands* lying against its roof. Behind the buccal mass the *oesophagus* receives the slit-like openings of the *oesophageal pouches*. It dilates then into a food-storing *crop*, and narrows a little to enter the *stomach*. Find also—close to the oesophagus— the loop of the intestine which runs forward to the head, returns to stomach level, and then turns forward to traverse the ventricle and opens by the anus into the pallial cavity.

Figure 56 The Ormer, *Haliotis*

A, External structure, in dorsal view, as shown after removal of shell, without further dissection; B, The alimentary canal complete, with its relation to the nervous system shown by a broken line; C, Diagram of the nervous system; D, Transverse slice of the head and buccal mass; E, Dissection of the pericardial and renal organs.

4. The *stomach* is a complex and characteristic organ in early gastropods (see Fretter and Graham, 1962). First note its extent before opening, embraced by the massive brown digestive gland. Find the spirally coiled *sorting caecum*, emerging behind the oesophageal opening and embedded in the digestive gland. Note the narrowly conical *style sac*, tapering forward to become continuous with the intestine.

5. Open the stomach by a longitudinal incision. Remove its contents, consisting of a *protostyle*, formed of a massive rod of faeces, continuous behind with mucous strings from oesophagus or caecum. Note the cuticle-lined part of the stomach and the *gastric shield*; and the two sets of multiple openings of the digestive gland. Locate the close-set ridges and grooves of the *ciliary sorting area*, and their relation to the style sac and caecum.

6. Observe the structure of the *digestive gland*, made up of tiny follicles from repeated branching of the paired diverticula. Examine stained sections of the digestive epithelium (ref. Owen, 1966).

7. Identify the single gonad, *ovary* or *testis*, forming the terminal part of the spire behind the digestive gland. Find its opening into the adjacent right renal organ. This is the larger of the two kidneys and acts as a gonoduct. Open both renal organs with fine scissors. Note their internal folding, *reno-pericardial openings* and *renal pores* into the pallial cavity.

8. Now complete the dissection of the *nervous system*. Using the diagram (Fig. 56, C) trace the *visceral loop*, noting its post-torsional figure-of-8 disposition. Clean up the half-loop of the *cerebral commissure*, finding its nerves from the structures of the head. Note the closely parallel *cerebropleural* and *cerebropedal commissures*. Cut away the oesophagus behind the nerve ring and remove the gut to show the *pleural ganglia* set close together and the origins of the visceral loop. In the superficial muscle layer of the haemocoele floor, trace out the primitive ladder-like pedal nervous system. Locate the parietal ganglia with adjacent *osphradial ganglia* at the base of either gill. Follow the *parietovisceral connectives* backwards to the *visceral ganglia*, at the base of the pericardium.

9. Finally, remove the buccal mass, and either lay it open from above, noting the odontophore, radula and jaws, removing the radula for mounting; or divide it into halves by a sagittal cut. Compare with Fig. 55 A, for a slit limpet.

(d) *Ciliary Systems*

1. Remove the shell of a living specimen and open the pallial cavity as previously directed. Mucus from the hypobranchial glands may at first be troublesome.

2. Refer to a thick section for the natural disposition of the *ctenidia*; and locate the currents along the filaments (frontal and abfrontal) and along the gill axes by gently adding coloured suspension. Which way does rejection proceed? Note too the currents on the hypobranchial gland folds and rejection currents on the pallial floor.

3. Cut a few filaments from the living gill, and in seawater in a cavity slide observe their cilia, including the current-driving lateral cilia. Reconstruct in a diagram the movements of water and sediment in the intact mantle cavity.

4. Study prepared sections of gill filaments, noting the ciliary tracts, skeletal rods, blood space and respiratory surface.

5. The internal ciliation of the *stomach* is harder to observe. With patience this can be seen, after opening the stomach and allowing time for ciliary action to resume. Do not over-smother with particles. With a good hand lens or a binocular microscope, observe the sorting area and style sac cilia.

11.5.2. *Patella*: AN ALTERNATIVE EARLY GASTROPOD

The true limpets are the most successful and numerous archaeogastropods. Though basically primitive they have become specialized for firm attachment on wave-beaten shores (Fig. 57). *Patella*—(or in the Pacific Ocean, *Cellana*)—species are good-sized limpets, offering a comparison with the more generalized zeugobranchs.

(a) *External study*

1. As with the ormer, attach to a glass plate, and observe the action of the foot, tentacles and radula.

2. Using particle suspensions if necessary, examine from below the pallial cavity and its openings. Find the site of inhalant and exhalant openings. Note the action of the marginal *branchial lappets* and their associated *pallial tentacles*.

3. In a living limpet detach with a scalpel point the *shell muscle* around its horseshoe-shaped course (See Fig. 57 D). Remove the animal and allow it to attach by the foot to a dish. Cover with water and view its external topography.

4. Locate the shell muscles, and the circle of branchial lappets, with the more spacious pallial cavity overlying the head. In an acmeid limpet, note the single bipectinate *ctenidium*. Note the *pericardium*, with the heart. The ventricle is not transversed by the rectum and there is but one auricle.

5. Trace the outlines of the very unequal *renal organs*, and find their pallial openings at either side of the anus. The larger *right kidney* serves as a gonoduct. It partly encircles the visceral mass, but leaves revealed at the centre part of the digestive gland and a superficial loop of the intestine.

6. Raise the flap of the mantle and trace with particles the ciliation of the branchial lappets, and (in Acmeidae) the gill, as well as rejectory currents on the side of the foot. Locate the small paired *osphradia*.

(b) *Internal structure*

1. Divide a fixed specimen into two clean sagittal halves and survey the arrangement of the viscera. The plan of the head-foot, pallio-pericardial organs and visceral mass may be compared with that of a keyhole limpet (p. 154).

2. Note the *ovary* or *testis* lying postero-ventrally in the concave bowl of the shell muscles. Identify the dark superficial layer of the two renal organs, and the light brown digestive gland compactly surrounding the coils of the gut.

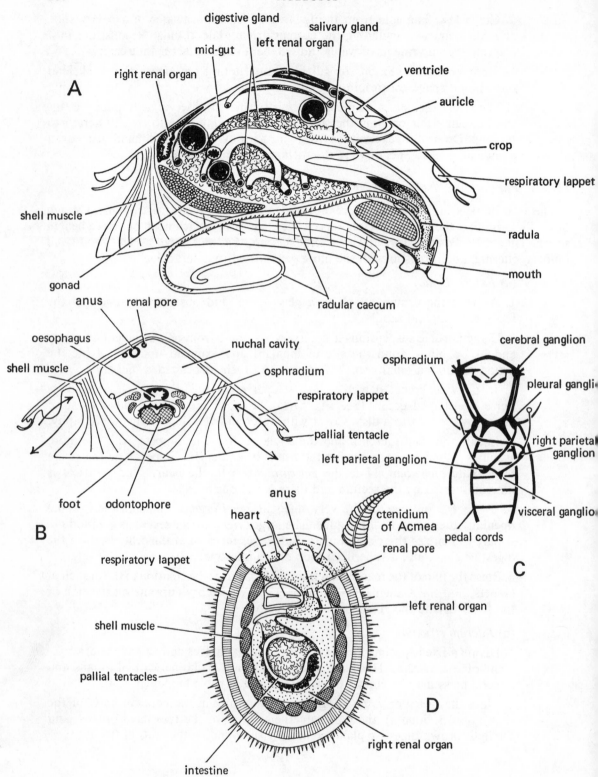

3. Observe the *buccal mass* in sagittal section. Compare it with that of *Haliotis*; note particularly the extreme length of the *radular caecum,* loosely coiled on the floor of the haemocoele above the foot. Take a portion of the radula and mount it. Note the narrowness of the ribbon and the reduced number of strong teeth, tools for abrading rather than sweeping. Note the *salivary glands* and their openings; and the *oesophagus* with lateral glandular pouches.

4. By minimal dissecting within the two sagittal halves the coils of the gut can be reconstructed. The gut structure is highly simplified. The widest part is the digestive and storage chamber, made up by the confluence of *crop* and *mid-intestine.* The *stomach*—lying between these—is marked simply as the place of entry of the *digestive diverticula.* The coils of the *intestine* are very extensive, related to the high volume of faeces produced.

5. The *nervous system* is basically as in *Haliotis*. Dissect it separately by pinning down a fixed animal through the edges of the foot, and approaching the haemocoele through the floor of the pallial cavity. Remove the renal organs, digestive gland and gut coils, but leave in situ the buccal mass and first part of the oesophagus. Identify the *cerebral ganglia,* separate in the limpet from the transverse commissure, the *pleural* and *pedal ganglia* and their connectives with the cerebrals. Trace the *visceral loop* and the *pedal ladder* as in the diagram.

11.5.3. *TROCHUS*: A Transitional Prosobranch (Fig. 59)

These prosobranchs stand intermediately between the Zeugobranchia and the Monotocardia. Since they form an instructive link in the series and are—especially in warmer seas—of ideal size for dissection, a generalized diagram of trochid structure is given here.

Like the ormers top-shells have a tentacled epipodial fringe; the snout and radula are used in the same algal-browsing manner, with a sweeping action of the marginal teeth.

The left ctenidium alone remains, but this is feather-like and still bipectinate; along most of its afferent side it is slung to the mantle by a suspensory membrane. Part of the hypobranchial gland, generally held to be the right member of a pair, lies to the right side of the rectum. There is neither penis nor pallial genital duct.

There are two kidneys, of which the left is the smaller and the genital products pass through the fully functional right member. The circulatory system (see Fig. 58) is transitional in type. The heart retains a right auricle and though the right ctenidium is lacking, its vascular arrangements largely survive to serve the right hypobranchial gland.

The digestive system is similar in general to *Haliotis*, with large oesophageal pouches and crop, a spiral sorting caecum, and a forward loop of the intestine alongside the crop.

The nervous system is unconcentrated but differs from *Haliotis* in having well-marked cerebral ganglia and the pedal centres not ladder-like.

Figure 57 The Limpet, *Patella*

A, Internal structure revealed by a simple longitudinal slice; B, Transverse slice through the anterior region; C, Diagram of the nervous system; D, The animal in external dorsal view after removal of the shell

11.6. Monotocardian Gastropods

11.6.1. *BUCCINUM*: A Carnivorous Prosobranch

Buccinum undatum is one of the largest British gastropods; it gives a good introduction to a monotocardian prosobranch, with the ctenidium and whole pallio-pericardial complex single, reduced to the organs of the left side (Fig. 58). The basic structure of higher prosobranchs is far more uniform than in the lower, and any of a number of whelks (or carnivorous Mesogastropoda such as the trumpet shells, Cymatiidae) will conform—with some variations—to these instructions (Fig. 60).

(a) *The living animal*

1. Examine living whelks in a seawater tank. Note the wide sole, the heavy operculum at the metapodial end of the foot, the cephalic tentacles, mouth and the vertically held inhalant siphon.

2. Chemodetection of distant prey can often be demonstrated, with dead or moribund animal tissue placed behind a narrowly opening screen, or in a sandwich between glass sheets. Note in such conditions the eversion of the slender proboscis, and the action of the radula.

3. Smaller carrion-feeding whelks, as of the genus *Nassarius*, may well illustrate chemodetection of food, if allowed to 'home' on a piece of fish-head. Note the side-to-side ranging of the inhalant siphon, a 'movable nostril', the mobility of the foot and the action of the proboscis in contact with food.

4. A shell-boring stenoglossan of the Muricacea, e.g. *Urosalpinx*, *Nucella* or *Ocenebra* can sometimes be demonstrated with the radula in action upon a mussel, oyster or large barnacle. The entry of the capilliform proboscis can be observed and the form of the radula-chiselled hole should afterwards be examined with a hand-lens.

(b) *Preparation of material*

Narcotized whelks should be fixed, after ceasing to respond to stimuli. Removal from the shell requires care, especially if injection is intended, to avoid rupture of blood spaces near the columella. The whole shell may be put under pressure in a bench vice, until the body and visceral whorls crack without injury to underlying parts. Fragments of the shell can be removed piecemeal with forceps. The visceral spire slips away freely, and the shell attachment is by the strong columellar muscle that should be dealt with last. Bouin's fixative or phenoxetol is best. Keep the fresh animal away from aqueous fluids; water will massively swell the hypobranchial or oviducal mucous glands if ruptured.

(c) *The pallial cavity*

1. Affix the whelk to a board or dish with strong pins through the foot, to leave the mantle uppermost. Identify the *ctenidium* and *hypobranchial gland* externally.

Figure 58 The Circulation of Prosobranch Gastropods

The vascular plan of *Haliotis*, *Trochus* and *Buccinum* is shown diagrammatically, to reveal homologies and to assist the interpretation of blood vessels in dissection.

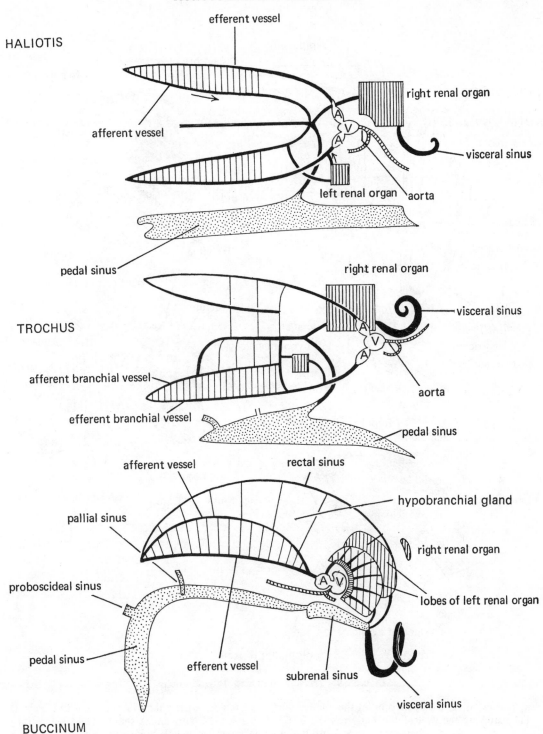

HALIOTIS

efferent vessel

right renal organ

afferent vessel

visceral sinus

left renal organ

aorta

pedal sinus

TROCHUS

right renal organ

visceral sinus

afferent branchial vessel

efferent branchial vessel

aorta

pedal sinus

afferent vessel

rectal sinus

hypobranchial gland

pallial sinus

right renal organ

proboscideal sinus

lobes of left renal organ

pedal sinus

efferent vessel

subrenal sinus

visceral sinus

BUCCINUM

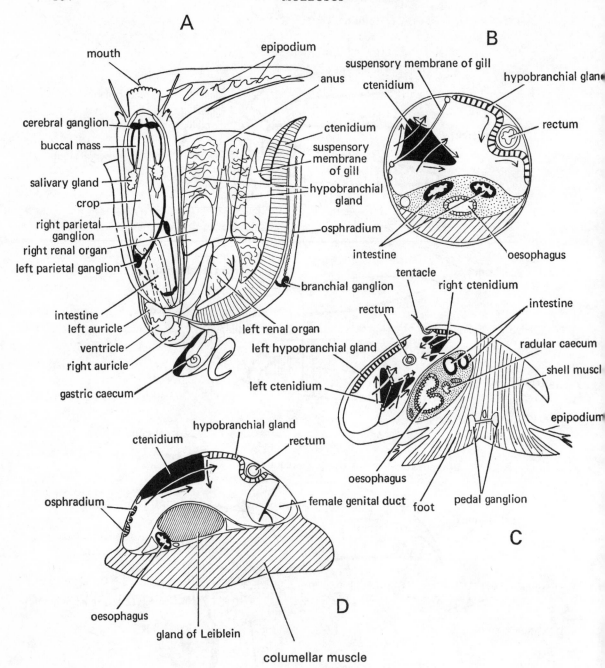

Figure 59 *Trochus* and Other Prosobranchs

A, *Trochus*, dissected by incision of the mantle along the left side of the gill and reflecting it to the right, and removing the wall of the body cavity; B, C, D, *Trochus, Haliotis* and a thaid whelk (see p. 168) showing basic structure of trunk and pallial cavity as revealed by thick slices

2. Open the pallial cavity with a long straight mid-line incision, running from the mantle edge. along the hypobranchial gland, just to the right of the gill. Pin out the two side flaps to display clear the pallial organs.

3. Note the *inhalant siphon*; the brown-pigmented *osphradium*, like a small bipectinate gill; the large *ctenidium* with a single row of triangular filaments, and the heavy folds of the *hypobranchial gland*. Find the *anus* and the course of the rectum, immediately to the right of the hypobranchial gland.

4. In a living specimen, opened in seawater, trace the course of the ciliary currents of the pallial cavity. Supplement your observations with a thick section of the pallial organs *in situ*.

5. The pallial parts of the genital ducts should be noted at this stage. In the male note the large muscular *penis*, on the right, reflected back into the pallial cavity from its origin behind the head. Trace back the superficial course of the *vas deferens*, a shallow duct beneath the body wall, and note the narrow *prostate gland* running forward from the vas deferens at the right posterior floor of the mantle cavity.

6. The pallial part of the female genital duct bulges massively into the mantle cavity, just ventrally and to the right of the much narrower rectum. Locate the cream-coloured *capsule gland*, and find the forwardly placed female genital pore, with a small *bursa copulatrix* incorporated in the oviduct wall at this level. Note at the posterior end, before the capsule gland begins, the U-shaped tract of the smaller *albumen gland*, receiving the narrow ovarian duct; and the small *ingesting gland*, in the oviducal wall between albumen and capsule gland.

(d) *Gonad and pericardial organs: Circulation*

1. Examine the intact visceral mass behind the mantle cavity. Note first the massive *gonad* (yellow *ovary* or reddish brown *testis*) overlying much of the brown digestive gland on the surface of the visceral spire. Trace its duct forward to the pallial genital tract, as it runs along the concave (axial) part of the spire. In the male the *gonadial duct* is thrown into strong convolutions and forms a white *vesicula seminalis*.

2. By gently detaching the first part of the visceral spire from the columellar muscle, many details can be seen externally: the base of the *pericardium* and *renal organ*, the inception of the *anterior aorta*, the course of the *posterior oesophagus* to the stomach, and the *visceral ganglia* at the posterior end of the visceral nerve loop. In an injected specimen this region will be occupied by a wide sub-renal blood sinus.

3. Locate the *pericardium* from the upper (convex) surface, just behind the mantle cavity, and the sole-surviving (left) *renal organ* lying behind and indented by the pericardium. Note the *pericardial gland* running around the kidney at its pericardial edge.

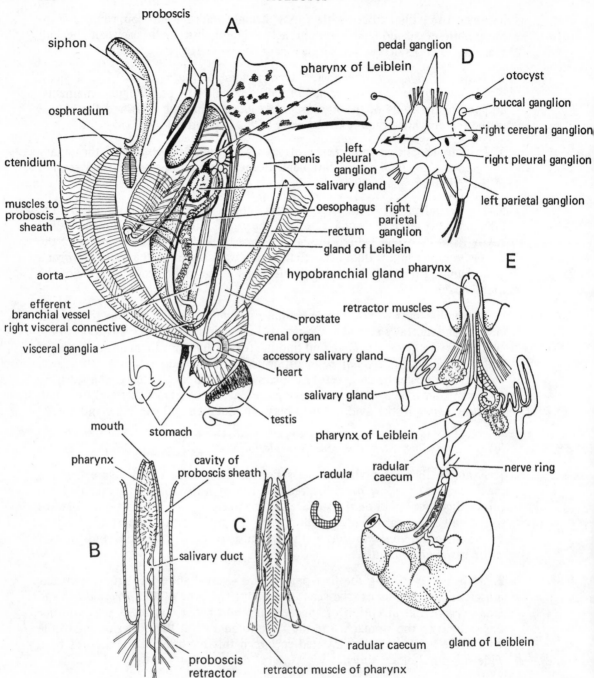

Figure 60 *Buccinum*, the whelk

A, Dissection of the whole animal. The mantle cavity has been opened in the mid-line and the roof of the body cavity and part of the wall of the proboscis sheath removed; B, The proboscis in its sheath opened to show the pharynx and oesophagus within it; C, The pharynx opened dorsally to show the radula (odontophoral cartilage in section, shown inset); D, The nerve ring, opened in the direction of the arrows by cutting the cerebral commissure in the mid-line; E, The foregut of a muricid whelk, such as *Thais*, to show differences in the glandular apparatus

4. Open the pericardium with fine scissors. Note the single *auricle* and *ventricle*. Into the auricle open the *efferent branchial vessel* and a small *efferent renal vessel* from the dorsal lobe of the kidney, emerging from the anterior end of the renal gland. From the ventricle departs the *aorta*, dividing immediately into wide *cephalic* and *visceral arteries*.

5. Open the renal organ, and note the complex folds of renal epithelium that hang down from its roof and reach across the lumen. These receive blood from the floor of the kidney by vessels which pass through the lumen. Locate the *reno-pericardial duct* and the slit-like *renal pore* to the mantle cavity.

6. The blood circulation can be better followed after injection of a fresh specimen. The arterial system is accessible through the ventricle. The venous system can be approached in two ways:

(i) through the efferent branchial artery with the syringe pointed towards the heart: injects the branchial, hypobranchial and pallial afferent vessels, as well as the auricle and the renal efferents.

(ii) through the foot, with the syringe reaching the trunk haemocoele where the oesophagus lies: injects the spacious cephalopedal returning blood to the renal organ.

7. See Fig. 58, for a diagram of the circulation. Blood from the head and foot collects in the *cephalopedal sinus*, and passes with blood also from the viscera into a *sub-renal sinus* beneath the kidney. From here it is distributed:

(i) through the internal folds of the kidney, by vessels passing from the floor across the lumen to the edges of the folds; it is thence carried by *renal efferent vessels*, visible beneath the kidney roof, through the renal gland direct to the auricle.

(ii) through vessels of the floor of the kidney, or directly into the pallial afferent vessel to the renal organ and gills.

(e) *Head and Trunk: The foregut and central nervous system.*

1. Make a long incision from the mouth along the floor of the mantle cavity, back as far as the pericardium. Underlying structures should be carefully protected, as they lie tightly against the thin but muscular body wall.

2. Follow the oesophagus and anterior aorta forward to the nerve ring. Draw aside the proboscis sheath to show clearly the nerve ring and the double-bent course of the oesophagus.

3. Before dissecting the gut further, examine the *nerve ring*. Note the *cerebral ganglia*, close together in the dorsal mid-line. Divide these ganglia, and open out the nerve ring to free the oesophagus from it. Examine with a good hand-lens the nerve ring from within. Note the relations of cerebral ganglia with *pedal* and *pleural ganglia*, the smaller *buccal ganglia* and the minute *otocysts*. The left and right *parietal ganglia* (*supra-oesophageal* and *sub-oesophageal*) are both drawn close into the nerve ring.

4. Follow back from the parietal ganglia the connectives to the *visceral ganglia* far back beneath the posterior floor of the mantle cavity. Where is the site of *chiastoneury*, or crossing-over?

5. Continue the dissection of the gut. Note the retractor muscles mooring the *proboscis sheath* to the haemocoele floor. Follow the oesophagus from the end of the proboscis sheath back to its hemispherical expansion called the *pharynx of Leiblein*, then through the nerve ring. Locate the *gland of Leiblein*, a long tubular diverticulum alongside the oesophagus, opening by a valve at the pharynx of Leiblein. Find the compact salivary glands and trace their ducts forward alongside the oesophagus.

6. Follow the oesophagus backward to the small bag-like *stomach*. Note the openings of the *digestive diverticula* and the course of the *intestine* and *rectum*.

7. Finally, complete the dissection of the foregut. Open the proboscis sheath down its whole length. Note the muscular *proboscis* within, the mouth at its tip. By stretching the proboscis from the anterior end, note its mode of eversion.

8. Open the proboscis in turn, to reveal the buccal mass within it, with the oesophagus, the salivary duct, and the short radular sac.

9. Open the buccal mass from above. Note the odontophore, and its cartilage and muscles. Remove and mount the radula.

(f) *The foregut of Muricacea*

Fig. 60, E illustrates the major differences in the foregut of a muricid whelk. In association with the boring habit, the radular sac is longer; and there are two sorts of salivary gland, acinose masses as in *Buccinum*, and convoluted tubular accessory glands. The gland of Leiblein attains a large size, forming a greyish-brown mass almost filling the trunk haemocoele.

11.6.2. *CREPIDULA*: A Specialized Ciliary Feeder

The slipper limpet, *Crepidula fornicata*, or any related calyptraeacean, offers an example of an advanced mesogastropod that has in its feeding and pallial organs evolved some way parallel to the bivalves (Fig. 61). It is a protandrous hermaphrodite, forming a sex pile with the oldest (female) individual beneath. *Crepidula* is studied to best advantage alive.

1. Examine the under-surface of a specimen attached to a glass plate, or viewed from below in seawater. Note the *foot*, in front of which may (at the female stage) lie a cluster of egg capsules. Locate the mouth and the oral lips, flanked by short cephalic tentacles. Behind the head identify the semicircular *nuchal lobes*, and on the right side the vestigial *penis*, from the previous male stage.

2. With carmine suspension, find the inhalant pallial opening on the left, with a filter of mucous strands excluding coarse particles. On the right lies the exhalant opening and the flap of the *food groove*.

3. Remove the living animal from the shell by passing a scalpel blade carefully under the foot, to detach it from the shell septum. Separate the shell muscle at its narrow attachment.

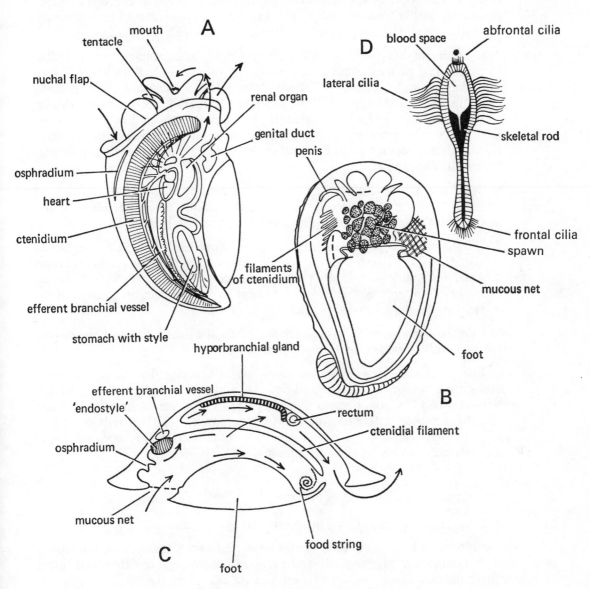

Figure 61 *Crepidula*, the slipper limpet

A, External structure as shown in dorsal view by removal from shell; B, The animal in ventral view with spawn mass; C, Diagram of the mantle cavity in transverse section; D, A single gill filament in transverse section

4. Pin out the slipper limpet mantle cavity uppermost, securing it at several points of the foot. In the intact animal, observe the general changes of topography, especially of the heart and gill, in relation to the enlarged mantle cavity. Note that the stomach has a well-developed *style sac* with a crystalline style.

5. Open the mantle cavity down the extreme right side, swinging the whole flap to the left and pinning it out under seawater. Note the enlarged *ctenidium* with long, finger-shaped filaments, the glandular strip of the "*endostyle*" running along its axis, and the short *osphradium* to its left. Identify the mucus-producing pouch in the left border of the mantle skirt. Locate the shallow gutter of the *food groove* along the right floor of the mantle cavity. Observe the forward passage of a mucous rope in the food groove to the head, intercepted from time to time by the radula.

6. With the aid of a thick transverse slice note the disposition of the gill filaments as an oblique lamella, crossing the mantle cavity, dividing it into inhalant and exhalant spaces, and with their tips resting in the food groove.

7. Using suspended particles study the ciliary currents of the gill filaments and the food groove. Remove several living filaments and observe them flat in seawater on a slide with a low power of the compound microscope. Note the action of the *frontal, abfrontal* and *lateral cilia*. Reconstruct on a diagram the working of the pallial cavity in life.

8. Examine a prepared transverse section of a gill filament to compare its proportions with a filament of *Haliotis* or *Buccinum*.

9. Living *veliger larvae*, extracted from spawn capsules that have attained a dirty-brown colour, are an excellent object of study in a watch-glass of seawater. Viewed with a low power monocular, they show the action of the ciliated *velum*, and the complete internal structure by transparency. The provisional *larval heart* pulsates beneath the pallial cavity. A minute *crystalline style* rotates in the style sac at some 30–50 turns a minute.

11.7.   Opisthobranch Gastropods

11.7.1.   *APLYSIA*: An Early Opisthobranch (Figs. 62 and 63)

With all opisthobranchs, live material should be studied first. Shape, colour and lability of the body are important characteristics of every species; even more than with other molluscs, fixed material alone presents a travesty of the real animal.

(a) *External study*

1. With small *Aplysia* on algae in a vessel of seawater, note the extensible body, and the stream-lined contour offered with the *parapodia* closed. On the head, see the *cephalic tentacles* and *rhinophores*, the *seminal groove* and *male opening*. The narrow soled foot takes a many-point adhesion to unstable surface: note the foot in action upon algal fronds (Fig. 62).

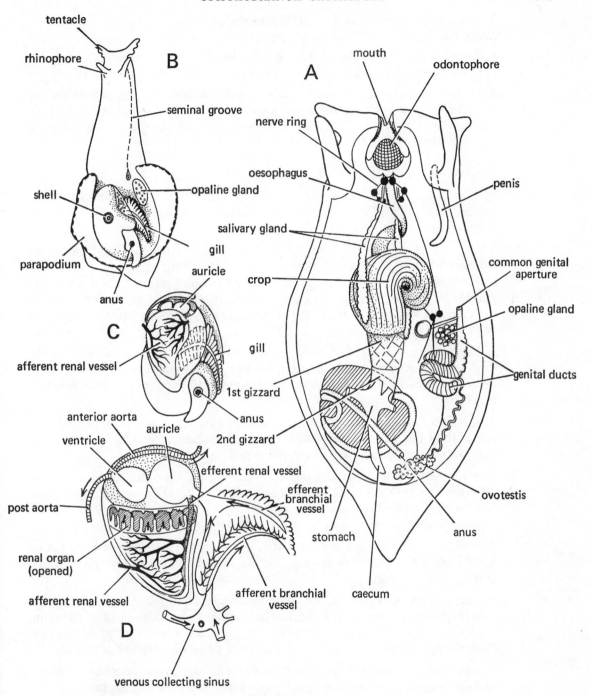

Figure 62 *Aplysia*, the sea hare

A, A general dissection of the internal organs after removal of the pallial cavity and pericardial organs, and with the dorsal body wall cut away (with modification, after Eales); B, A sea hare showing external structure after drawing back the parapodia to reveal the pallial organs; C, The shell removed to show the organs lying beneath it in the pallial roof; D, The renal organ partly dissected to show its internal structure and to reveal fully the heart in the pericardium, with the branchial circulation

2. Locate the visceral hump, mantle and shell (partly covered by integument), mantle cavity, gill and anus, as the parapodia open.

3. Observe the faster movement possible by the backward wave along the parapodia, or by their lifting free of the body to form a swimming skirt.

4. In the sea-hares *Bursatella*, *Dolabella* and *Dolabrifera*, the parapodia are fused in the mid-line save for a short slit; they enclose a subparapodial space that by their pulsations can be filled with water. *Dolabrifera* lives on wave exposed rocks, and has a broad adhesive sole.

(b) *Pallio-pericardial organs*

Specimens for dissection should be carefully narcotized and fixed as well expanded as possible. Fixative should be injected through the body wall into the haemocoele.

1. Cut away the parapodia around their horseshoe-shaped base and locate the entrance to the pallial cavity. Note the partly projecting fleshy *gill*, and behind it the *anus*, in a back-directed channel of the pallial floor. Lift the edge of the pallial roof to show the *purple gland* beneath it and the clear *opaline gland* on the pallial cavity floor. Locate the *common genital aperture* and the beginning of the seminal groove.

2. Locate the chitinous *shell* plate, remove its covering integument, and lift the shell away with forceps. The *pericardium* and *renal organ* lie beneath the transparent mantle. Note on the roof of the kidney the afferent renal vessels bringing blood from the haemocoele and mantle.

3. Cut away the anterior part of the renal organ, to show its interior, and to reveal fully the pericardium and heart. Remove the rest of the mantle roof to show the crescentic shape of the gill.

4. Note the pendent lamellae from the roof of the renal organ, and the passage of blood vessels through the lamellae to an efferent renal vein beneath the floor. Locate the *renal pore* opening into the mantle cavity. The *reno-pericardial opening* is difficult to see with the naked eye.

5. Find the position of the so-called 'ventral abdominal sinus' beneath the pallial floor behind the gill. This receives venous blood from outlying parts of the body, passing it chiefly to the gill by the afferent branchial vessel.

6. Examine the *gill* in detail, noting its differences from a ctenidium; it forms a closely plicate, double-walled sheet. The characteristic ciliation of a ctenidium is lacking. Blood spaces run through the gill from afferent to efferent sides. Note the *efferent branchial vessel*, and its close point of entry to the auricle.

7. Open the pericardium to display the *heart*. The auricle receives from the *efferent branchial* and *efferent renal veins*. From the ventricle the main arteries arise (i) by an arterial trunk passing beneath the heart across the front of the pericardium, giving rise to a large anterior aorta and a smaller gastro-oesophageal artery. (ii) a posteriorly directed visceral aorta.

(c) *Transverse section*

An important adjunct to dissection is to appreciate the organs *in situ* as shown in a thick slice (Fig. 63). Or a stained and mounted section of a young animal (c. 10 mm) may be examined with a low power microscope.

(d) *Internal dissection*

Opening the body wall:

1. Remove the mantle and the gill completely, and expose the viscera by removing the floor of the pallial cavity and opening the dorsal body wall.

2. Locate the site of the *opaline gland* and the *common genital aperture*. Begin the incision between these, keeping the opaline gland to the right of the cut. To the left of the cut, immediately beneath the body wall, lie the *genital ducts* and the *parietal* and *visceral ganglia*.

3. Continue the cut backwards, to the right of the anus, and around the horseshoe of the parapodial base. Cut forwards from the common genital aperture obliquely towards the mid-line between the tentacles. Expose the underlying viscera as in Fig. 62, A.
Note the compact mass of the digestive gland and its contained organs, loosely held by connective tissue within the haemocoele.

Digestive system

1. Examine the course of the gut from the mouth backwards. Identify the *buccal mass*, the paired *salivary glands*, the spacious *crop*, the two triturating *gizzards*. Within the digestive gland, expose by dissection the *stomach* and its *caecum*, and the course of the *intestine* and *rectum*.

2. Having noted the cerebral ganglia on the upper surface of the oesophagus, divide the nerve ring by cutting between them, and open the buccal mass from above. Note the spinose lateral *jaws*, the *odontophore, radula* and *radular sac*.

3. Open the first and second gizzard, noting the form of the chitinous lining teeth. The first is a mincer with heavier teeth, the second a strainer with the teeth more slender. Slice away the roof of the stomach to show the several openings of the *digestive diverticula*, and the caecum in its relation to the intestine. Examine the compact arrangement of the digestive tubules, and study their histology in a prepared stained section (for details of digestive system see Howells 1942).

Nervous system

1. After cutting the commissure between the *cerebral ganglia*, sever the oesophagus behind the nerve ring and lift it out from the ring. Note the sensory nerves from the cerebral ganglia, and the *buccal ganglia* beneath the buccal mass attached by cerebrobuccal connectives. Find the *pleural* and *pedal ganglia*, linked with the cerebrals by *cerebropleural* and *cerebropedal connectives*. Note the *pedal commissure* and slender *parapedal commissure*, the first dorsal, the second ventral to the aorta.

2. Trace back the *visceral loop*, which may vary in degree of concentration in different aplysioids (the condition shown is for *Aplysia punctata*). Locate the pair of ganglia terminating the loop just beneath the pallial floor. To the right is the *parietal ganglion* (before detorsion the *left* or supraoesophageal member. The infraoesophageal or *right parietal ganglion* of prosobranchs is incorporated in the *visceral ganglion*. Note the *branchial* and *genital ganglia*.

Reproductive system

1. Find the *ovostestis* applied to the digestive gland and trace the *little hermaphrodite duct* forward to the genital complex.

2. Carefully unravel this glandular complex. Contact of its fresh tissues with aqueous fixatives, or formalin, will cause swelling of its mucous areas. Encircled by the thick *mucous gland* are the *winding gland* and the compact *albumen gland*. Show the mutual relations of these parts. (see Fig. 65).

3. Note that the little hermaphrodite duct bifurcates on entering the complex. The female path continues through the fertilization chamber, winding gland and mucous gland. The male path proceeds directly forward, its lumen confluent with the straight oviduct.

4. Make a thick slice across the oviduct-spermduct to show its three compartments. (A) the oviduct carrying the eggs forward, (B) the sperm-grooves carrying internal sperm outward, and (C) the copulatory path bringing in outside sperms and receiving the penis of the partner. There are two sperm-containing diverticula, a *spermatheca* from (B) and a *receptaculum seminis* from (C).

5. Follow the ciliated seminal groove along the right side to the head. Note the *penis* which is a simple tube, everted at copulation.

6. If possible examine one of the copulatory chains of a number of individuals, sometimes to be observed in aquarium tanks. The foremost member (A) acts as a female only, while behind and superimposed above A is an individual (B) acting as male to A and as female to C, lying in turn behind.

11.7.2. A SHELLED OPISTHOBRANCH (Fig. 63, A)

One of a number of species of Bullomorpha will provide an alternative to *Aplysia*, with the retention of an internal spiral shell (*Haminoea* or *Scaphander* or *Actaeon*) or with a still spiral internal shell (*Philine*). These bullomorphs are sand-dwellers with a broad foot and parapodial flanges tending to enclose the shell from the sides. The body is smooth, with none of the excrescences typical of sea hares, and the head forms a broad, depressed *cephalic shield*. Within this lies the hard-walled oesophageal gizzard, lined with chitinous or calcified plates. There is no crop, and the *gizzard* is single; *Philine* and *Scaphander* swallow and crush whole molluscan food, *Haminoea* grazes on *Zostera* and other plant food.

The mantle cavity agrees generally in its arrangement with that of *Aplysia*. There is a plicate gill, but the pallial glands are not specifically purple or opaline in type. The water current through the mantle cavity is maintained by strong cilia, not upon the gill, but on two folds, a dorsal and ventral raphe, behind the gill and—in *Scaphander* and *Actaeon*—reaching beyond the mantle cavity in a spiral caecum.

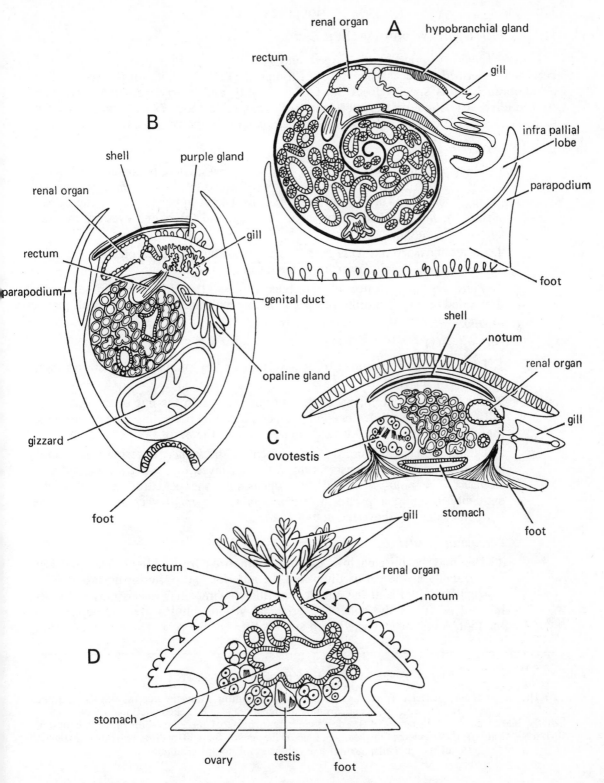

Figure 63 Transverse slices showing the organization of four opisthobranchs
A, A bullomorph such as *Haminoea*; B, *Aplysia* (with modification, after Eales); C, A pleurobranchoid, such as *Oscanius*; D, A dorid nudibranch

11.7.3. *TRITONIA*: A Nudibranch Slug

Tritonia hombergi is one of the largest and best known British nudibranchs. It has been well-described in recent papers by Thompson (1961–2). Like all the soft-bodied opisthobranchs, this species loses most of its beauty and instructive value if not first studied alive, if possible on the substratum of the specific food. *Tritonia* grazes on the living tissues of the coelenterate *Alcyonium digitatum* (dead man's fingers).

(a) *External study*

1. Observe the form of body from above (Fig. 64). There is no shell or mantle cavity and the gill is replaced by a row of dorso-lateral *branchial appendages* along either side. The *rhinophores* are club-shaped appendages, on the dorsal surface, or *notum*. Head tentacles are lacking, and the paired eyes are minute, embedded in integument.

2. Locate the *anus* and *renal pore*, two thirds of the way back on the right side, and the *common genital aperture* some distance in front.

3. Note the prominence of the buccal mass, the distensible mouth with fimbriated *oral veil*, and the sharp *jaw*, acting with the radula to chop off pieces of *Alcyonium*.

(b) *Dissection*

1. Display a fixed specimen, dorsal surface up, by pinning obliquely through the edges of the foot.

2. Cut around the periphery of the notum, just within the line of the branchial tufts. Remove the whole dorsal integument, carefully protecting the pericardium, that lies just beneath the surface, mid-way along the body.

3. Behind the pericardium, trace the extent of the *kidney*, covering the visceral mass superficially, an extensive, thin walled sac investing the digestive gland and stomach. The *reno-pericardial* duct opens into the pericardium at its right extremity, forming a pyriform '*syrinx*'. Note the *renal duct* and *renal pore* lying immediately above the anus.

Circulatory system

4. This may be injected in a carefully narcotized fresh animal. The ventricle lies medianly in the pericardium and the anterior and posterior aortae may be reached from it. From the auricles the efferent branchial vessels may be back-injected as far as the gills. The principal venous sinuses are large spaces in the foot, sometimes injected with success through the sole. These communicate

---

Figure 64 *Tritonia*, a nudibranch Sea-Slug

A, The whole animal dissected from the dorsal surface by cutting away the notum, and showing the underlying organs after removal of the pericardium and renal organs; B, Schematic longitudinal section of the buccal mass; C, Diagram of the nerve ring; D, Schema of the circulation (after Thompson, with modification); E, Side view of the head; F, Plan of the genital ducts, after dissection and unravelling of the glandular genital mass (based on Thompson's figure)

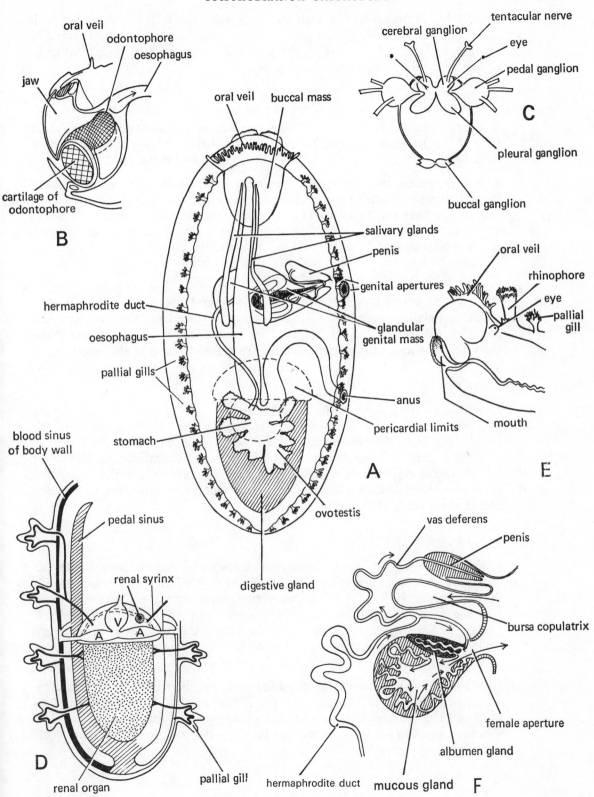

B
oral veil
odontophore
oesophagus
jaw
cartilage of odontophore

A
oral veil
buccal mass
salivary glands
penis
genital apertures
glandular genital mass
anus
pericardial limits
ovotestis
digestive gland
stomach
pallial gills
oesophagus
hermaphrodite duct

C
cerebral ganglion
tentacular nerve
eye
pedal ganglion
pleural ganglion
buccal ganglion

E
oral veil
rhinophore
eye
pallial gill
mouth

D
blood sinus of body wall
pedal sinus
renal syrinx
renal organ
pallial gill

F
vas deferens
penis
bursa copulatrix
female aperture
albumen gland
mucous gland
hermaphrodite duct

directly by ascending trunks with the body wall sinus near its connexion with the auricle. An alternative route for venous blood is through the portal system of the kidney, and thence through the gills before return by a peripheral sinus in the body wall to the heart (see Fig. 64).

Digestive system

5. Remove the pericardium and kidney to disclose the alimentary canal more fully. Note the large spherical *buccal mass*, the *oesophagus* arising from it dorsally, and the ducts of the narrow *salivary glands* entering close to it.

6. Follow the oesophagus back to its entry into the massive *digestive gland*. Note the simple course of the intestine as it emerges from the gland and curves over to the anus on the right. The *stomach* is simple, chiefly an atrium where the digestive diverticula converge. Locate it by opening the digestive complex just below its surface by horizontal slicing with a broad scalpel blade. In doing so, note the irregularly lobed *ovotestis* lying just beneath the kidney sac on the surface of the digestive gland, and retain this for later study.

7. Open the stomach to note within it the several openings of the digestive gland, containing tongues of ciliated tissue drawn out from the stomach wall. (The original paired lobes of the gland have fused into a common mass). Note the typhlosole and wide channel leading from the stomach to the intestine.

Nerve ring

8. With the enlargement of the buccal mass, the *pleural* and *pedal ganglia* are concentrated alongside the *cerebral ganglia*, on the dorsal surface of the oesophagus. Note their form and relations as shown in Fig. 64.
The nerve rings of *Tritonia* and *Aplysia* have recently acquired an important interest to the experimental physiologist. Their centres contain the pigmented cell bodies of giant axons, large enough to be found with dissecting microscope and to allow entry of recording electrodes.

Reproductive system

9. From the *ovotestis* follow the narrow *hermaphrodite duct* forward into the complex glandular genital mass. This lies on the floor of the haemocoele beneath the oesophagus, with its openings converging on the right side. Find the *albumen* and *mucous glands*, packed closely alongside the pyriform *bursa copulatrix*. Locate the opening of the oviduct and the bursa. A little in front of these the penis also traverses the body wall. The coiled *vas deferens* enters the penis at its broader end.

10. Carefully remove the entire glandular mass and determine its structure by unravelling, or making thin razor slices. The hermaphrodite gland appears to enter the mucous gland; it actually runs along a deep furrow beneath it, to bifurcate into (a) *vas deferens* leading to the *penis*, and (b) the female tract,

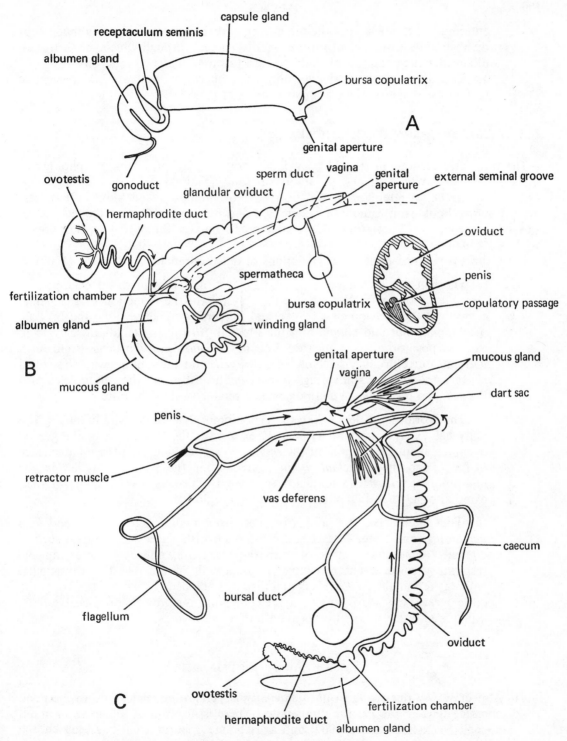

Figure 65 Comparative diagrams of gastropod genital ducts

A, A female *Buccinum*; B, Hermaphrodite genital tract of *Aplysia*; a second through the distal part of the oviduct and sperm duct is shown inset, with the penis of the partner *in situ*; C, Hermaphrodite genital tract of the land-snail, *Helix*. For *Tritonia*, see Fig. 64

entering the complex near the albumen gland. The mucous gland is much subdivided, with a number of outpushed diverticula, through which the fertilized,
albumen-coated egg string winds its path towards the female aperture.

11. Examine some fresh spawn, noting the successive nutritive and protective
secretions. What are the glandular sources of these?

11.7.4. THE DIVERSITY OF OPISTHOBRANCHS

The transverse sections in Fig. 63 provide a ready comparison of the morphology of
the shell, pallial cavity, foot and parapodia in four opisthobranchs. For ecology and
anatomy of *Oscanius*, reference should be made to Thompson and Slinn (1959). The
leading source book on opisthobranch structure and function, reviewing the rich literature of the group, is Thompson's recent monograph on the 'Biology of Opisthobranch
Molluscs' (1976).

For a short general account of the radiations of opisthobranchs, see Morton (1967).

(a) The Pleurobranchomorpha or Notaspidea, as represented, for example, by
Oscanius, show a further stage after aplysiids in the reduction of the pallial
cavity and the development of external bilateral symmetry. Note the broadly
overlapping *foot* and *notum*, and the plumed *gill* (not a ctenidium) carried in
exposed position at the right side. Locate the *anus, genital apertures* and *renal
pore*, all on the right side. Note the *oral veil*, and the mouth, with the rolled
rhinophoral tentacles still springing from the head beneath the notum. See also
the vestigial *shell plate*, a chitinous shield retained below the notum.

(b) The Doridacea belong—like the Tritoniomorpha—to the Nudibranchia. The
body has external bilateral symmetry, except for the lateral site of the genital
openings. Note the position of the *rhinophores*, springing from the notal surface
in front, and the *branchial plumes* surrounding the anus, which lies in the
mid-point behind. Head tentacles are absent. There is a wide range of diet,
external ornamentation and colour among the dorids.

(c) The Aeolidiacea are a highly specialized nudibranch group. *Aeolidiella
papillosa* is one of the largest and common British species. The dorsal surface
contains numerous clusters of club-shaped *cerata*, not only increasing the
respiratory surface, but receiving an outgrowth of the digestive diverticula,
with a terminal '*cnidosac*' containing unexploded nematocysts from the
coelenterate prey incorporated into the epithelium as a protective device. Examine
a prepared transverse section of the tip of a single ceras.

11.8. Lymnea: An aquatic pulmonate

The still or slow-water snails *Lymnea stagnalis, L. peregra* or related species are good
sized representatives of the aquatic pulmonates. There is no want of information about
the land-snail *Helix* (see Rowett 1962) and a basommatophoran snail has been chosen
here as illustrating in many ways the transition towards the higher pulmonates.

(a) *External study*

1. Large living specimens should be examined first in a dish of pond-water. Let the animal emerge until the sole of the foot is extended. Note the *mouth, oral lappets* and the divergent, triangular *head tentacles* with the minute eyes at their outer bases (Fig. 66).

2. The most informative view of the animal is in its upside-down position with the sole outspread and hanging by surface tension immediately beneath the water. Note the shape of the sole, its glistening appearance with mucus; observe the direction of ciliary currents, by which a mucous sheet spreading over the surrounding water film is drawn backwards in the vicinity of the mouth. The radula regularly plucks at the mucous sheet with its adherent food particles.

3. Observe the more general mode of feeding on water plants, or lettuce, cropped between the dorsal chitinous jaw and the protracted odontophore.

4. Locate the slit-like aperture or *pneumatostome* of the mantle cavity, elsewhere closed by a thin septum running from mantle skirt to body wall. Detect the pulsating movements of the body, and the rhythmic opening and closure of the pneumatostome, regulating entry and exit of air into the mantle cavity. *L. stagnalis* breathes generally from the atmosphere, *L. peregra* can use the mantle cavity either as a lung, or a water-filled gill space.

5. Find the rudimentary external *osphradium* on the exposed skirt of the mantle. Note the *anus* and the *renal pore*, both shifted forwards to lie outside the pallial cavity. The *female genital aperture* lies in the same region, but the *male aperture* is far forward (in the position of the common aperture in *Helix*), just behind the right tentacle. Trace the line of the *vas deferens* just beneath the integument on the right side.

(b) *Pallial cavity*

1. Narcotize animals in the extended state, as can be done also with land-snails, by placing them in a jar of cooled, boiled-out water, and leaving for 12 hours or more with air space excluded and the surface covered by a glass plate. After all responses cease, clip away the thin shell with scissors or forceps, slipping out the visceral spire intact, and avoiding damage to the animal where the columella attaches.

2. Dissect narcotized animals fresh, after covering with clean water, or after fixation and hardening.

3. Insert the points of fine scissors at the opening of the pneumatostome and open the mantle cavity by cutting carefully round the thin septum that attaches the skirt to the muscular body wall.

4. When the edge of the mantle is freed, cut it longitudinally parallel to the rectum, as shown in Fig. 11.1. Reflect the mantle, as a flap containing the respiratory area, renal organ and rectum, and pin it out for display under water.

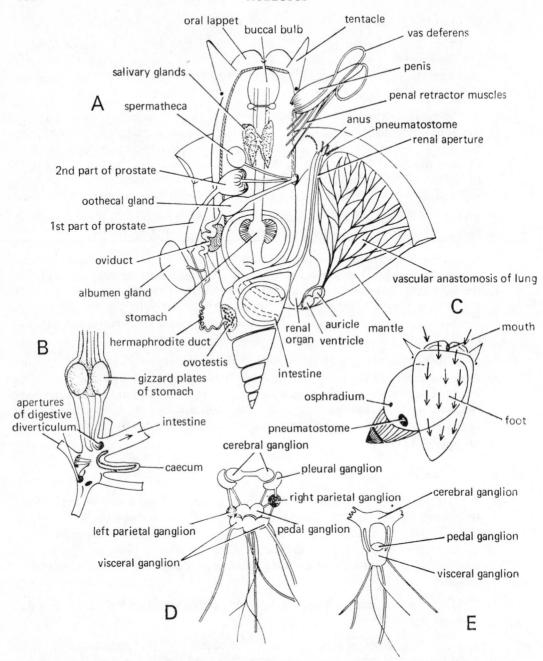

Figure 66 *Lymnea*, A Pulmonate Pond-Snail

A, General dissection, with the mantle cavity opened along the left side, and the body wall removed from the head backwards; B, Detail of the stomach with its gizzard plates laid open; C, External view of living snail, with foot extended beneath a water film, showing the course of ciliary currents drawing a mucous sheet back past the mouth. The mantle wall over the shell aperture shows the pneumatostome and the small external osphradium; D, Nerve ring of *Lymnea*, with (E) nerve ring of *Helix* for comparison

5. Find the extent of the vascular area of the *lung*, noting the direction of the principal *afferent* and *efferent vessels*. Locate the pericardium, with the *heart* consisting of a single auricle and ventricle. Trace the extent of the *renal organ* and its duct, noting its relations with the pericardium.

(c) *Internal structure*

1. Identify the positions of the digestive and reproductive systems through the thin, transparent floor of the pallial cavity. With fine scissors open the haemocoele to the full extent shown in Fig. 66, A, keeping the male and female apertures intact.

2. Now dissociate and display, with little further dissection, the systems laid open to view.

3. Trace back the alimentary canal from the large *buccal mass*, through the wide tube of the *oesophagus* to the *stomach*. Note the paired *salivary glands* and their ducts and keep the nerve ring for the present intact. Find the *stomach* at the base of the visceral mass, with its division into a *gizzard*, with two prominent muscular lobes for its walls, and a thinner-walled part receiving the *caecum* and *digestive diverticula*.

4. Dissociate the soft mass of the digestive gland to disclose the loop of the *intestine* and follow it as the *rectum* along the mantle wall. Note the *hermaphrodite gland*, forming a small yellowish patch on the surface of the digestive gland. Keep it intact, with its duct, for later study.

5. The working of the stomach can best be appreciated after feeding coloured or other recognizable experimental food-stuffs. Remove it, with the ducts of the digestive diverticula, the caecum and the first part of the intestine. Carefully open it with fine scissors for viewing in water under a dissecting binocular. The caecum elaborates a mucus-bound string of fine faecal waste. This is wound around a thicker mucous rope, which is composed of alternate sections of coarse non-nutritive particles and of material extruded by the digestive gland.

6. Open the buccal mass from the dorsal surface and note within it the broad odontophore, and the crescentic jaw. Remove the radula and mount it for microscopic study.

Nervous system

1. Cut through the oesophagus just behind the buccal mass, drawing its cut end through the nerve ring and turn the buccal mass forward so as to expose the ring fully.

2. Note the extreme concentration of the eleven ganglia, with the visceral loop closely drawn into the ring. Identify the *cerebral ganglia*, receiving nerves from the sense organs of the head, and the *pleural ganglia* lying close to them. Long connectives run from both to the ventral part of the nerve ring. Find the pedal centres in close apposition in the ventral mid-line, and the *parietal ganglia*. The *visceral loop* behind these comprises three ganglia, sending out important nerve trunks, as illustrated (Fig. 66, D).

3. The nerve ring of *Helix* is illustrated as a final stage in concentration, with the pleural, parietal and visceral centres fused into one ventral mass. Note the comparable nerve trunks to those of *Lymnea* (Fig. 66, E).

Reproductive system

1. Identify the *hermaphrodite gland* (or *ovotestis*) on the surface of the digestive gland and follow the convoluted *hermaphrodite duct*, to its bifurcation into male and female tracts.

2. Draw out the male and female glandular tracts and display them by pinning to the left. Similarly draw out and pin the *penis* and *vas deferens* to the right.

3. Identify the separate structures. In the female tract note the solid *albumen gland*, joining by a narrow duct, the convoluted *oviducal gland* and its attached *mucous gland*, the bulb-like *oothecal gland* and the narrow *vagina*. A stalked *bursa copulatrix* opens close to the female aperture.

4. Study with a binocular microscope a portion of *Lymnea* spawn mass, noting the jelly coat around the separate eggs, each with its own capsule and albumen supply. Comparison of stained sections (azan or mucicarmine) of egg mass and genital duct shows much about the origin and fate of the successive secretions.

5. In the male genital tract, note the wide *upper prostate* (a cylindrical gland made up of minute acini) and the pyriform *lower prostate*. Trace from this the vas deferens, running just beneath the body wall to the level of the male aperture. Here it has a length of slack, with one or two broad convolutions before entering the penis. Open the penis sheath to show the invaginate penis inside it. Note the strips of retractor muscle securing the penis sheath to the body wall.

11.9.   A Chiton or Coat-of-Mail Shell

Along with the Monoplacophora, the chitons or coat of mail shells have the most primitive pattern of any of the molluscs. Nonetheless they are—as in their multiplication of shell valves and ctenidia—somewhat peculiarly specialized, and have hence been postponed for treatment to this point.

The chitons form only a modest element in the British Mollusca, and are rather small for easy dissection (Fig. 67). Living material of *Lepidochitona cinereus* and *Lepidopleurus asellus* affords a good study of the mantle cavity and its disposition of currents. For internal dissection, some of the western N. American species, *Cryptoconchus stelleri* or *Ischnochiton magdalenensis* are ideal. The present dissection guide is based on a New Zealand species, *Sypharochiton pelliserpentis*; but will apply with slight modifications to any of the Chitonidae.

(a) *External features and pallial cavity*

1. Specimens should be attached to glass plates for viewing from the lower surface. Observe the dorsal surface of the valves and the girdle, minutely scaled in most of the chitonids. Disarticulated valves show division into two regions,

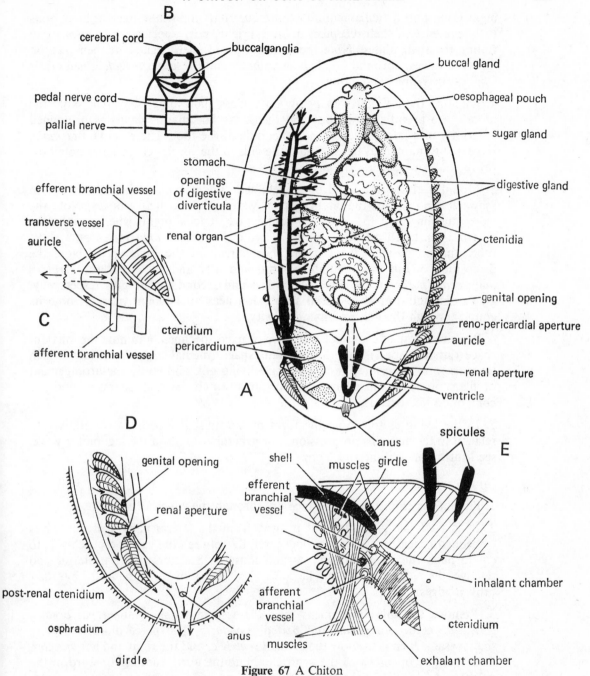

Figure 67 A Chiton

A, General dissection after removing the shell plates (outlines shown by broken lines) and with the pericardium opened to display the heart; B, Detail of the nerve ring; C, Arrangement of the blood vessels in relation to one of the gills; D, View of the left side of the posterior ventral surface showing the gills lying in their natural positions in the pallial groove; E, Transverse section through the right pallial groove and girdle showing one of the gills, semidiagrammatically. (See Fig. 55, C, for filaments and ciliation of the gills)

tegmentum and articulamentum. A unique and important feature of chitons is the presence of shell receptors, minute bright spots appearing like eyes penetrating the shell plates. Note the distribution of shell organs, or their pits, on different valves, both the simple *micraesthetes* and the larger *megalaesthetes* (ref: Charles, in Wilbur and Yonge, 1966).

2. Examine the animal from below on a glass slide under seawater. Note the mouth, the broad ventral foot and the anus. The mantle cavity is prolonged anteriorly into lateral grooves at either side of the foot. These contain *ctenidia*, increased to numerous pairs. Count the gills in the species chosen and note how far the gill row reaches forwards.

3. Mount a chiton mounted on a glass slide under a binocular microscope in seawater, and record the circulation of water in the pallial grooves. Note the close attachment of the girdle all round save at local points where an inhalant aperture can be formed by lifting it. Locate the median *anus* raised on a short tube, and on either side of it the short streak of the *osphradium*. Note the disposition of the final ctenidium, fore and aft; and the separate *excretory* and *genital pores* at the bases of the last and second to last gills respectively. By careful addition of coloured suspension near an inhalant passage, note its course through the gills and mantle cavity.

4. Remove a row of ctenidia from a live or fixed specimen, and note the inhalant space (lateral aspect) and the exhalant space (medial aspect) between which the gills form a curtain. Isolate a single living gill, and study the arrangement of filaments, and the ciliary currents in a watch glass of seawater under a compound microscope.

5. In the living chiton, note the form of the muscular pedal wave and try to relate it to the animal's progression. For general account of the locomotor wave, see Morton, in Wilbur and Yonge, 1964.

(b) *Internal dissection*

1. Pin out the chiton dorsal surface up, through the edge of the girdle.

2. Bend back the girdle to split it away from the lateral flanges of the valves embedded beneath it. Slide a scalpel carefully between the valves on each side to cut through the joining ligaments and muscles. Starting at the posterior end remove the valves, avoiding laceration of the structures beneath by their broken edges.

3. Remove the thin integument over the viscera to show the *renal organs*, *pericardium*, median *gonad* and paired *gonoducts*. The percardium is a wide shallow sac. Open it to show the median *ventricle*, and the right and left *auricles*. Each has two openings to the ventricle. Note the aorta running forward in the dorsal mid-line.

4. Cut away the muscles at the right mantle edge, to uncover the gills from above. Note the longitudinal *efferent branchial vessel*, with several openings to the auricle. Deeper down lies the longitudinal *afferent branchial vessel*,

receiving blood by a transverse vessel from a median collecting sinus in the haemocoele floor.

5. Trace the forward limb of one renal organ with its complex side branches, beginning from the renopericardial opening into the anterior corner of the pericardium. Trace back the returning limb, enlarging into a bladder and discharging by the *renal pore*. From the median dorsal gonad, locate the *gonoduct* passing obliquely outwards to the *gonopore* slightly in front of the renal pore.

6. Remove the heart, renal organs and gonad to display the *digestive system*. Note the large *buccal mass*, with its glandular appendages, the small anterior *salivary glands* and the *pharyngeal pouches*. At the base of the oesophagus open the two large, bilobed 'sugar glands'. The oesophagus crosses obliquely to the left to open into the *stomach*, which expands into a wide ventral storage sac. The shape of the stomach and the coils of the intestine may be revealed by dissecting away the yellowish brown *digestive gland*. Note the openings of its two lobes into the stomach. Cut through the rectum and the oesophagus to lift the alimentary canal clear of the body.

7. Open the buccal mass from above and remove and mount the *radula*. Alternatively, make a sagittal section of the whole buccal mass. Note the considerable length of the radular sac. The radula is supported on a pair of tensely fluid-filled sacs, the functional equivalent of the odontophore.

8. Examining the floor of the perivisceral cavity with the gut removed. Look for the large *longitudinal ventral blood vessels*, collecting blood from head and viscera, and the pair of *transverse vessels* passing blood to the branchial afferent vessels. Note the branch renal tubules surrounding these vessels, and their openings into the renal bladder upon either side. Find more clearly the position of the renal pore.

9. After the removal of the buccal mass, the circum-buccal *nerve ring* may be located, running as a semicircle (cerebral half-loop) in front of it. Note the stout nerve trunks, inner *pedal* and outer *palliovisceral*, running back from it: these are often difficult to see since obscured by muscle strands on the haemocoele floor. Find the small *buccal* and *subradular ganglia* on the buccal mass and their connection with the ring.

11.10. *Nucula*: An Early Bivalve

A protobranch such as *Nucula* or *Nuculana* gives an appropriate beginning to a study of bivalves. They show the most ancient surviving pattern of bivalve organization. The mouth is already raised from the ground, the buccal mass and radula lost, and the pallial cavity re-organized as an all-inclusive space, with anterior and posterior adductor muscles. The labial palps with their large 'proboscides' have more importance than in later bivalves. The ctenidia are confined to the posterior part of the mantle cavity. They have simple triangular leaflets, neither linear nor reflected, and they do not (significantly) collect food. The water current enters at the anterior end of the mantle cavity, passing between the gill filaments from the ventral (inhalant) to the dorsal (exhalant) chamber of the mantle cavity, and leaves posteriorly.

Live nuculids if available should be studied burrowing in surface sand. The plug-like foot when fully spread has a flat surface comparable with a sole. The foot is rhythmically plunged forward or downward into the substrate, serving as an anchor on which the body is pulled down by the shortening of the pedal retractors. The palp proboscides can often be seen in action, with an animal at rest near a glass partition.

Examine the pallial cavity in a fixed animal by removing one valve and fastening the other valve on its side. Identify the *foot, palps* and *palp proboscides, ctenidia, adductor* and *pedal retractor muscles*. A sagittal section stained and mounted is also instructive; note the simplicity of the stomach, with its short, thimble-shaped style sac.

Study also a mounted transverse section through the posterior part of the mantle cavity, showing the arrangement of the two gills in an inverted W. (in an actual section, the filaments may be cut somewhat obliquely rather than seen in conventional face view) (Fig. 68).

11.11.   The Sea Mussel: *Mytilus*

The mussels show well the organization of the mantle cavity in a filibranch bivalve without siphons and with the pallial edges still unfused. They are rather specialized in their attachment by the byssus to a hard surface, but unlike most such attached forms they retain both adductor muscles. The anterior end is however small, pointed and reduced. This trend is carried further in the monomyarian bivalves (Fig. 71) (see p. 195).

(a) *External study*

1. Note the attached mussel and the mode of planting of byssus threads. The tongue-like foot may be protruded, and a thread secreted down its groove, attached and hardened. Or the mussel may become mobile by the purchase of the foot against the substratum to break the threads. Observe the mobility of small specimens, less than a centimeter long: the shell is held upright, and pulled forward by the rhythmic forward extension and contraction of the foot.

2. Locate the inhalant region of the mantle, at the rounded posterior end, noting its crenulated margin, and dorsally to it the exhalant region, a short incipient siphon with plain margin.

(b) *Pallial cavity*

1. Lay the mussel on its side, and insert a strong scalpel blade at the postero-ventral end, between the mantle and the shell lining, cutting through the posterior adductor muscle close to the shell. Carry the blade forward to cut through the pedal muscles and the small anterior adductor at their shell insertions. Take the valve so freed with strong forceps and, with sufficient force to break the ligament, remove it.

2. With the mantle lobe still intact and the animal fixed by the other valve, identify the *adductor muscles*, the six or more *pedal retractor muscles* and the anteriorly placed *pedal protractors*. Note their scars, and the line of pallial attachment to the shell. Observe the form of the inhalant and exhalant openings.

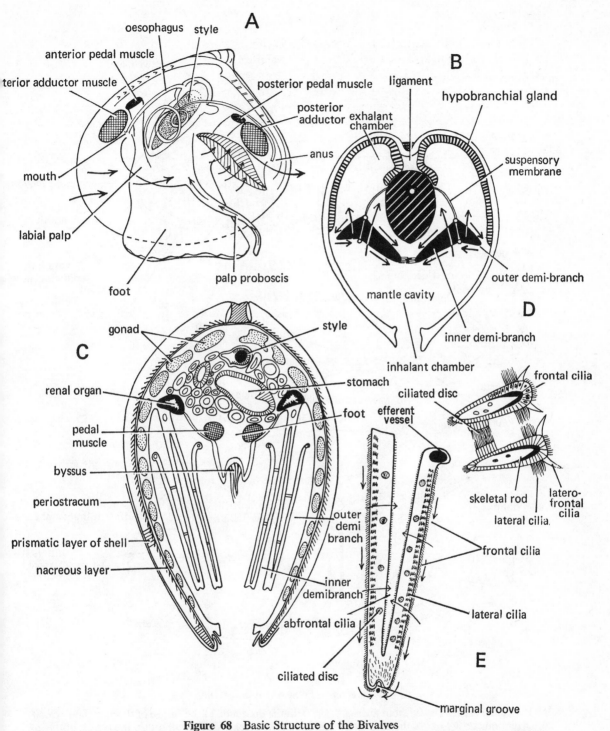

Figure 68 Basic Structure of the Bivalves

A, *Nucula*, with the mantle cavity opened from the left side, showing the pallial organs intact and the course of the gut as by transparency; B, Schematic cross section of *Nucula* through the gills (with modification, after Yonge); C, Schematic cross section through the mantle cavity of *Mytilus* (with modification, after White); D, Transverse section of two gill filaments of *Mytilus*; E, A single gill filament of *Mytilus* in face view, its breadth exaggerated to show the arrangement of the ciliary tracts.

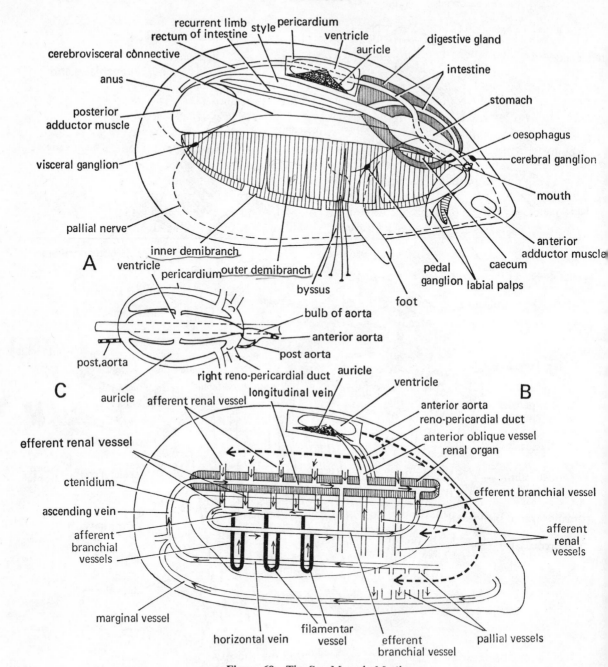

Figure 69 The Sea Mussel, *Mytilus*

A, The pallial organs viewed from the right side, with the dissection of the course of the alimentary canal; B, Diagrammatic lay-out of the circulatory system, with the veins shown colourless, the vessels of the gill filaments black, and the anterior and posterior aortae cross-hatched; C, The heart in the pericardium, opened from the dorsal surface

Make a razor section through the thickened mantle border, noting the *inner* and *outer lobes*, the *middle fold* and the periostracal groove. Examine a stained section of the mantle edge with the microscope.

3. Raise the mantle flap and cut it off near the line of attachment. The pallial organs are displayed beneath. Each gill is composed of an *inner* and *outer demibranch*, corresponding to the two rows of filaments in the bipectinate ctenidium. A demibranch is in itself a double sheet, formed by the folding backwards upon itself of each row of linear filaments to make a *descending* and an *ascending lamella*. Compare the diagrams of the archaeogastropod gill (Fig. 55) and those of *Nucula* and *Mytilus* to understand fully the morphology of the bivalve gill. Note that the lamellae of the *Mytilus* gill fray apart easily into sections by the detachment of some of the filaments from those lying next to them. This is a feature of the filibranch condition, with the filaments united not by vascular connections but only by hairbrush-like ciliary tufts.
Note the parts of the gill, and the labial palps, of which a pair on either side encloses the narrow anterior ends of the two demibranchs.

4. Examine stained sections of the gill running transversely through the filaments. Note the blood spaces and chitinous *skeletal rods*, the *ciliary discs* connecting adjacent filaments, and the *frontal, lateral,* and *laterofrontal cilia* (see for details Atkins 1936). The laterofrontals—not seen at all in gastropods—form immobile filtering combs as water is drawn between them by the action of the lateral cilia. The frontal cilia transport intercepted particles along the length of the filament. Abfrontal cilia, important in gastropods, now have only a subsidiary role, on the 'inner' edges of the reflected filaments. Higher lamellibranchs lose them.

5. Open a fresh specimen and study under clean seawater the action of the pallial cavity. With particle suspensions (see p. 148) carefully investigate all the surfaces of the gill as well as the grooves at the demibranch bases and the currents at the edge. Enter the results of a ciliary survey of the gill on a diagram as in Fig. 71. Note also the course of the rejection currents on the inner wall of the mantle.

6. A clearer view of the gill cilia in action can be obtained from a small square of living gill with part of the demibranch edge, examined first under a binocular and then with a compound microscope.

7. Cut off a *labial palp* at its base and pin out to show its structure and ciliation. The outer surface of both palps is smooth, while of the inner side facing the gill, half the area is transversely ridged. Note the groove bounding the ridges, the four grooves convering one from each palp upon the mouth in the mid-line. Note the passage of suspended particles along the grooved and smooth areas. Large particles are rejected at the tip. Fine material is kept in suspension and passed over the ridges towards the mouth.

(c) *Internal structure*

(In ventral view)

1. Before opening the body wall, much can be learnt of the internal structure by viewing from below, through the roof of the mantle cavity. Divaricate the valves of the fresh specimen, so far as possible without cutting the adductors. Or, with a fixed specimen, the posterior adductor can be severed at its middle length (avoiding damage to the visceral ganglia).

2. Cut away the gills to their bases. (Fig. 70). Note the *renal organs* extending from the labial palps to the posterior adductor. They are reddish brown, and run dorsal to the gill bases, though spreading further sideways. *Mytilus* lacks the second, reflected and parallel, limb of the kidney, that forms in many bivalves a non-glandular 'ureter'. The *renal pore*, mid-way along, shares a common papilla with the genital opening.

3. Follow the widely ramifying *gonadial tubules* in the mantle, white in the male, orange in the female. In a young specimen, not congested with gametes, they can be traced to their convergence near the mantle attachment, whence the main duct leads to the genital opening. Most bivalves, with a foot larger than in *Mytilus*, store most of the gonad in the pedal connective tissue spaces.

4. Trace the *nervous system* in transparency from outside. The *visceral ganglia* and their commissure lie on the posterior adductor muscle. The *visceral connectives* run forward superficially, in the wall of the renal sac, laterally to the external papilla. The *cerebral ganglia* and their commissure, (all that is left of the nerve ring) lie in front of the oesophagus, deep to the mouth. *Pedal commissures* run back from there with the visceral trunks to diverge into the foot. Dissect the base of the foot to expose the pedal ganglia. From the cerebral and visceral ganglia, trace the origin of the *circum-pallial nerves*.

5. After removal of the kidney, the *pedal muscles* may be seen radiating to their shell attachments. Identify first the two *pedal retractors*, springing from the body of the foot itself. Locate the *byssus gland* just behind the foot, with six pairs of *posterior byssus retractors* and one pair of *anterior retractors* inserted upon it.

(d) *Dissection*

1. The animal may now be removed from its shell, by cutting the muscle attachments. On the dorsal surface, open the rectangular *pericardium* to display the heart. The median *ventricle* is traversed by the *rectum*. The *aorta* opens from the anterior end, with a dilated *aortic bulb* immediately in front of the pericardium. The *auricles* are thin sacs, opening at mid-length into the ventricle, and invested by brown *pericardial glands*. Note the arrival of the *afferent oblique vein* from the mantle, near the anterior end of the auricle. Follow with a seeker the downward course of the *renopericardial duct*, from the antero-lateral angle of the pericardium.

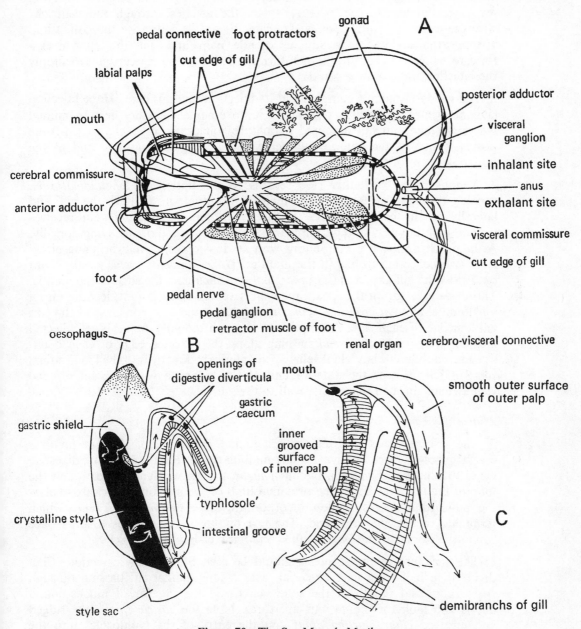

Figure 70 The Sea Mussel, *Mytilus*

A, Dissection of the animal after opening the valves and removal of the gills, with nervous system and renal organs visible by transparency through the roof of the pallial cavity; B, The stomach and its caecum opened from the right side, showing the crystalline style *in situ*; C, The ciliary currents of the labial palps of the right side, with the outer palp drawn aside from its natural position

2. The *circulatory system* may be studied at this stage, and requires injection for successful attack. (see p. 149). Inject the arteries through the ventricle, with the cannula pointing forward. From the aortic bulb arise the pallial and visceral arteries. The former supplies mantle, palps and foot; the latter gives a forward trunk to the stomach, then turns back below the pericardium to supply the intestine and digestive gland.

3. The venous system is complex and its injection requires care. Three injection sites are generally necessary to show it up completely. Choosing an animal where the mantle is not obscured by mature gonads, inject (i) the foot, (ii) the posterior side of the anterior adductor muscle (iii) the posterior end of the horizontal vein of the mantle.

4. Venous blood on either side returns to the important *longitudinal vein*, running through the tissues of the renal organ, dorso-medially to the gill, and laterally to the cerebro-visceral connective. Each longitudinal vein receives visceral blood by way of renal capillaries from the visceral organs lying dorsally. At its posterior end it is continuous with an anastomosing vein bringing blood from two collecting trunks in the mantle: (i) the *horizontal vein*, parallel with but below the gill axis, and (ii) a *marginal vein* encircling the edge of the mantle. Through the length of the renal organ short vessels open from it, leading into a longitudinal *afferent branchial vein*, running along the common base of the two gill lamellae of each side. After passage through the gills, blood is re-collected into an *efferent branchial vein* running along the reflected edge of each demi-branch. The efferent branchial veins open anteriorly directly into the longitudinal vein; and from near the anterior end of the kidney a wide afferent oblique vein ascends through the mantle wall to the auricle.

Alimentary canal

5. The visceral mass is firmly compact, and dissection of the gut involves careful piece-by-piece removal of surrounding tissue. First remove the digestive gland to expose the *stomach*, the *oesophagus*, the *caecum* lying beneath and the loop of the *intestine*. Trace the proximal limb of the intestine backwards below the pericardium. This section incorporates the *crystalline style sac*, which terminates at the posterior end. The rest of the intestine loops to pass once round the stomach, and back as the *rectum* traversing the ventricle.

6. The structure of the stomach should be seen by internal dissection. First divide it in half from fixed material, with a longitudinal cut passing through the caecum and mouth of the style sac. Clean the interior of mucus-bound debris and look for the various apertures. Note the *gastric shield*, the ridged and grooved *ciliary sorting areas* and the entry of the typhlosoles into the caecum. The right series of openings from the digestive diverticula also lie in the caecum.

7. The stomach can be studied with advantage from freshly collected living material. Laying it open with a longitudinal incision, note the flexible *crystalline style* projecting into the stomach. A mucus food string may remain attached to

it. The amylolytic enzyme of the style may be identified by testing it upon starch and titrating for reducing sugar with Benedict's solution. Finally the stomach and style sac cilia can be demonstrated in action with carmine or fine carborundum. For detailed account of the working bivalve stomach, refer to Graham (1949); see also Owen (1966).

8. Stained sections of the digestive gland, style sac and sorting area should be studied with the compound microscope.

11.12 Monomyarian Bivalves

The scallop and the oyster may be dissected as alternative examples to the mussel, or be used for a shorter examination of the pallial cavity and organization plan. They illustrate the extremes of the 'monomyarian' condition reached by bivalves evolving from forms with a byssus-attached habit. (see for a general account Morton (1967) and for details Yonge (1953).

11.12.1. *PECTEN* OR *CHLAMYS*

(a) *External study*

The scallops (Pectinidae) have representatives still byssus-attached (*Chlamys*), *Pecten* is characteristically free-moving. It has developed a specialised mode of locomotion, by expulsion of water from the shell valves, after extreme reduction of the foot. (Fig. 71).

1. First examine specimens alive if possible. The smaller *Chlamys opercularis* shows the escape reaction, by swimming vigorously in the presence of starfish predators. Note the duration and direction of movement, and the site of expulsion of the water current. Live specimens should between experiments be removed from starfish or starfish water: they will otherwise become unresponsive.

(b) *Pallial cavity*

1. Open the valves carefully slicing through the (central) adductor muscle with a scalpel blade. Compare with *Mytilus* in the orientation of the pallial organs and the disposition of the viscera. Much of the latter can be traced by external form and colour with little dissection required. Note the single central adductor, which is the original posterior one. It is divided into 'fast' and 'slow' portions, the latter responsible for prolonged shell-closing contractions, the former for the rapid opening and closing movements expelling sediment in monomyarian surface bivalves and later brought into use for swimming. The gill is broadly crescentic in form, reaching concentrically round the adductor. The foot is vestigial in the extreme anterior position. The visceral mass incorporates the brown digestive gland, white testis and—(forming a distal keel) the coral-red ovary. (The scallop is hermaphrodite). The mouth lies at the bases of the labial palps, anteriorly to the foot, and the stomach—embedded in the digestive gland— can be laid open by tangential slicing. Note the course of the intestine, running crescentically, above the adductor muscle. The rectum pierces the ventricle; and behind the pericardium lies the dark paired tube of the renal organ, embracing the anterior half of the adductor muscle.

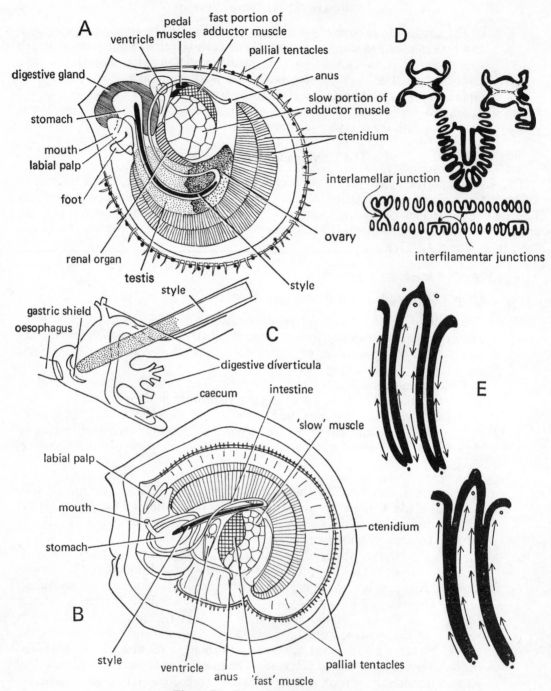

Figure 71 Two Monomyarian Bivalves

A, *Pecten* opened from the left side, showing the pallial organs intact and the viscera as visible externally; B, *Ostrea* opened in the same view, for comparison of the pallial organs and viscera with those of *Pecten*; C, The stomach and the proximal part of the style sac of *Ostrea* (based on Yonge); D, Comparison of a heterorhabdic gill (above) and a homorhabdic gill (below) in horizontal section cutting the filaments transversely; E, Diagram of the inner and outer demibranchs of a heterorhabdic gill, showing ordinary filaments (above) and principal filaments (below) with the course of ciliary currents

2. The ciliary currents of the ctenidia and labial palps should be investigated as in *Mytilus*. Note that the gill, while filibranch, is heterorhabdic, that is composed of normal and principal filaments, and appears thrown into folds or plicae (for details see Atkins 1936). Observe the inhalant currents around most of the circumference, and the restricted exhalant site at the posterior side of the hinge.

3. The edge of the mantle in *Pecten* or *Chlamys* deserves particular attention Note the development of the three lobes in comparison with *Mytilus*. The inner lobe forms a prominent velum, controlling water entry and exit. The middle lobe, which in many bivalves has sensory functions, is prolonged in *Pecten* into a dense fringe of tactile tentacles interspersed with longer oculiferous tentacles. Note the appearance of the eyes, and study a mounted stained section of a *Pecten* eye.

11.12.2. *OSTREA*

1. The true oysters (Ostreidae) illustrate a stage of the sedimentary habit in surface-dwelling bivalves. As in *Pecten* the single adductor muscle is central, divided into 'quick' and 'slow' areas; the gill forms a crescent around it and has—as in the scallop—a plicate structure, with principal and normal filaments. The foot is however entirely lost and the adult has no byssus; the shell is directly cemented to the substratum by the right valve. The mantle margin, with its inhalant currents, is open nearly all round. It is beset with fine tactile tentacles, but never pallial eyes as in *Pecten*. Find the point, just opposite the posterior end of the gill, where the mantle lobes are fused, separating on the one side an exhalant site, distinct from the inhalant sector. Carefully plot the ciliary currents of both demibranchs, as with previous bivalves.

2. The labial palps of *Ostrea* are broad and triangular and may be spread out flat to show their ciliary action. The inner and outer palp on either side enclose the gill, as shown in Fig. 69. Ciliated grooves are confined to the inner surface, and may be examined in a palp cut off at the base, or pinned back against the visceral mass. By coloured particle suspensions the action of the grooves can be seen, carrying coarser material towards the edge of the palp, where it enters a marginal rejection groove, for passage to the tip. Here it spills off. Across the summits of the ridges, lighter particles are carried towards the mouth. The exposed (outer) face of the palp is smooth, and appears to reject material alighting on it. For internal structure and function of the oyster, especially the gut see the account by Yonge (1926), also Yonge (1950).

3. Oysters of the genus *Ostrea* are larviparous as distinct from warmer water oviparous oysters, *Crassostrea*. Larvae are retained between the lamellae of the maternal gills, to an advanced stage with eye-spots are velum. These afford an excellent study of living larval morphology. As well as the velum and gill, the rotation of the style and the movements of food between the stomach and diverticula can be studied by transparency (see Yonge (1926) and Millar (1955)).

11.13. Some Higher Bivalves

The bivalves broadly known as the 'Eulamellibranchia' have as a whole avoided the modifications shown by the monomyarians for surface attachment. They have kept the original anterior and posterior adductor muscles, and the foot is active and unreduced. Evolution has been chiefly in the direction of fusion of the mantle margins, with the production of the posterior edges into inhalant and exhalant siphons.

The following comparative notes deal with three representative eulamellibranchs, the cockle, *Cardium*; the gaper *Mya*, and one of the wafer shells, *Tellina*. Others can be dissected with relatively slight modifications. The best account of bivalve diversity is to be found from the many papers of Yonge (see Wilbur and Yonge (1962) for summarized bibliography).

(a) *Study in life*

Ideally each of these bivalves should be examined *in situ* in the soft substrate. The depth of burrowing and the nature of the expanded siphonal tips noted. *Cardium* embeds only shallowly in the muddy sand of estuaries. The posterior end of the shell is frequently exposed, with the two short siphons extended. Species of related genera, *Laevicardium* and *Acanthocardium*, show a mobility that is in some measure found in all the cockles. Note the firm, strongly compressed foot, oval in section, able to be extended beneath the shell and abruptly straightened to give a propulsive 'leap'.

Tellinids are chiefly found on cleaner sands of exposed beaches, but the small British *Tellina tenuis* and *Macoma balthica* live along with *Cardium edule* (and sometimes *Scrobicularia plana*) in sheltered flats. All the tellinaceans are deep burrowers: note the broad muscular blade of the foot and the swift embedding movements in clean sand. These animals are deposit feeders; the two long siphons are independent, the inhalant one sucking in surface detritus from a radius of two or three centimetres.

Mya arenaria, the gaper, is in the adult much less mobile. Note its deep permanently buried situation in muddy sand, or clay, the long fused siphons encased in horny periostractum, and the vestigial foot, generally not protruded in the adult.

(b) *The pallial cavity*

1. Inspect the pallial organs wherever possible in living material, by removing one valve and affixing the other to a dissecting dish (Fig. 72). Allow time for the gills and retracted palps to re-expand before study. An added appreciation of the topography of the pallial cavity can be gained from studying thick sections.

2. Note the position of the retracted siphons, and the extent of fusion of the mantle edges. Find the *siphonal retractor muscles* inserted in an embayment of the pallial attachment line on the shell. In fixed material, slit open the *siphons* longitudinally to see their interior. Note the *anus* opening into the base of the

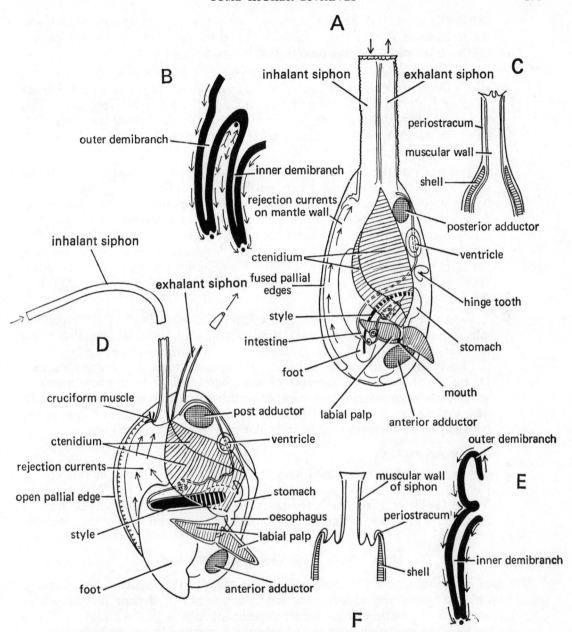

Figure 72 Two Eulamellibranch Bivalves

A, *Mya arenaria* opened from the right side with the pallial organs intact and the alimentary canal shown as if by transparency; B, Diagram of the gill of *Mya*, in cross section showing the course of ciliary currents; C, Structure of the siphon of *Mya* from the ventral aspect; D, *Tellina tenuis* opened from the right side with the pallial organs intact and the course of the alimentary canal represented by broken lines; E, Structure of the siphon of a tellinid; F, The gill of *Tellina* in cross section, showing the course of ciliary currents

exhalant siphon, and the continuity with the suprabranchial spaces, sending out water received from filtering through the gills. See also the visceral ganglia on the lower surface of the posterior adductor muscle.

3. Observe the sizes and proportions of gills and palps, especially (from thick slices) the relative extent of inner and outer demibranchs. In tellinids the outer member is dorsally reflected and reduced, and the palps are very large. This can be related to the large volume of coarse deposits to be sorted before ingestion. Find the directions of the various ciliary currents on gills, palps and mantle lining, as for *Mytilus*.

4. The study of the eulamellibranch gill is complicated in many species by its plicate structure. An originally flat lamella becomes folded and "*heterorhabdic*" by the development of "*principal*" and "*ordinary*" filaments. A tellinid species may be selected to show the flat structure of a "*homorhabdic*" gill. Cut out a small area of a demibranch, including its edge, to show its net-like structure under the binocular microscope. Vascular *interfilamentar junctions* alternate with slit-like openings or '*ostia*' leading into the interlamellar space. *Interlamellar vascular junctions* pass between the adjacent ascending and descending lamellae.

5. Compare transverse sections of a tellinid gill, with those of heterorhabdic gills (e.g. Cardiidae and Myidae); and refer also to Ostreidae and Pectinidae for examples of heterorhabdic filibranch and pseudolamellibranch gills).

6. Inspect the line of fusion of the mantle margins. Begin, in a living specimen, by tracing the rejectory currents of the pallial lining. Find where particles accumulate ventrally and pass back in a median track along the fusion site. In *Mya* such material reaches the base of the *inhalant* siphon, where it is extruded by backflow of current at the periodic contraction of the adductors.

(c) *Internal anatomy*

This is not dealt with in detail here. A detailed and still sound monograph of *Cardium*, though an old one, is that of Chadwick (1900). For Tellinacea see Yonge (1949) and for *Mya arenaria* Yonge (1923). The diagrams in Fig. 72 show the general disposition of the stomach, style sac and coils of the intestine.

11.14. Some Experiments with Bivalves

The bivalves are sedentary molluscs combining, in many of their systems, high working efficiency with simplification of structural components. Many of their functions lend themselves well to observation by laboratory experiment (Fig. 73).

Some modes of investigation, with useful references, are briefly set out here.

11.14.1. LOCOMOTION AND BURROWING

Use for experiments active lamellibranchs such as *Tellina*, *Donax*, or *Macoma*. Place a living specimen in good condition of the sand surface and note the cycle of action of the foot. This expands to a sharply pointed muscular strip, able by its pulsations to be thrust into the sand. It there becomes dilated so as to anchor the animal by its outstretched,

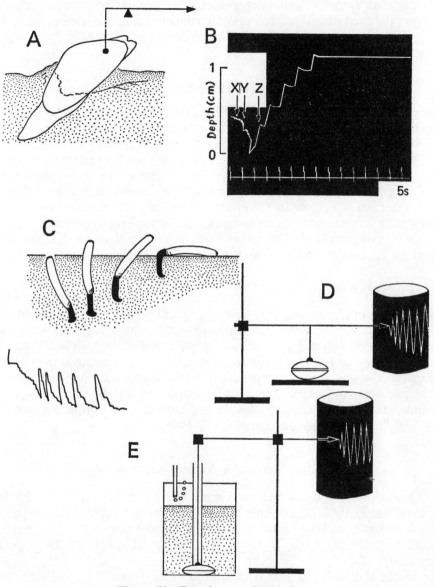

Figure 73 Experiments with bivalves

A, *Donax* showing angle of digging and method of recording by attaching a thread to the valve; B, Kymograph record of digging by *Donax*: *x*, probing by foot; *y* attempts to erect the shell, followed by normal digging cycles beginning at *z* (after Trueman, Brand and Davis); C, Stages in the digging of *Ensis*, with trace showing successive phases of movement; D, Kymograph recording of adductor muscle rhythms in *Anodonta*; Kymograph recording of adductor muscle rhythms; E, In a sand-burrowing tellinid, with thread attached by way of a glass tube through the sand

distended end. The foot of the razor-shell *Ensis* acts similarly, expanding from a pointed probe to a swollen anchor. The retractor muscles of the foot are then strongly contracted so as to pull the shell down into the sand. Trueman and his colleagues have, in several papers, described apparatus for kymographic recording of the bivalve burrowing cycle. See Fig. 73.

11.14.2. ADDUCTOR MUSCLE: ACTIVITY AND QUIESCENCE

A general feature of many bivalves is the regular alternation of bursts of activity and quiescence by the adductor muscles. In *Anodonta*—as investigated by Barnes (1955) the valves are for a short period rapidly opened and closed every six hours; this is by the action of fibres producing rapid or *phasic* contractions (and represents a mechanism for expelling collected sediment—see in *Pecten* its adaptation to swimming by sustained bursts of clapping the valves). Between phasic contractions the valves are held tightly closed, by a portion of each adductor producing prolonged contractions of *tonus*. In Fig. 73 is shown the equipment for recording this activity. With a burrowing bivalve such as a tellinid it is sometimes possible to get more natural results with the same experiment performed on the animal buried in sand. For further details see Barnes (1955).

11.14.3. FILTERING RATE

Bivalves placed in a suspension of a living phytoflagellate readily begin to filter these particles and remove them from the suspension. Alternative filterable suspensions may be water made pink with red corpuscles of dogfish blood, or with a colloidal graphite suspension ("Aquadag") of an appropriate grade. The rate of removal may be calculated by reading the changes in the optical density of the suspension by placing samples in a light absorptiometer or colorimeter. Optical density should be plotted logarithmically against time. The rate of filtering may be established by a simple formula from the clearance rate. For details see Jørgensen (1955, 1966).

11.15. *Loligo*: A Modern Cephalopod

The Cephalopoda are the largest and most complexly organized of the Mollusca, and the most widely available example in most parts of the world is one of the squids of the genus *Loligo* or a related group. The following account has been generalized to include most squids. On the cuttlefish, *Sepia*, an excellent monograph has been provided by Tompsett (1939).

Living squids should be examined in an aquarium if available, and their swimming and resting posture, with the various uses of the funnel, in forward or back propulsion or steering, noted and sketched. Cuttlefish, or better still *Octopus*, are more adaptable to aquaria. For *Sepia*, observe the submerged posture in sand, and the undulatory motion of the marginal fin, with its effects in steering. The elaborate colour changes of the cuttlefish have been described by Holmes (1949). With *Octopus* note the extent of the inter-arm web, and if possible the action of apprehending food by pouncing upon it with the whole arm circlet. The small squids *Sepiola*, of an inch or so long, are sometimes taken living in a fine seine net at the edge of sandy beaches.

It is often possible, near the coast, to obtain the egg masses of a squid or *Octopus* from which live embryos can at an advanced stage be removed. Placed in a watch glass of seawater under a binocular or compound microscope, these will not only show the working of the pallial organs by transparency, but also the expansion and contraction of the three types of chromatophores, black, golden yellow and wine red.

(a) *External features and pallial cavity*

1. Observe the stream-lined shape of *Loligo*, and the arrow-tipped strength of its posterior end, stiffened by the axis of the internal *pen*. Note also the triangular *fins*, the disposition of the *funnel* and its ability to be variously directed in performing steering movements. Examine the circlet of eight short arms around the mouth and the nature of the suckers, pedunculate and with horny rings; also the two long prehensile arms and the sheaths into which they normally retract in life (Figs. 74, 75, 76).

2. Open the mantle cavity by cutting its wall in the median longitudinal aspect along its whole length. Pin the animal with pallial organs and funnel uppermost in a dissecting dish, and pin out the two pallial flaps to either side. The pallial cavity when intact forms a crescentic space extending round the fusiform visceral mass. Protecting the viscera on the upper surface lies the internal skeleton or chitinous '*pen*'. This should be reserved for examination until the viscera have been removed. Note the long axial shaft and the expanded chitinous shield, the *pro-ostracum*, to which the mantle tissues are attached anteriorly.

3. Locate the principal landmarks in the mantle cavity (Fig. 74). The *funnel* is a dorsoventrally flattened tube lying in front. Note the "*resisting apparatus*", a pair of cartilaginous ridges and sockets, effectively closing the cavity to expulsion of water except by the funnel.

On either side of the mantle cavity, suspended from the body wall, lies a long bipectinate *gill*. This is a true ctenidium, but in relation to the cephalopod's increase in size its filaments have become more complex than in a gastropod. Note that they are no longer ciliated, but thrown into primary and secondary folds to increase surface area. An accessory *branchial heart* is visible at the base of each gill. Note the *afferent* and *efferent vessel* traversing each gill. In contrast with gastropods, the afferent vessel runs along the suspension edge of the gill, and the efferent along the free edge.

The *rectum* opens in the median line by a short tubular spout carrying the anus. The *ink-sac* discharges just inside the anus and is a densely black ovoid sac, immediately beneath the integument. Its narrow duct runs along the rectum to a point just behind the anus. Take care not to puncture the ink-sac, for its contents will obscure the whole dissection.

In a female specimen, the ink-sac and rectum may be partially overlapped by the *nidamental glands*, two elongate whitish glandular sacs, in contact along the middle line and with paired anterior openings. A smaller pair of *accessory nidamental glands* lie in front of them, oval and reddish-brown mottled. Note also the paired openings of the *renal sacs*, to either side of the anus, and—in either

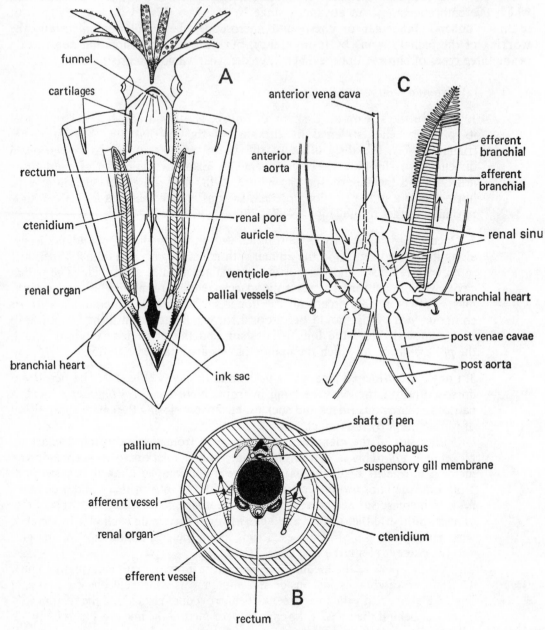

Figure 74 The Squid, *Loligo*

A, Dissection of the pallial organs of the male, after opening the mantle with a mid-line ventral incision, and slitting open the funnel; B, Arrangement of the pallial organs as seen in a transverse slice; C, The heart, venous system and branchial circulation

sex—the single *genital opening*, a spout-like tube on the left side just behind the renal openings.

Find the *stellate ganglia* near the dorso-lateral mantle attachment at either side with their large trunks (carrying third order giant fibres) innervating the musculature of the mantle. The stellate ganglia are connected with the brain by stout pallial nerves carrying second order giant fibres.

(b) *Internal structure*

1. The viscera should be displayed by continuing the dissection from the pallial cavity. Carefully remove (from the female) the nidamental and accessory nidamental glands and after incising the transparent body wall free the ink-sac from its connective tissue attachments and ligate it before removing or reflecting to one side. Continue the median line incision from the base of the funnel back to the apex of the visceral mass. Remove the integument as far as possible to either side.

2. Upon opening the body wall the first structure revealed will be the *renal sac*, with its important contained organs. Follow the external renal pores back into the sac, which has arisen from two originally separate renal organs now grown together by the breaking down of the contiguous walls.

The body of the renal sac occupies the greater part of the visceral mass between the heart and gastric caecum behind and the 'liver' and ink-sac in front. Its walls are wrapped round and intimately applied to the surface of the pancreas, intestine and great veins, so that it appears to contain these organs. The renal sac may be likened to a flask with a neck projecting backwards, vertically flattened and bilobed. The neck is directed posteriorly to overlie the heart and the gastric caecum, and in the female lies between these organs and the nidamental glands which cover it superficially. The neck of the renal sac contains the two side arms of the renal blood sinus, separated dorsally by the remains of the original septum between the two sacs. The total confines of the renal sac can be determined by flooding it with a coloured fluid introduced through one of the external renal pores.

Identify the *reno-pericardial canals*, which are tubes passing obliquely for a short distance in the wall between the renal sac and just above the branchial veins and at the junction of the body with the neck.

3. The *circulatory system* is best traced out after the injection of a fresh specimen. Use a coloured gelatine injecting mass, administered warm, or an oil paint mixed with turpentine. The two longitudinal vessels of the ctenidium, the afferent and efferent branchial veins, should now be located. The venous system can be injected through the afferent vessel or through the branchial heart which lies just inside the body wall at the gill base. The arterial system is best approached through the ventricle direct.

4. Open the renal sac to inspect the contents. Most conspicuous structures are the *renal sinuses* produced by the branching of the *anterior vena cava*. These pass freely through the sac being simply attached at either end to the wall. Within

the flask-like body of the sac the vena cava expands into a large median *renal sinus*. In the neck of the renal sac this divides into two *lateral renal sinuses*. Note the vascular renal appendages covering these sinuses and projecting freely into the renal lumen.

Into the posterior end of each lateral renal sinus opens a spherical sinus from the nidamental gland. This receives—on either side—the long wide blood space of a *posterior vena cava*, one of a pair of convergent blood spaces from the posterior visceral mass. From the lower end of the lateral renal sinus openings, at the outer side, the spherical *branchial heart*, with its glandular appendage. This is joined by a vein from the mantle, and serves as a booster heart for the supply of blood to the gill by the afferent branchial vessel.

5. Note the pancreas and part of the intestine contained in the renal sac. Then remove the venous system to display the underlying *pericardium*, which is a compartment open to the spacious visceral coelom. At the same time the parts of the digestive and genital systems will be revealed. The muscular *ventricle* lies transversely across the mid-line of the pericardium. It receives on either side a slender *auricle*. This continues from the efferent branchial vein, and the heart is thus entirely arterial, receiving and distributing re-oxygenated blood.

Trace the origins of the *arterial system* as illustrated in Fig. 74, C. The *anterior aorta* runs forward from the ventricle as a single trunk to the head. The two *posterior aortae* branch soon after leaving the ventricle. Before they branch a dorsal artery ascends directly to the mantle.

6. For a full dissection of the alimentary system, the rest of the renal and pericardial organs should be removed, and the body wall opened right to the mouth. Carry forward the median incision through the funnel past the level of the eyes. The visceral cavity is in its anterior part filled by the smooth, massive '*liver*'.

Alternatively, in a small animal, the gut may be seen to great advantage by making a sagittal section along the entire body to show the digestive tubes medianly opened *in situ*.

7. The most superficial part of the gut is the ink-sac, already seen attached to the end of the rectum. Follow back the *intestine* from the rectum, to the part in the renal sac and finally to its connection with the *stomach*. This is a stout sac composed of two different structures, both part of the generalized molluscan stomach. Lying to the left is a muscular *gizzard*, lined with cuticle. This receives the narrow *oesophagus*, which can be seen running back from the head in a deep longitudinal cleft of the 'liver'.

The gizzard opens by a narrow isthmus across the median line into the *caecum*. Near this point it receives the wide duct of the 'liver' to which the pale follicles of the *pancreas* are applied. Nearby is the exit of the intestine. Beyond the isthmus the large caecum expands. This contains the gastric sorting area, the '*ciliated organ*' made up of convergent leaflets visible by transparency from outside. Trace it back into the spacious thin-walled *caecal sac*, not seen in the cuttlefish or octopus as such. The gonad lies deep in the caecal sac and the genital duct is superficial to it.

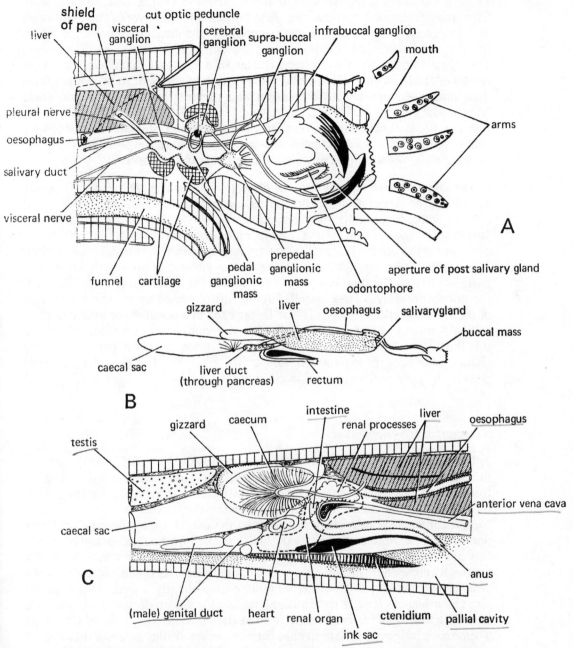

Figure 75 The Squid, *Loligo*: General structure after cutting a longitudinal slice

A, The head with the buccal mass and brain in section; B, Diagram of the course of the alimentary canal (after Bidder); C, The central part of the pallial and visceral region, passing through the stomach, heart and renal organ

Trace towards from the isthmus of the stomach the single duct formed from tributaries from each of the two fused lobes of the 'liver'. The molluscan *digestive gland* is represented in cephalopods by two distinct portions, for which the only common names in use are the inaccurate terms '*liver*' and '*pancreas*'. The white follicles of the pancreas cluster in a compact mass over the main course of the liver duct. Expose the whole length of the pinkish brown fusiform *liver*, noting the oesophagus cleaving it and the anterior vena cava and aorta lying below it.

For the finest functional account of the cephalopod stomach and digestive gland the student should consult the papers of Bidder, especially (1950) on *Loligo* and in Wilbur and Yonge, (1966) on comparative aspects.

8. The alimentary canal may now be dissected forward to the head. The *mouth* lies within the circlet of eight short arms, surrounded by a frilled oral veil, and opening immediately into an ovoid, muscular *buccal bulb. Dorsal* and *ventral jaws* border the mouth, like a strong chitinous parrot-beak, but the lower overlapping the upper. Trace their insertions deep within the buccal musculature. The *odontophore* or tongue carries the *radula* ribbon, relatively slight in proportions. Remove and mount part of the radula for microscopic study. Beneath the odontophore is a large papilla, at the tip of which opens the single duct of the *posterior salivary glands*. These lie far back in a notch at the antero-ventral end of the liver. Their two ducts join in a common median channel. Above this duct is a short papilla with the opening from a second pair of salivary glands, embedded in the musculature of the pharynx. The oesophagus opens postero-dorsally and continues back through the brain and its cartilaginous capsule.

Reproductive System

9. Locate on the left part of the reflected body wall the opening of the genital duct into the pallial cavity. This is single in both sexes and lies just in front of the left renal pore. The nidamental glands, structurally separate from the genital duct, will have been already noted in the female.

Next identify the *gonad* which lies deep to the caecal sac against the posterior part of the pen. Both testis and ovary are single and median, the ovary considerably the larger. From the lining wall, ova are shed into the spacious *genital coelom* forming the ovarian cavity; from this opens the single remaining (left) *oviduct*, a beginning with white, glandular tube slightly coiled. This expands into the *oviducal gland*, a flattened enlargement made up of two glandular pads bounding a slit-like lumen, leading to the external orifice.

The *testis* is longer and more slender than the ovary. From it runs a rather wide *vas deferens* to the glandular male genital duct which lie just beneath the wall of the mantle cavity. The glands form a rather intricate complex consisting of a long receptacle for the spermatophores, called *Needham's Sac*, opening by the external male orifice, and a membranous *genital sac* containing within it the structures that secrete and elaborate the separate parts of the spermatophore.

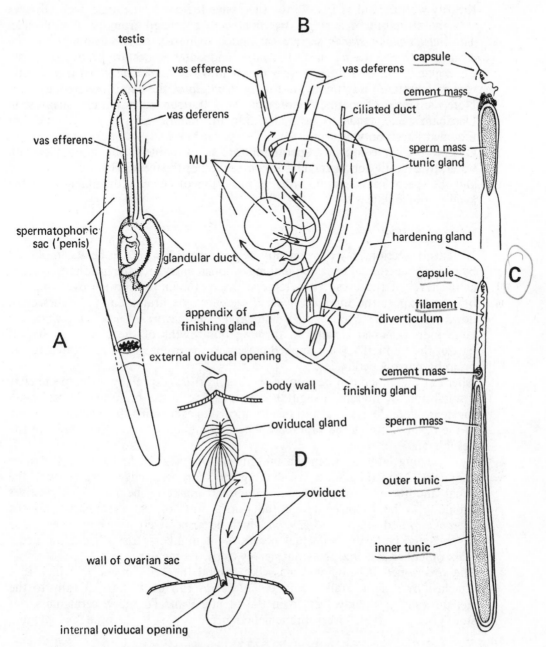

Figure 76   The Squid, *Loligo*: Genital Organs

A, The male genital ducts without dissection; B, Dissection of the glandular mass, showing the course of the ducts and accessory glands. MU indicates the several mucous glands; C, The structure of a spermatophore, with (above) the tip of an exploded spermatophore; D, The female genital duct

Identify with the aid of Fig. 76 the successive regions of the male duct. Observe also the structure of a single *spermatophore* extracted from Needham's Sac. The *mucilaginous glands* secrete the mucous matrix of the *sperm mass*, the *cement body* and the *filament*. The inner and outer tunics are produced by the tunic gland. The structure is then thrust into the *hardening gland* (incorrectly termed 'prostate') and finally, in the *finishing gland*, the cap is secreted.

Note the stores of spermatophores, up to 400 arranged in parallel sheaves, in Needham's Sac, ready to be pulled into the mantle cavity by the somewhat modified "hectocotylus arm".

Observe the explosion of a spermatophore (if available fresh) on contact with water, by osmotic action and elastic contraction of the tunic. The cap is dislodged and the sperm mass everted with its provision of cement for attaching to the head of the female.

Nervous System

10. This dissection may conveniently be performed last. It can be accomplished by turning the animal over and making a dorsal entry into the head musculature at the level of the eyes. Then remove the cartilaginous capsule overlying the brain, to expose the *supraoesophageal region* of the brain. Carry the dissection downwards to the *viscero-pallial* and *pedal nerve centres*. The *optic ganglia* are large kidney-shaped structures, forming outgrowths of the supraoesophageal part of the brain. They are applied closely to the eyeballs and connected to the brain by narrow peduncles.

Alternatively, much can be seen of the brain *in situ*, especially if the specimen is a small one, by making a sagittal section taking in the brain and buccal mass together. Note the clear, semitranslucent cartilages forming the investing capsule of the brain. Observe the supraoesophageal ganglionic mass, lying above to the oesophagus.

By dissecting laterally from the mid-line in one sagittal half expose the optic peduncles. Note the connective linking the supraoesophageal centres with the pedal and the separate prepedal ganglionic masses. The buccal connectives should also be followed along the dorsal wall of the oesophagus, to the *suprabuccal* and *infrabuccal ganglia* which lie respectively above and below the oesophagus. Note the branchial nerves (4 in number) arising from the pedal mass on either side and the stout nerves to the tentacular arms.

The *viscero-pallial* ganglionic mass lies behind the pedal mass, to which it is attached by a short stalk. Stout *pallial nerves* run back dorsolaterally to the stellate ganglia already located in the mantle wall. To the visceral mass run stout visceral nerves. Viscerobranchial connectives pass to the branchial ganglia.

12. LOPHOPHORATES

Three Phyla of typically sessile animals, the Phoronida, the Bryozoa and the Brachiopoda, are similar in having a ciliated lophophore with which they catch their food. The numerous tentacles which form the lophophore are hollow and contain extensions of the coelom.

12.1. Phoronids

Exclusively marine, small colonies of *Phoronis* spp. may be found on stones or shells. *Phoronopsis*, which occurs on the Pacific coast of North America from California to British Columbia, occurs in muddy sand flats.

These animals are best examined alive. They may be narcotized with MS 222 or with menthol. Bouin, Helly or Zenker all fix well. Paraffin sections stain well with Azan. Vertical (sagittal) sections are the most informative.

Feeding currents may be demonstrated with particles of carmine or carbon. Very little should be added. If too much is given the animal reacts by throwing off mucus when feeding stops. For details, see Grassé, Vol. 5(1) and Hyman, Vol. 5.

12.2. Bryozoans

12.2.1. INTRODUCTION

Most bryozoans are marine, superficially hydroid-like colonial animals. There is a great variety of species, some of which are very common. Dead colonies in which only the skeleton remains may be confused with plumularian hydroids, but living colonies are unlikely to be confused. The characteristic form of the lophophore and the enclosure of the whole zooid within a zoecium into which the lophophore can be retracted is quite distinct from the hydrothecal arrangements of calyptoblasts.

Bryozoans are best studied alive. *Membranipora* spp. forming mats on *Laminaria*, *Fucus* or other weeds is easily handled and most suitable for examination under a binocular microscope. Cyphonautes larvae of this genus may be found in the plankton at various times of the year, and settlement and colony formation may be observed.

Various other forms, locally common, may be maintained in aquaria providing that there is adequate circulation, oxygen-supply and food. *Phaeodactylum, Isochrysis* or other algal culture should be supplied as food.

Reference should be made to specialist monographs for details. In Britain, see Ryland (1962).

12.2.2. CLASSIFICATION

There seems to be general agreement that the Phylum may be divided into two Classes, and that of these, the Gymnolaemata, may be further subdivided into three Orders.

Class 1. GYMNOLAEMATA
Almost entirely marine. Colonies polymorphic. Lophophore a circle of tentacles.
Order 1. CTENOSTOMATA
Colonies not calcareous ('chitinous', membranous or even gelatinous)
Bowerbankia, Paludicella

Order 2. CHEILOSTOMATA
Orifice of zoecium usually with operculum. Zoecia calcareous boxes.
Bugula, Membranipora

Order 3. CYCLOSTOMATA
Orifice of zoecium round. Zoecia calcareous, fused together.
Crisia

Class 2. PHYLACTOLAEMATA
Entirely fresh water. Not polymorphic. Lophophore with distinct horseshoe-shaped arrangement of tentacles.

Plumatella, Cristatella

12.3. Brachiopods

12.3.1. INTRODUCTION

The brachiopods are entirely marine and very locally distributed. Some are superficially like bivalve molluscs, but these are readily distinguished by the flexible stalk which protrudes from one valve near the hinge. The internal anatomy may be seen by removal of the upper (dorsal) valve. To do this the adductor muscles have to be cut through. Ease the valves apart and cut the muscles in the same way as if the animal were a bivalve mollusc, using a small scalpel and keeping the blade as near the shell as possible. For details, see Hyman, Vol. 5, and Cambridge Natural History, Vol. 3.

Ciliary feeding may be demonstrated in *Lingula* with carmine or carbon particles. Very little should be added, otherwise the animal throws off copious mucus and feeding stops. See Chuang (1956).

12.3.2. CLASSIFICATION

The brachiopods are simply divided into two Classes.

Phylum BRACHIOPODA

Class 1. INARTICULATA (ECARDINES)
Valves held together by muscles only.
Lingula, Glottidia

Class 2. ARTICULATA (TESTICARDINES)
Valves held together by muscles and by a posterior tooth and socket arrangement. (lamp shells)
Terebratella

13. ECHINODERMS

13.1. Introduction

The Phylum Echinodermata is clearly separated from all others. Its members are coelomate, and the coelom is of great importance in the economy of the animal; not only does it play a part in carrying and distributing nutrients, respiratory gases, excretory and storage products, but a part of it, the water-vascular system, operates the hydraulic and extensible tube-feet, which are diagnostic of the phylum and which play an essential part in the life processes of the animal and its interplay with the environment.

Most members have a calcite skeleton. In some, such as crinoids and ophiuroids, this is in the form of rods supporting arms and other projections; in others, such as starfishes and sea-urchins, it forms a more-or-less rigid capsule or theca; and in others, such as holothuroids, it forms isolated spicules in the integument. The skeleton is generally not solid, but is built as a crystalline meshwork, with living material permeating the crystalline phase.

The better-known members, such as sea-urchins, brittle-stars and starfishes, have spines, to which the name Echinodermata refers (Greek: *echinos*, a hedgehog + *dermatos*, skin, or 'spiny skinned'). But possession of spines is not diagnostic of the phylum.

Curiously, this phylum has almost universally adopted and retained pentamerous symmetry. The reason is obscure, but it may well be that a 5-rayed arrangement of initial plates is demanded at a critical stage in the post-larval development of all echinoderms for mechanical reasons (Nichols, 1967a & b). The pattern of symmetry has particularly interesting implications in some aspects of the animals' behaviour, such as the manner in which the leading arm of a walking starfish is decided, and how the directional swing of the locomotory organs of this arm is reflected by those of other radii, as shown on p. 227.

General accounts of echinoderm biology are Hyman Vol. 4 (1955), Clark (1962), Boolootian (1966), Millott (1967) and Nichols (1967, 1975).

13.2. Classification

The earliest echinoderms were probably cup-shaped animals with mouth uppermost and arms held out radially to catch food. The crinoids, alone among the recent groups, retain this primitive form. Subsequently, two main lines evolved: the one (Asterozoa) showed radial growth patterns, while the other (Echinozoa) showed meridional growth patterns. While the Asterozoa universally retain radial symmetry, there is a tendency within the Echinozoa to assume secondary bilateral symmetry, which is reflected in the worm-like body of the holothuroids and in the heart shape of some so-called 'irregular' echinoids.

Phylum ECHINODERMATA

Sub-phylum CRINOZOA
Mouth uppermost; ambulacra restricted to oral surface.

Class CRINOIDEA
Well developed, movable arms.

Antedon

Sub-phylum ASTEROZOA
Star-shaped; mouth on under surface.

Class OPHIUROIDEA
Arms cylindrical and sharply demarcated from central disk.

Ophiothrix, Ophiocomina

Class ASTEROIDEA
Arms not sharply demarcated from central disk.

Asterias, Asterina

Sub-phylum ECHINOZOA
Globoid or worm-like, without radiating arms.

Class ECHINOIDEA
Plated echinozoans; body enclosed in rigid test; ambulacra meridional.

Order ECHINOIDA
Mouth in centre of under surface, anus in centre of upper; teeth operated by complex 'Aristotles lantern'.

Echinus

Order SPATANGOIDA
Not radially symmetrical; mouth not in centre of under surface; no Aristotle's lantern; anus on posterior side.

Echinocardium

Class HOLOTHUROIDEA
Echinozoans elongated in the oro-anal axis; skeleton reduced to ossicles in the integument.

Holothuria, Leptopentacta

13.3. Crinoids: *Antedon*

13.3.1. INTRODUCTION

Crinoids are the most primitive of the living classes of echinoderms. Early crinoids had stems, by which they were attached to the sea-floor, but subsequently the stem was lost at least once, so that the animal could swim. The common British crinoid, *Antedon bifida*, is one of these free-living, or *comatulid*, forms, and though it spends most of its life clinging to hard surfaces, such as rocks, wrecks and pier-piles, it can swim by downward strokes of five of its ten arms, alternating with upward strokes of the other five. When it is stationary, its arms are held out for feeding.

13.3.2. EXTERNAL STRUCTURE

Specimens are best killed by immersion in fresh water, and then preserved in alcohol. This ensures that the tube-feet, lining the food grooves, are extended. Note that the almost spherical body is largely composed of the *tegmen* (see Fig. 77) which is not supported by ossicles. The arms emerge from near the base of the tegmen, at a structure called the *calyx*, formed of a series of plates representing the original whorls of plates which surrounded the primitive crinoid body. All that can be seen externally of the calyx is its aboral base, the *centrodorsal*, which bears between 17 and 40 hooked *cirri* for attachment.

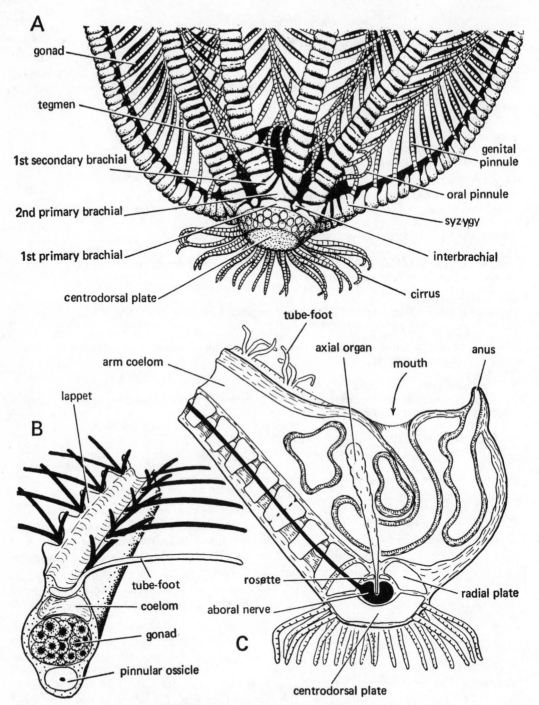

Figure 77 A typical stemless crinoid, *Antedon*

A, External view of the main part of the body, showing the arrangement of the principal ossicles. About two-thirds of the cirri have been removed from thier bosses on the centro-dorsal plate, and the two pairs of arms on the far side of the specimen have been cut off close to their origins; B, Perspective view of part of a genital pinnule, showing diagrammatically the arrangement of the groups of tube-feet and the lappets when the animal is feeding; C, Internal organs visible after the cut shown in Fig. 78 has been made, looking at the cut surface of the right-hand piece

The arms are supported by *brachial ossicles* which can be clearly seen in the inverted specimen, even before dissolving away the soft parts. Remove the cirri from around the base of one pair of arms, so that the arm ossicles can be seen in position. There are two *primary brachials*, the second of which forms the axil of the arm-division into two; after the division the ossicles are called *secondary brachials*. Examine the secondary brachials carefully, and note that the joint between secondary brachials 3 and 4, and again between 9 and 10, and 14 and 15, are different from the other joints. These joints (*syzygies*) are thought to be places where autotomy can occur fairly easily.

In some specimens of *Antedon* there may be tiny *interradial* plates between the axillary primary brachials.

From each arm ossicle, on alternate sides, a *pinnule* arises, and there is a division of labour along each arm. The first pair arises from the second (on one side) and third (on the other) of the secondary brachial ossicles, and each of these *oral pinnules* curls up around the sides of the tegmen, attached to it by soft tissue, to lie over the oral surface of the disk. There are no food-grooves or tube-feet on the oral pinnules. Their function is to sweep the oral surface of the tegmen free of foreign matter.

The pinnules arising from the 4th brachial segment to well over half the way down each arm bear the *gonads* as swellings on the oral side of the pinnular ossicles. Sexes are difficult to distinguish. The genital products are shed by rupture of the pinnule wall.

Distally, the pinnules are solely food-catching and respiratory in function.

The pinnular *food grooves* lead into the grooves of the arms, and these in turn lead into those on the oral surface of the tegmen and thence to the central *mouth*. The *anus*, on a short spire, opens in one interradius a short distance from the mouth, and breaks the radial symmetry of the animal. The muscular anal spire can ensure that faecal pellets are directed away from the mouth and food grooves.

The food grooves are avenued by muscular, papillate *tube-feet*, but unless the animal was killed by immersion in fresh water, it is likely that the tube-feet will have been withdrawn into the groove and covered by the tiny protective *lappets*, which are triangular flaps, hinged along the base, by which the groove can be covered when danger threatens.

If you have a living specimen for examination, the feeding process is easily seen under a medium-powered binocular microscope, with good illumination. Focus first on the oral surface of a *pinnule* and note that the tube-feet are in groups of three (see Fig. 77B), the groups lying alternately on each side of the food groove (Nichols, 1960). Each group consists of a long foot, held out nearly horizontally and acting as the main food catching organ, a short foot held nearly vertically, and a medium-sized foot at an angle between the other two. There are mucous glands on papillae over all the tube-feet. The medium and short tube-feet of each group help thrust the food-laden mucous strings into the food groove. Ciliary currents carry them to the mouth. Focus now on the oral surface of an *arm* and note that here the tube-feet of each group of three are roughly the same size. As the pinnules are the main catchment organs, the tube-feet of the arms are principally for helping the mucous strings into the grooves.

In addition to their feeding function, the tube-feet are important in respiratory exchange; when they are extended, their walls are thin enough for the coelomic fluid they contain to be brought within easy diffusion distance of the surrounding sea-water.

Dark-coloured *saccules* form a dotted line on each side of the food grooves and are scattered like freckles over the tegmen. They are colourless in life, but apparently take up the pigment from the rest of the animal as it dies. The function of the saccules is concerned with excretion, though little is known of the process.

13.3.3. INTERNAL ANATOMY OF CENTRAL DISK

Make a clean sagittal cut through the disk with a sharp scalpel, making sure that the cut passes through the anal spire and the right-hand member of the anterior pair of arms (as shown in Fig. 78). In this way it should pass through the edge of the mouth as well as the anus, and on the smaller of the two pieces (the right-hand one) you should see the axial complex, which passes from the centre of the calyx to end blindly in the right anterior radius (see Fig. 77,C). The function of the axial complex is not at all clear, particularly in crinoids, but is thought to be concerned mainly with excretion, and possibly also cytopoiesis (production of cells, in this case coelomocytes).

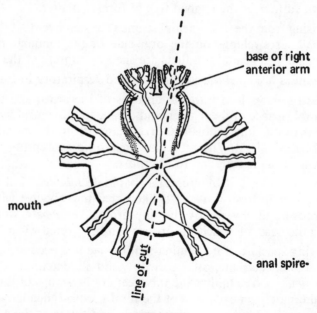

Figure 78

Diagram of the disk and arm bases of the crinoid *Antedon*, showing the suggested line of cut to expose the internal organs shown in Fig. 77C. The bases of one pair of arms, with their oral pinnules, are drawn in more detail

The main structure visible at the cut surface is the gut, with heavily convoluted walls, lying in the coelomic space. Note also the calyx, cut in section. Identify the ossicles of the calyx. In addition to the centrodorsal basally, which was seen externally (p. 215), five *radial* plates in a circlet form the 'sides' of the calyx and enclose a tiny cavity. The 'roof' of the cavity is formed by a single *rosette*, made up of five fused *basal plates*; the rosette does not sit like a lid on the ring of radials, but is wholly internal. If your section has passed through the mid-line of an arm base, you will see the brachial nerve passing through the brachial ossicles. The other arm structures, such as the coelomic spaces, and the canals concerned with tube-foot operation, are not easy to see without wax-embedding, sectioning and staining.

13.3.4. OSSICLES

Older techniques for freeing echinoderm ossicles from their surrounding and permeating tissues include the use of hot caustic soda or potash; these techniques are highly dangerous in a crowded laboratory. A better technique involves the use of cold sodium hypochlorite solution ('Chlorox'). Wash in running water after treatment, decant as much as possible and allow to dry, or examine a wet- preparation.

To separate the ossicles, place a small segment of an arm and its attached pinnules in neat sodium hypochlorite solution. Most of the organic material will have been removed after an hour or so, but the blocks of muscle which move the arms may still be present. Note the flexors orally and the extensors aborally. If they have not freed completely, force two adjacent ossicles apart and examine the transverse face of one of them; note the canal running longitudinally through the ossicle for the brachial nerve, and the shallow pits for the attachment of the arm muscles.

If you have a second specimen which has not been bisected, gently detach the tegmen from the calyx, cut down the arms to about the axillary ossicle and leave the calyx in sodium hypochlorite solution overnight. Note the following:

1. Attachment of the cirri to the centrodorsal.
2. Close attachment of the radials to the centrodorsal; these plates separate only after some hours in NaOCl.
3. Cavity beneath and between the radials, covered by the rosette (in pulling away the tegmen you may break and destroy the rosette).

13.4. Ophiuroids: *Ophiothrix*

13.4.1. INTRODUCTION

The phyletic relationship between the various classes of the echinoderms is difficult to assess; but judged on the criterion of the probable course of the evolution of the coelom and particularly that part of it concerned with the operation of the all-important tube-feet, it seems likely that the ophiuroids, or brittle-stars, are the group next most primitive to the crinoids. The British ophiuroid *Ophiothrix* and its relative *Ophiocomina* are both quite common animals of the shore and shallow sea, and are generally readily available for dissection, and alive for experimentation. The account that follows is based on *Ophiothrix*, but mention is made at the end of the differences between this and *Ophiocomina*, in case this genus is available instead.

Like *Ophiothrix*, *Ophiocomina* (which sometimes occurs with it) is highly variable in colour and spination. The main difference on the aboral surface is that the disk of *Ophiocomina* bears a covering of tiny spines, usually called '*granules*', over its entire surface, even covering the radial shields, On the oral surface of the disk the main difference is in the spine complement of the jaws: spines occur on the sides of the jaw plates as well as on the tips. On the oral side of the arms, the difference is that *Ophiocomina* has two tentacle scales at the base of each tube-foot, where *Ophiothrix* has one.

Like crinoids, ophiuroids are not suitable for gross dissection, but a good knowledge of their organization and activities can be obtained by examining the external structure and isolated ossicles, by single cuts across arm and disk and by watching the activity of the living animal.

13.4.2. EXTERNAL STRUCTURE

Ophiothrix is extremely variable in spine armature and colour pattern: no two individuals are alike. The dorsal (aboral) surface of the disk (see Fig. 79A) has short spines in a stellate pattern, and adjacent to the origin of each arm is a pair of plates, the *radial shields*, forming a support for the aboral integument of the disk, and having a function in respiration, as will be seen later. The shape of the radial shields in various ophiuroids is used extensively in the group's classification.

The spines on the disk continue round its sides, interradially, and on to the edge of the oral surface. The stellate mouth (see Fig. 79B), has a complicated system of plates, spines and tube-feet. Interradially are the *buccal shields* and beneath them (looking down on the oral surface) are the *lateral buccal shields* and beneath them again the *jaws*, made up of two half-jaws. The jaws bear short, closely-packed spines, pointing in towards the centre of the mouth. In each mouth angle are two pairs of *buccal tube-feet* which help to manipulate food into the mouth.

On each side of the arm bases, where they run beneath the disk, are slit-like openings each leading into a *bursal sac* (see Fig. 79B, C). There is a pair of sacs on each side of each arm, making 20 in all (Smith, 1940); *Ophiothrix* is probably unusual in having two pairs of sacs in each radius, where most ophiuroids have only one pair. The actual openings into the sacs will be seen when the disk is cut in two, under 'Internal Anatomy'.

The functions of the bursae are not entirely clear. Some writers call them 'genital-bursae' for two reasons: first, young post-metamorphic specimens of the brittle star have been found nestling in the bursae; and, secondly, the gonads of some ophiuroids have been said to discharge their products directly into the bursae. But in *Ophiothrix* at least it appears that the main function is respiration, since currents are drawn into the sacs by ciliary action and expelled by a pumping action of the disk roof. This action is brought about by muscles originating orally at ossicles in the disk wall and inserting aborally at the radial shields, seen externally on top of the disk. On *Ophiothrix* the gonoducts open separately from the bursae, though the gonopores cannot normally be seen without sectioning.

Examine the oral side of the arm, about mid-way along. The near-rectangular *ventral plate* can be seen axially (see Fig. 79B), and arising from close to its lateral edges, a pair of *tube-feet*. Notice the tiny *tentacle scale* at the base of each tube-foot. On the aboral side, there is a single *dorsal plate*, diamond shaped with a keel.

Make a few rough transverse sections of the arm, about mid-way along and examine the cut faces (see Fig. 79C). The arms are mainly supported by strong *vertebrae*, representing a pair of plates fused in the mid-line. Surrounding a vertebra in each 'segment' of the arm are the dorsal and ventral plates, already seen externally, and a pair of *lateral plates* each bearing 6 spines. These plates will all be seen when you examine the isolated ossicles, below. In some of your cuts you should see the two pairs of *intervertebral muscles* which move the arm.

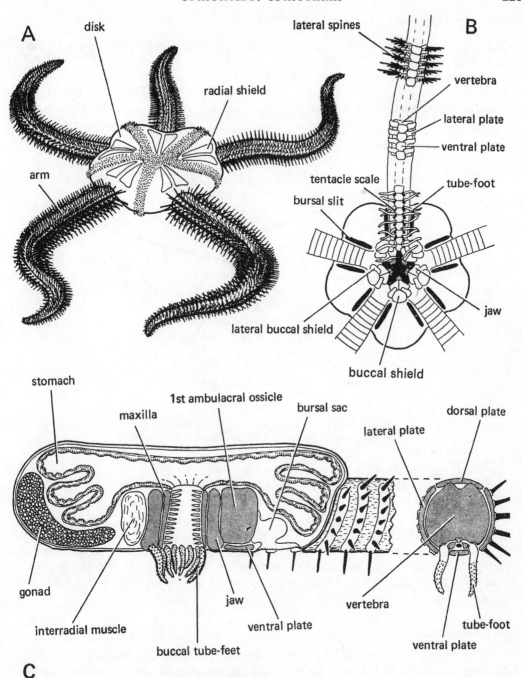

Figure 79 The brittle-star, *Ophiothrix fragilis*

A, Perspective view of complete specimen, showing particularly the aboral surface of the disk; B, Oral surface of disk and arms. One arm is drawn in more detail to show, proximally, the arrangement of the tube-feet, centrally the ossicles visible dorsally, and distally the lateral spines; C, Internal organs visible after a sagittal cut across the disk, such as that shown in Fig. 80, has been made. On the right, an arm is shown in transverse section in the region of a vertebral ossicle

13.4.3. INTERNAL ANATOMY OF THE DISK

Make a single cut across the disk along a line running through one of the bursal slits on one side and through the inter-radius opposite (see Fig. 80). Where you have cut through an inter-radius, you will see a lobe of the stomach, and where you have cut through the bursal slit, you will see the distal of the two bursal sacs. The other bursal sac lies within the inter-radial muscle alongside the jaw, and is generally difficult to see with certainty.

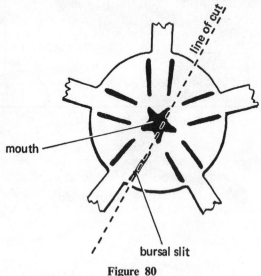

Figure 80

Diagram of the disk and arm bases of a brittle-star, such as *Ophiothrix*, showing the suggested line of cut to expose the internal organs shown in Fig. 79C

On the larger of the two pieces produced by your cut, note the construction of the jaw apparatus and the placing of the buccal armature of spines and the tube-feet. Cut round the edge of the aboral face of the disk and remove the semicircle of dorsal body wall. Open the stomach, if you have not already removed its aboral wall with your cut, and note its lobulation and single opening from the mouth. Push aside the stomach wall inter-radially and if your specimen is mature you may be able to see the gonads lying beside the bursae. Looking at the origin of one arm, feel with a seeker between the dorsal wall of the stomach and the dorsal body wall and note that an extension of the perivisceral coelom passes down the dorsal side of the arm.

13.4.4. OSSICLES

Place a small segment of arm from mid-way along in NaOCl (see p. 219). After the ossicles have fallen apart, you should be able to identify spines, vertebrae, dorsal, ventral and lateral plates, and even the tiny tentacle scales. Fit two vertebrae together with needles or fine forceps, and notice how they articulate. Now examine the transverse face of a single ossicle and note the groove ventrally which carries the radial water-vascular canal, radial nerve and other ambulacral structures.

13.4.5. Feeding

Brittle-stars show both microphagous and macrophagous methods (Fontaine, 1965) as follows:

1. *Microphagous*
   (i) Mucous net method. The animal assumes a feeding posture with arm tips upraised and spines spread as widely as possible. Mucous glands in the integument and spines secrete strings of mucus which come to lie on the arm surface and to form festoons between the spines. One or more arms moves laterally to 'sweep' the segment of the catchment area it commands. The food trapped by the net is removed either by ciliary activity or by the tube-feet.
   (ii) Surface film feeding. The animal will move its arms parallel to the surface of the water to pick up neuston from the air/water interface. Because the animal is mainly sub-tidal, this method must be relatively unimportant to it.
2. *Macrophagous*
   (i) Arm-loop capture. The arms may be flexed round a large (1 to 2 cm) particle and moved so as to bring the particle to the mouth.
   (ii) Browsing. Sometimes, though seldom, an animal moves over the substratum using its teeth and oral tube-feet to remove encrusting matter.
   (iii) Tube-foot capture. When faced with a particle of intermediate size, or a bolus of food entrapped by the mucous net, the animal will seize it with one of its tube-feet and pass it down the arm to the mouth from tube-foot to tube-foot.

Clearly, the availability of these mechanisms opens for the animal a wide diversity of food resources.

If you have living material available, starve a few ophiuroids for a few days by keeping them in filtered sea-water. Present them with pieces of chopped mussel of the following sizes to observe their various modes of feeding:
   (i) 10 mm × 5 mm;
   (ii) About 2 mm square or less;
   (iii) Finely-ground mussel. To observe the mucous net, introduce into the sea-water of the dish a weak solution of toluidine blue so that the water is coloured a faint blue. The mucus will show up metachromatically in this solution, and the net should be visible as pinkish strands.

13.5. Asteroids: *Asterias*

13.5.1. Introduction

Asteroids share common ancestry with ophiuroids (Fell, 1963). Some of the arm plates and plates round the mouth are homologous in the two groups. Whereas the ophiuroids use their arms for propulsion over the sea floor, the asteroids rely on movements of the tube-feet for locomotion, and whereas the ophiuroid arms are extremely fragile, those of asteroids are strong enough for such activities as pulling apart bivalve mollusc shells to get at the meat inside.

Asteroids are excellent animals for dissection, to show the structure of a typical advanced echinoderm, and are also particularly useful to demonstrate the basic features of the echinoderm nervous system.

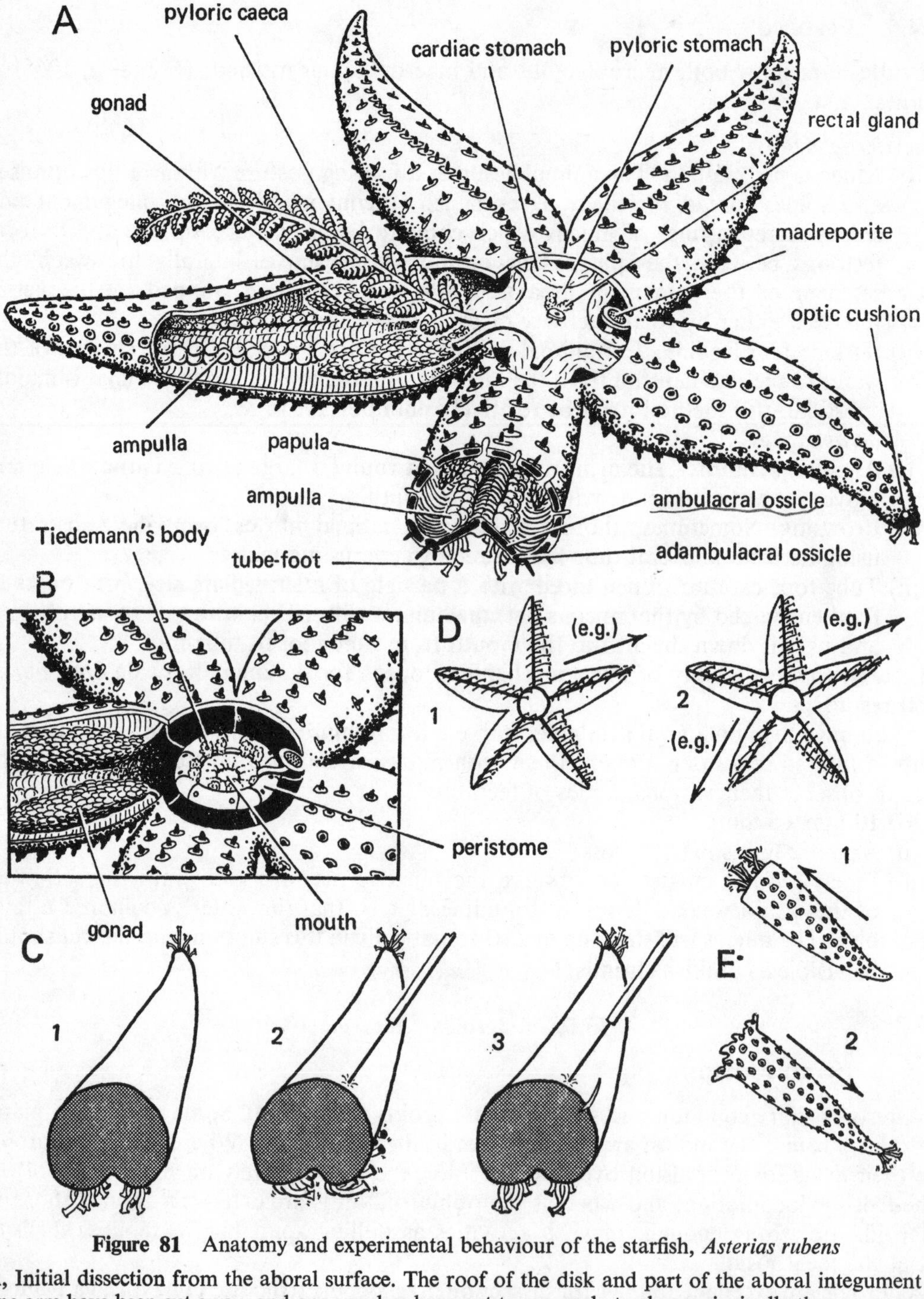

Figure 81 Anatomy and experimental behaviour of the starfish, *Asterias rubens*

A, Initial dissection from the aboral surface. The roof of the disk and part of the aboral integument of one arm have been cut away, and one arm has been cut transversely to show principally the arrangement of the arm ossicles and the soft parts that protrude between them. In the dissected arm, one of the two pyloric canals, with its caeca, has been removed; in the sectioned arm both canals have been removed;

13.5.2. EXTERNAL FEATURES

On the oral surface, note:

1. Central *mouth*.

2. *Ambulacral grooves* down each arm; the roof of each groove is formed by the *radial nerve* which you may be able to distinguish from other tissue.

3. *Tube-feet* within the grooves, originating lateral to the radial nerve.

4. *Optic cushion* at the end of each arm (see Fig. 81A). You may need to probe among the terminal spines and tube-feet to see it in a preserved specimen.

5. *Spines* bordering each groove and round the mouth.

6. *Pedicellariae*, pincer-like structures surrounding some spines. These organs, also present in sea-urchins, are modified spines that help to protect the animal. A more detailed study of their activity is given under 'Echinoidea' (see p. 235).

On the aboral surface, note:

7. *Madreporite*, the only external structure breaking the radial symmetry of the animal.

The function of this perforated plate is uncertain (Nichols, 1966). Although it opens into a part of the water-vascular system, it does not appear to act either as a safety-valve or a topping-up point. It may possibly function either as a device for ensuring equal pressure between the water-vascular system and the external environment, or as a pressure receptor to provide information to the tube-feet about the changing head of water above the animal as the tide rises and falls.

8. *Anus* at the centre of the disk, or slightly eccentric; it may be tightly shut and difficult to find.

9. *Papulae* emerging between the ossicles. These, with the tube-feet, constitute the animal's respiratory organs. They are mere blisters of the external and coelomic epithelia and contain coelomic fluid. If possible examine one or two on a living animal under a binocular microscope: you may be able to see

---

B, Further dissection of the specimen shown in A, in which the alimentary system has been removed, showing the circum oral water-ring and its accessory structures; C, Diagrammatic representation of the experiment described on p. 227 showing the role of the basi-epithelial nerve plexus in tube-foot response to tactile stimulation. Only part of one arm of an intact animal is shown, (1) Normal behaviour; tube-feet pointing as in normal locomotion, (2) Stimulation of the arm exterior by a probe; tube-feet respond by pointing towards the area of stimulation, (3) After a cut through the arm integument, including the nerve plexus, has been made between the point of stimulation and the oral surface, the tube-feet no longer respond; D, Diagrammatic representation of an experiment to show the role of the circum-oral nerve ring in coordinated pointing. Starfishes viewed from the oral side, as though the nerve ring and radial nerve cords were visible as black lines, (1) Normal starfish, with tube-feet pointing as if the upper right arm were leading, (2) After severing the nerve in two places, the two parts thus produced assume independent pointing directions; E, Diagrammatic representation of an experiment suggesting that in a severed arm the centres responsible for determining the direction of pointing are located along the length of the complete arm, (1) When an arm is severed so that the nerve cord is not complete, the tube-feet point as though it is a trailing arm, that is, with the cut end leading, (2) When severed as a complete arm, the tube-feet point as though it is a leading arm.

coelomocytes in the lumen; sometimes you can see a clump of them caught in an eddy of the internal ciliary current. At intervals, papulae may be autotomized, and thus the clot of coelomocytes, with ingested waste matter, is removed from the body and a new papula grows. Coelomocytes move out through the epithelium over the entire body, too.

13.5.3. INTERNAL ANATOMY

Remove the dorsal body wall from the centre of the animal and one arm (see Fig. 81A); take care in cutting round the apical disk and madreporite, so that the structures leading to the orifices on the dorsal side are not damaged. It is convenient to remove the body wall from the arm opposite the interradius in which the madreporite is situated. Note the following parts of the alimentary canal.

1. Large *stomach* divided into lower *cardiac* and upper *pyloric* portions.

2. Canals from the pyloric part to each arm, which give off the lumina to the *two caeca*; currents take food to the caeca, and digestive enzymes from the caeca to the stomach.

3. *Rectal sac*, a gland looking like a bunch of grapes, concerned with the consolidation of the faeces and with ridding the coelom of some substances. stances.

Remove the alimentary system by lifting out the pyloric caeca from each arm and cutting through the 10 retractor muscles of the stomach, two to each arm.

Note the *gonads* in the proximal part of each arm. The gonoducts will probably not be visible; they run to the sides of the arms and open in the angle between two adjacent arms.

Find the *water-vascular ring* round the mouth, where the peristomial membrane joins the circum-oral ring of ossicles (see Fig. 81B), with its nine *Tiedemann's bodies*, two to each radius, except where the stone canal arises. Find the *stone canal*, leading from the ring upwards to the madreporite, and note the *axial organ* attached to its outer side. Cut through these two structures and examine the axial complex in section. Note the scroll-like partition in the stone canal, to increase the surface area of ciliated epithelium.

The stone canal, so called because it is heavily endowed with ossicles, links the central parts of the water-vascular system, operating the tube-feet, with the external opening, the madreporite, seen externally. The axial organ has a function of decontaminating the animal of foreign matter.

Note the tightly-packed rows of *ampullae* in each arm. These assist in tube-foot protraction, as mentioned below on p. 228.

Make a transverse cut across one of the four arms from which you have not stripped the dorsal body wall, and examine the cut surface (see Fig. 81A). Note the arm coelom, which is a radial extension of the body coelom of the disk, as in ophiuroids, and the hepatic caeca (which are not firmly attached and may fall out) and possibly also the gonads lying within it. The arm is supported by rather loosely connected ossicles, the largest of which are the two *ambulacral ossicles* forming support for the ambulacral groove. These ossicles are homologous with the two ossicles that have fused to form the ophiuroid vertebrae (p. 220, above). Attached to the ambulacrals are the tube-feet, in two rows on each side of the groove, and there is a canal from each tube-foot to its ampulla between adjacent ambulacrals. Attached to the outer ends of each ambulacral ossicle is an adambulacral ossicle which bears spines and pedicellariae to protect the groove. Other ossicles lie in the lateral and dorsal walls of the arm.

The *radial water-vascular canal*, supplying fluid from the central parts of the system for operation of the tube-feet, lies in the angle between the two ambulacral ossicles. You will probably not be able to see it in your rough section. Beneath it is the V-shaped *radial nerve*, with a coelomic space, the *perihaemal canal*, on one side of it and the surrounding sea-water on the other.

13.5.4. Experiments on Nervous Organization

The following simple experiments, based on Smith (1966), show the presence in star-fishes (as typical members of the phylum) of a decrementally conducting integumental nerve plexus, of through-conducting paths both peripherally and centrally and of co-ordinating centres which enforce upon this radially-symmetrical animal a pattern of activity resulting in movement in one direction.

1.   Observe the activity of a straight pedicellaria. Stimulate the body wall beside one and observe the opening and closing movements of the jaws. Increase the strength of stimulation gradually and observe the effect on pedicellariae further away. Make repeated and stronger stimulation on the outside of a pedicellaria, to bring neighbouring pedicellariae into activity. Note carefully the range of influence following stimuli of varying strengths. The fact that more pedicellariae are brought into activity the stronger the stimulus shows that there is a sensory connexion to the general basi-epithelial plexus of the skin surrounding the stimulated organ, and thence to other organs, and that the plexus conducts decrementally. This plexus represents the *peripheral nervous system* of the animal, which allows local patterns of activity in response to small environmental stimuli. These can, as we shall see, also excite the 'central' parts of the nervous system to bring about more general reaction by the animal. (Note that more detailed experiments on pedicellaria response are given under Echinoids, on p. 234.)

2.   Observe the reaction of the tube-feet to stimuli (see Fig. 81C). While gently holding the starfish above the floor of the dish, probe one side of an arm and notice that the tube-feet on both sides of the ambulacrum tend to bend towards the point of stimulation. Make a cut between the probe point and the *ventral*, i.e., *oral*, surface of the arm. Note that this destroys the response of the tube-feet of *both* sides of the ambulacrum. This experiment shows that directional transmission occurs. In this case, the excitation passes towards the radial nerve, and the experiment indicates that excitation does not pass over the dorsal mid-line of the arm from one side to the other: the transverse circumferential pathways are dorsoventrally polarized and the mid-dorsal line is an excitation barrier.

3.   (a) Invert a starfish on a glass ring in a dish of seawater so that its arms do not droop on to the floor of the dish, and observe the normal 'pointing' and 'stepping' activity of its tube-feet. Note how one arm

tends to lead, as it would in normal locomotion (see Fig. 81D, 1), its tube-feet swinging parallel to the long axis of the arm and the tube-feet of other arms taking up a swing in the same direction. This indicates the presence of a *central nervous control*, generating co-ordinated activity. In this experiment, has anything influenced which arm leads? Is the chosen direction (i) the resultant of excitation interaction between all five radial nerve cords *via* the nerve ring, or (ii) an imposition by one arm of dominance over all the others? The next experiment suggests which.

(b) Cut the circum-oral ring in two places to produce 3-arm and 2-arm segments (see Fig. 81D,2). Observe the effect on tube-foot pointing in all the arms. Note how the tube-feet of any one arm of each piece assume dominance of direction, so that *two* arms of the starfish now react as though they are leading arms. This excludes the possibility of co-ordination through a general interaction and resultant, and supports the alternative proposal of one-arm dominance.

(c) (i) Cut off one arm by severing it on the distal side of the nerve ring (see Fig. 81E,1). Note that it steps as though it were a *trailing* arm.

(ii) Now cut off a second arm, leaving part of the disk and nerve-ring attached (see Fig. 81,E,2). Eventually this arm will move as though it were a *leading* arm. This is because the centres capable of imposing direction, which lie along the length of each arm, are complete, and so the normal assumption of dominance by this arm can occur. If this preparation is observed for some time, its tube-feet may change direction, which is what is required in the arms of a radially-symmetrical animal with no persistent dominant radius.

13.5.5. ACTIVITY OF THE TUBE-FEET

Direct observation of the protraction activity of the tube-feet is not easy, since a knife-cut into the body wall aimed at opening a window to observe the ampullae will pierce the basi-epithelial plexus and disturb the sensory input to the region under observation, causing the tube-feet to remain retracted, or, at best, to behave abnormally. Minimal disturbance to normal activity can be caused by cutting very rapidly through the body wall. The best way to do this is mechanically, using a circular saw made from a 12 mm diameter metal washer spun by a dental drill. If you have access to such a device, make a rapid cut between the ambulacral and adambulacral ossicles lengthwise down one arm of an active animal while holding it gently in one hand. Continue to cut up the sides and along the dorsal surface to remove an appreciable-sized window such as that drawn in Fig. 81A, though not continuing to the disk. Immediately replace the animal in seawater and watch a region within your window where the tube-feet are active. Note that the first part of tube-foot extension is not brought about by the ampulla: it is only after the tube-foot has extended some way that the ampulla contracts to drive its tube-foot to the maximum extended length. On retraction of the tube-foot, the ampulla can be reinflated either by the first part of the column of fluid to leave the tube-foot or at any time subsequently, even after the tube-foot has completely retracted. This shows that parts of the water-vascular system additional to the ampullae are involved in tube-foot protraction and it also indicates that the valve which can close the canal leading to the tube-foot and its ampulla can be opened against a build-up in pressure on the tube-foot side of it.

13.5.6. Demonstration of Geotaxis in *ASTERINA*

Some echinoderms are negatively geotactic. In the case of *Asterina gibbosa*, the tiny British rock-dwelling starlet, the response is particularly strong. It appears to be a response to the pull of the body on the tube-feet. What could be its function?

1. Place a living *Asterina* on the bottom of a dish of sea-water. Notice how it moves to the sides and climbs the wall.

2. Attach a cork to the animal by means of a thread tied round the disk. Support the buoyancy of the cork in the water as you replace the starlet on the aquarium wall until the tube-feet have gripped. Does the asteroid still move upwards?

3. Dissect off one arm with part of the disk attached, and repeat (1). The response should still be shown by the isolated arm. Now remove the remaining segment of the central disk and repeat. The response should not now be shown.

4. Do *light* or *depth* affect this response?

13.6.   Echinoids: *Echinus* and *Echinocardium*

13.6.1. Introduction

As in asteroids, dissection of a typical echinoid such as *Echinus* is very helpful in understanding the organization of the echinoderm body. Though the origin of the echinozoans is obscure, it is likely that the earliest ones were flexible-bodied, living on the sea-bed. Of the two living classes, the ECHINOIDEA evolved a rigid test, while the HOLOTHUROIDEA retained the flexible body. Later in echinoid evolution the radial symmetry was broken by migration of the mouth and anus, giving rise to the so-called 'irregular' echinoids, such as *Echinocardium*, which are generally burrowers.

In this section, dissection of *Echinus* is described (pp. 229-237) followed by an account of the external morphology of *Echinocardium* (pp. 237-240).

13.6.2. Dissection of *ECHINUS ESCULENTUS*

(a) *External features*

On the aboral surface (see Fig. 82A), note the *apical disk*, with central *periproct* (the anus is too small to see), 5 *genital* and 5 *radial* plates (sometimes called 'oculars', though they are no more sensitive to light than other regions, so far

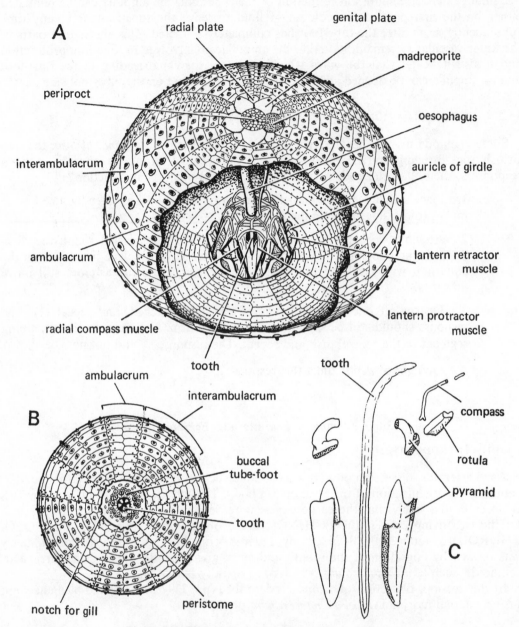

Figure 82 The regular sea-urchin, *Echinus esculentus*

A, General view of a denuded test which has had a window cut in its side, showing the arrangement of plates at the apex and the way the Aristotle's lantern is slung within the peristomial girdle. The oesophagus and axial system are shown emerging from the top of the lantern; B, Oral view of a denuded test, with the peristomial membrane still intact; C, Exploded view of one segment of the Aristotle's lantern, showing the component ossicles. The faces which abut other ossicles are shown stippled

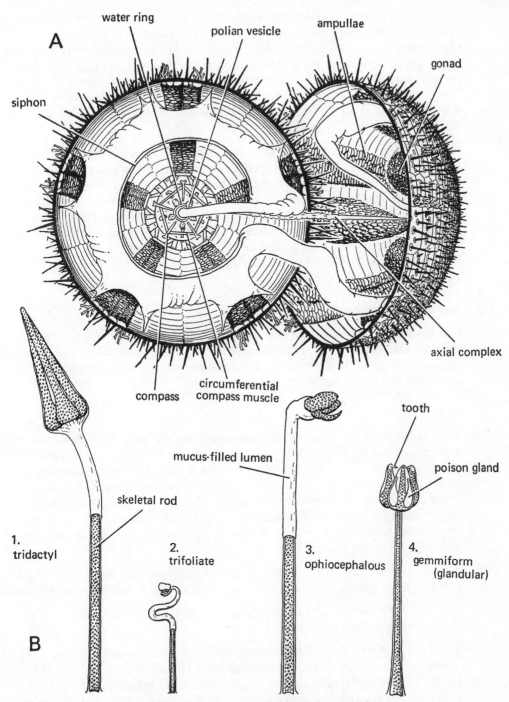

Figure 83 The regular sea-urchin, *Echinus esculentus*
A, The test has been cut round the ambitus and the halves gently pulled apart; B, Examples of the four types of pedicellaria, drawn to the same scale; (1) tridactyl, (2) trifoliate, (3) ophiocephalous, (4) glandular (or gemmiform)

as is known), one genital plate modified to form a sieve plate (*madreporite*) leading into the water-vascular system and axial sinus.

On the oral surface (see Fig. 82B), note the *mouth* and surrounding *peristomial membrane*, *teeth* protruding from mouth, the sensory *buccal tube-feet* on the peristomial membrane, *accessory gills* at the peristome edge for supplying the respiratory needs of the masticatory apparatus, *ambulacra* with *tube-feet*, *interambulacra*. Note the various sizes of *spines*, and the highly modified spines, the *pedicellariae*.

Echinus moves by contracting and bending the sucker-ended tube-feet, using the spines as fulcra. Food is scraped from the hard substratum over which the urchin is moving by means of the teeth, operated by the complex Aristotle's lantern (see p. 233). The body is protected from sediment and from the dangerous larvae which may settle on it by the pedicellariae. Respiratory exchange, except that of the masticatory apparatus, is brought about across the walls of the tube-feet, which may be extended and gently waved in the water; respiratory needs of the masticatory apparatus, as noted above, are supplied by the accessory gills round the peristome.

(b)   *Internal structure*

Cut through the shell round the ambitus (the widest part). As the two halves separate (see Fig. 83A), be careful not to pull them apart before you have seen the orientation of the soft parts within the perivisceral coelom, particularly the axial complex (see para. 4, below).

Note:

1.   *Alimentary canal* making one complete circuit, then turning on itself for another, before passing aborally to the *anus*. Find the *siphon*, an accessory tube running parallel to the gut along part of its length; the siphon originates close to the point where the oesophagus emerges from the lantern, it passes along the inner border of the gut to the point at which the gut doubles back, then opens again into the gut. The function of the siphon appears to be to act as a by-pass for water taken in with the food, so that the enzymes in the secretory part of the gut are not diluted.

2.   *Gonads*, hanging in the interambulacral areas on the aboral side. These are eaten in some places, a practice reflected in the urchin's specific name, *esculentus*.

3.   *Aristotle's lantern*, the masticatory apparatus surrounding the first part of the oesophagus—see pp. 233–234.

4.   *Axial complex*, consisting of *stone canal* and *axial organ*, running from the aboral side of the lantern to the madreporite. The stone canal (so-called because its walls are supported by large, loosely connected ossicles) serves to keep the tubes of the water-vascular system supplied with fluid; the axial organ is a complex structure, partly excretory, partly endocrine and probably partly cytopoietic.

Examine more closely to find:

5. Inner and outer *intestinal haemal vessels*, the inner lying close to the siphon, the outer on the opposite side. Trace the two vessels to the oesophagus, and note that they arise from:

6. The *circum-oral haemal ring*, on the aboral side of the lantern.

7. *Water-vascular ring*, adjacent to the haemal ring. Note that it has 5 *polian vesicles* in the inter-radii; these are low-pressure reservoirs, to maintain turgor in the water-vascular system. If you have difficulty in seeing the various circum-oral rings, wait until you dismember the lantern—see below.

8. *Radial water-vascular canals* in each ambulacrum, with branches leading to the ampullae of the tube-feet.

13.6.3. PREPARATION OF WHOLE MOUNT OF A TUBE-FOOT AND AMPULLA

Where your ambital cut crosses an ambulacrum, dissect out one tube-foot and its ampulla by breaking away the ambulacral plate until it cracks across a tube-foot pore. Since the external epithelium of the tube-foot is continuous with that covering the test, and the epithelium of the ampulla continuous with that lining the coelomic cavity, you may need to cut round the base of the tube-foot and the neck of the ampulla with a sharp scalpel to avoid tearing the epithelium. Place the preparation in a dye such as carmalum, purpural or acid haemalum for the appropriate time, dehydrate, clear and mount.

13.6.4. DISSECTION OF ARISTOTLE'S LANTERN

Echinoids feed on encrusting organic matter by holding the test close to the substratum with the tube-feet, then using five teeth to rasp off the organic matter. Food loosened in this way is wafted into the oesophagus by cilia.

The five teeth are manipulated by a complicated apparatus of ossicles, muscles and ligaments resembling a lantern in shape (see Fig. 82A), though Aristotle is said to have compared the whole urchin to a lantern, not merely its masticatory apparatus. The whole apparatus is enclosed in its own compartment of the coelom, and in *Echinus* receives its own oxygen supply from 5 accessory gills which protrude from the body in a ring round the peristome. The movement of coelomic fluid into and out of these gills is also performed by the lantern.

If time is limited, dissection of the lantern may be done while the tube-foot—ampulla preparation is staining (see 13.6.3. above).

The lantern is held in place by muscles slung between its ossicles and the perignathic girdle (see Fig. 84A), alternating arch-like *auricles* (radial), and low *apophyses* (inter-radial).

The lantern is made up of:

1. Large *pyramids*, consisting of a pair of 'V' shaped structures fused together, and a pair of projections aborally, also fused together, and forming an arch (see Fig. 82C).

2. *Teeth*, one to each pyramid, the soft (young) ends protruding aborally. The teeth are abraded away as the urchin feeds, and the dental sac, in which the soft end lies, adds to the tooth throughout the urchin's life.

3. Beam-like *rotulae* resting on the aboral surfaces of two adjacent pyramids.

4. 'Y'-shaped *compasses* on top of the rotulae.

5. *Muscles*. 5 pairs of *protractors* attaching the aboral arches of the pyramids to the apophyses of the perignathic girdle to protrude the teeth, 5 pairs of *retractors*, originating at the oral ends of the pyramids and inserting at the top of the auricles (see Fig. 84A); 5 *comminators*, between adjacent pyramids by which the teeth can be moved relative to one another; 5 pairs of *posturals*, continuous with the protractors, but innervated from a different source; 5 *circumferential compass muscles* joining adjacent compasses and forming a pentagonal pattern on the aboral face of the lantern (see Fig. 83A); and 5 pairs of *radial compass muscles* attaching each compass to the apophyses to either side of it (see Fig. 84B).

To remove the lantern, cut through the muscles supporting it, and make a circular cut round the periphery of the peristomial membrane so that the entire lantern can be removed.

Examine closely the outer surface of the comminator muscles, and note that 5 *radial water-vascular canals* pass down the lantern here. Follow one canal back to the water-vascular ring by carefully removing the compass and rotula which cover it aborally.

At the oral end of the comminator muscles it is usually possible to see the *radial nerves* emerging just oral to the water vascular canals. Follow one of these inwards, towards the central axis of the lantern, and find where it joins the *circum-oral nerve ring*, which lies above the oral end of the pyramids.

Mechanics of the lantern

Using the protractors, retractors, posturals and comminators, not only can the whole lantern be moved relative to the peristomial girdle, both laterally and axially, but also the teeth can be moved relative to one another. The radial compass muscles will pull the outer, bifid, ends of the compass orally, and the circumferential compass muscles, antagonistic to the radials, will pull the compasses up again (see Fig. 84B). This mechanism will pulsate the membrane which stretches across the aboral surface of the lantern, so that coelomic fluid is forced into and out of the accessory gills.

13.6.5. Pedicellariae of *ECHINUS*

Search the remains of your specimen for the following pedicellariae (see Fig. 83B):

(i) *tridactyl*, the largest type;

(ii) *trifoliate*, with broad flat blades, and flexible neck, the smallest type;

(iii) *ophiocephalous* ('snake-headed'), found chiefly round the peristome;

(iv) *glandular*, or *gemmiform*, with tiny teeth distally and poison sacs along the outer edge of the valves.

13.6.6. Nervous Organization of the Pedicellariae

Echinoid pedicellariae make interesting preparations for the study of nervous and muscular physiology, because of the persistence of their reflex activity hours after they are removed from the test of the urchin.

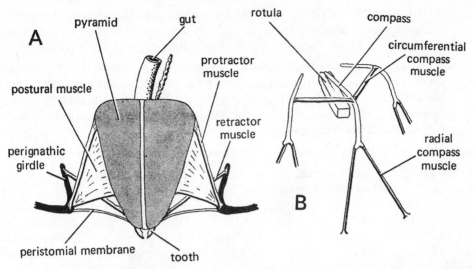

Figure 84

Schematic diagrams showing the main muscles responsible for the movements of the Aristotle's lantern of a regular sea-urchin, such as *Echinus*

A, Vertical section across the peristome (girdle in black) and pyramids of the lantern (stippled), showing the main muscles which operate the teeth; B, Perspective diagram of one compass, its underlying rotula and its two adjacent compasses. Each compass is pivoted at the inner end in a groove in the underlying rotula. Contraction of the *radial compass muscles* pulls down the outer part of the compass; contraction of the antagonistic *circumferential compass muscles* restores it. The perioesophageal membrane is stretched across the top of all five compasses so that their movement causes the membrane to be raised and lowered, thus creating an ebb-and-flow circulation in the accessory gills round the peristome.

Experiments on the whole animal

The experiments described first are all performed without removing the pedicellariae, though their removal does not seriously affect the reflex activity of the jaws, and so the experiments in (4) to (6) could, if more convenient, or if living material is scarce, be performed on excised pedicellariae. Whichever is done, allow a few minutes between experiments for the material to return to a background level of tone.

1. With a mounted needle, probe the test of the urchin and notice how the spines and pedicellariae 'focus' on to the point of irritation.

2. Choosing a tridactyl or ophiocephalous pedicellaria, stroke the side of the stalk with a mounted hair. Note that: (a) the jaws open (b) the stalk sways from its base (c) other nearby pedicellariae move towards the stimulated one.

3. Now make a cut about 1 cm. long in one of the interambulacra, making sure that the cut passes through the entire epithelium and into the skeleton beneath. Leave a short time for the urchin to recover from the cut. Stroke the stalk of a pedicellaria near the cut, as before. Note that while the nearby pedicellariae on the same side of the cut are again stimulated into activity, those on the other side of the cut do not respond. This shows that excitation is carried to other effectors by nerve elements in or near the epithelium, and that the spread of excitation from a point of stimulation is directional, and is interrupted by a cut.

4. Moving to a different part of the test (or on excised pedicellariae), select pedicellariae of the different sorts with their jaws closed. Stroke the outside of the closed jaws with a mounted hair. Notice that the jaws invariably open.

5. Using a mounted hair, stimulate the *inside* faces of open jaws of the different types of pedicellaria. Notice that in the case of the tridactyl and ophiocephalous ones the jaws snap shut; in the case of glandulars the jaws open slightly wider than before; and in the case of the trifoliates the jaws move independently and may close, though not for a sustained period.

An experiment using isolated glandular pedicellariae

Unlike the other pedicellariae, the glandulars exhibit a chemosensory response in addition to the usual mechanosensory one. The inside of each jaw of a glandular pedicellaria responds to a mechanical stimulation, as shown in (5) above, by opening wider than before rather than by snapping shut. Closure of the jaws in this type can be elicited by a chemical stimulus such as is provided by a piece of, or the juice from, a predatory starfish. The centre for chemoreception appears to be located in a tiny *sensory hillock* in roughly the centre of the inside surface of each jaw (Campbell, 1976).

6. Pluck off a few glandular pedicellariae and place them in sea-water in a watch glass or cavity slide. Select one with open jaws, or stimulate one to open by touching the outside of the jaws. You should be able to see the sensory hillocks under the highest power of a stereomicroscope. Touch one sensory hillock with a fine twist of filter-paper. This will elicit the 'further-opening' response. Now dip the twist of filter-paper in a suspension of mashed-up starfish tube-feet, and again touch the sensory hillock with it. This time, the jaws close. Continue observing the pedicellaria for a further 60 seconds. During this time (usually about 40–50 secs) you should be able to see the poison sacs on each jaw contract as the poison is squeezed from the duct near each jaw-tip and injected into, or squirted over, the twist of filter-paper on which the jaws have closed.

13.6.7. COELOMOCYTES

Floating in the coelomic fluid, both perivisceral and that of the tubular systems, are numerous free cells, the functions and activities of which have not yet been clearly worked out. There are so many different types of cell (Boolootian and Giese, 1958, have recognized 14 from 28 species of echinoderms) that it has been suggested that some are growth stages of others, and that some may not be coelomocytes at all but entoparasites.

The coelomocyte system is probably used for excretion (there are no separate excretory organs in echinoderms), regeneration and transport of essential substances, so it is hardly surprising that so varied a system exists.

The best way to observe living coelomocytes is by phase contrast microscopy, but failing this a reasonable picture is obtained by dark-ground illumination, or simply by cutting down the condenser iris below the usual limits. They must be observed within 5 minutes of being taken from the body cavity, that is, before clotting sets in (unless an anti-coagulant is used), though even in a clot many of the cell types can still be recognized.

Suck a *small* quantity of coelomic fluid into a capillary tube and seal the ends with vaseline. Place this tube on the stage of the microscope and rotate it to get different angles.

If possible, inject an urchin *via* the periproctal membrane with finely ground carmine in sea-water, to see which cells, if any, are phagocytic.

The following is a rough guide to the cell types you may see in *Echinus esculentus*.

TABLE 1. Coelomocytes of *Echinus esculentus*

	Size range	*Colour by phase*	*Granules*	*Vacuoles*	*Function*
Bladder amoebocytes	9–57 μm	Colourless	Many black	Several	Phagocytic
Filiform amoebocytes	8–55 μm	Grey	Some black	2 to many	Clotting; phagocytic
Fusiform corpuscles	2–12 μm by 6–30 μm	Grey	None	None	?
Colourless spherules	8–12 μm by 13–28 μm	Pale Yellow	Lobular	None	Lipid transport?
Vibratile corpuscles	3–12 μm by 9–44 μm	Grey	Many black	Many small	Circulation nutrients?
Eleocyte amoebocytes	11–30 μm by 7– 8 μm	Red	Small red	None	O_2 transport?
Hyaline haemocytes	9–13 μm	Pale yellow	None	Numerous	Clotting

13.6.8. *Echinocardium cordatum*

This is an advanced echinoid, a member of the spatangoids or heart-urchins. It has bilateral symmetry superimposed on the basic pentamery seen in *Echinus*. Little further information on echinoid organization can be gained by dissection, but points of considerable interest can be seen by examining the effector organs borne on its test. These effectors, based on those already seen in *Echinus*, have undergone modification to suit a habitat very different from that of the more primitive regular echinoids.

The animal lives in a snug burrow up to 150 mm below the surface of the sand (see Fig. 85). It moves through the sand by scraping away at the front end of the burrow and moving sand to occupy the space behind it. It builds a vertical tube to the surface of the sand, down which water is drawn by ciliary currents, and as it moves it leaves a tubular extension of the burrow into which the water seeps which has passed over the animal (Nichols, 1959). Both the vertical tube and the horizontal soak-away are built and maintained by tube-feet, while spines do most of the burrowing and sand-moving. The animal feeds by picking up 'handfuls' of sand in special tube-feet round the mouth and it respires by means of other special tube-feet on the dorsal surface. Currents are produced by cilia over the entire body, but special regions, the *fascioles*, have heavily ciliated spines to augment the general currents.

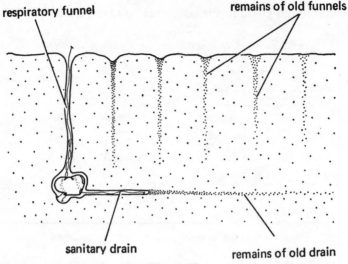

Figure 85

Diagram showing the common British sand-urchin *Echinocardium cordatum* in its burrow. The animal is drawn as though moving from right to left. Only a few tube-feet of each series are drawn (From *Syst. Assoc. Publ.* 4, (1962), 106)

Examine the effectors mentioned above, using the following check-list:

1. *Three fascioles* (see Fig. 86, A), usually appearing as darker bands among the spines:
   (i) Dorsal, producing a current down the vertical funnel;
   (ii) Anal, clearing the region of the anus;
   (iii) Sub-anal, creating a current along the sanitary soak-away.

2. *Five types of spine* (see Fig. 86, D and E):
   (a) Protective—particularly arching over the ambulacra; the anterior ambulacrum has a double arch of spines, one inside the other.
   (b) Burrow-building—the tuft on the dorsal surface and that within the sub-anal fasciole.
   (c) Burrowing—the 'scraping' spines of the anterior side, and the 'sand-shifting' spines of the lateral parts of the body. These may also help in plastering sand on to the lateral walls of the burrow, so that there is a space round the outside of the animal, between it and the sand, for essential circulation of water.
   (d) Locomotory—the paddle-shaped spines of the roughly triangular region on the ventral side.
   (e) Current-producing spines (*clavulae*) in
      (i) the fascioles;
      (ii) the anterior groove, for helping the mucous string down the groove to the mouth.

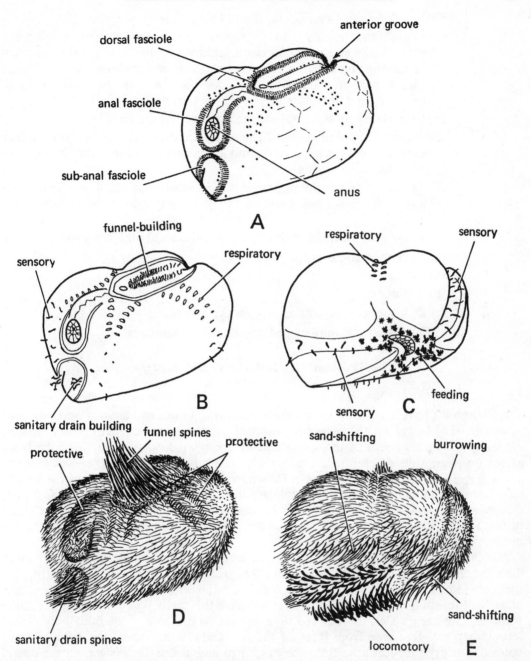

Figure 86. External features of an irregular echinoid, *Echinocardium cordatum*, the British sand-urchin A, Denuded test, drawn as though only the fascioles remained after removing the other spines and the tube-feet; B, C, Postero-lateral and oro-lateral views to show the division of labour in the tube-feet. The feet are drawn contracted, as in a preserved or handled specimen; D, E, Postero-lateral and oro-lateral views to show the division of labour in the spines

3. *Types of tube-feet* (see Fig. 86, B and C):

(a) Dorsal funnel-building in the top section of the anterior ambulacrum. These make and maintain the respiratory funnel, and are said to assist in feeding, by extending beyond the top of the funnel and drooping over to pick up material from the sea-bed. Note the papillate nature of the disk, and the arcuate spicule on the 'neck' of the foot.

(b) Respiratory, in the petal-shaped areas on the dorsal side.

(c) Feeding, round the mouth. These also are papillate, but here the papillae extend across the entire disk, and are not restricted to a fringe round the edge, as in the case of (a).

(d) Sanitary drain building, within the sub-anal fasciole. These keep open a drain so that waste currents can seep away into the sand behind the animal.

(e) Sensory, at the widest part of the test and around the locomotory plastron.

4. Examine the various types of *pedicellaria*.

(a) Tridactyle;

(b) Trifoliate;

(c) Ophiocephalous, particularly on the peristome.

Note: There are no glandular pedicellariae in spatangoids.

13.7.   Holothuroids: *Holothuria* and *Leptopentacta*

13.7.1. INTRODUCTION

Holothuroids have assumed a worm-like shape, with mouth and anus at opposite ends of a cylindrical body; but the basic pentamery of the typical echinoderm is still evident, especially the arrangement of the ambulacra in five rows down the length of the body of the burrowing dendrochirote *Leptopentacta*. There is an extrusible proboscis at the anterior end in *Leptopentacta*, but not in *Holothuria*; both have a ring of extensible feeding tube-feet surrounding the mouth. The integument is leathery and the calcite ossicles are reduced to mere spicules, except for a calcareous ring round the oesophagus, used principally for the attachment of body muscles. The body has longitudinal and circular muscles by which it changes shape.

Holothuria creeps over the sea bottom or into crevices, while *Leptopentacta* burrows into the substratum and lives in a 'U' shaped gallery. *Holothuria* protects itself by extruding sticky threads from its anus when it is molested, a habit which has gained it the common name 'Cotton-spinner'; *Leptopentacta* does not have this method of protection. As *Holothuria* is a surface dweller, its suckered tube-feet are on one side of the body only. These locomotory tube-feet arise from three of the five ambulacra, termed the *trivium*, while pointed, sensory tube-feet arise from the other two ambulacra, the *bivium*, on the dorsal side. In *Leptopentacta*, all the tube-feet are used to help build the burrow, and there is no marked division of labour between the ambulacra.

Both *Holothuria* and *Leptopentacta* exchange respiratory gases mainly across the walls of special internal folds of the cloaca, the *respiratory trees*. Sea water is pumped into and out of their lumina by muscles of the body wall and cloaca.

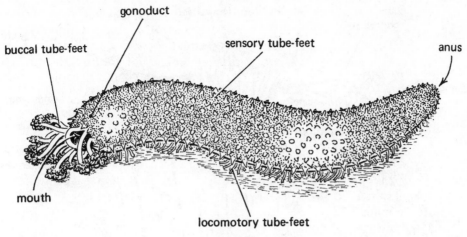

gonoduct

buccal tube-feet

sensory tube-feet

anus

mouth

locomotory tube-feet

Figure 87

Diagrammatic drawing of the British aspidochirote sea-cucumber *Holothuria forskali*. The animal is drawn as though feeding: one of the buccal tube-feet is inserted into the mouth, while the others are foraging

13.7.2. DISSECTION OF *HOLOTHURIA*

(a) *External features.* Orientate the body so that the mass of suckered tube-feet, representing the trivium, lies ventrally (see Fig. 87). Note:

1. Large anterior *mouth*, surrounded by *buccal tube-feet*. Unless the animal was killed in a relaxed condition, these will be hidden inside the mouth, but they may be extended by gently pinching the body just behind the mouth.

2. *Trivial locomotory* and *bivial sensory tube-feet*. Note that the ambulacra are hardly recognizable as separate columns.

3. *Gonoduct*, between the bivial ambulacra at the anterior end.

4. Terminal *anus*, surrounded by sensory tube-feet: locomotory tube-feet do not extend to the anus.

(b) *Internal anatomy.* Make a longitudinal dorsal incision from just in front of the anus to just behind the mouth. Make sure this cut extends to the anterior side of the calcareous ring (which can be felt through the body wall). Make transverse incisions so that the flaps of body wall can be pinned out to expose the viscera (see Fig. 88). Separate the viscera slightly with a seeker and identify:

5. *Alimentary canal*, with haemal *rete mirabile* (brown); it appears that amoebocytes carrying nutritive substances pass through the walls of the alimentary canal into the *rete*, and thence, *via* the haemal spaces, to the rest of the body.

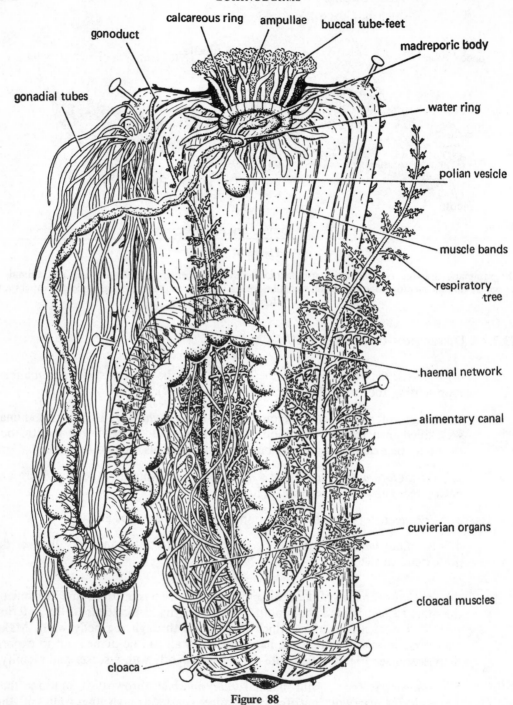

Figure 88

Dissection of the British aspidochirote sea-cucumber *Holothuria forskali* from the dorsal side. The cuvierian organs are drawn on one side only.

6. *Gonad*, slightly pink tubules, about 3 mm in diameter.

7. *Respiratory trees*, yellowish and diffuse.

8. *Cuvierian organs*, brownish-pink or white, slightly smaller than the gonadial tubules.

9. *Ampullae* of the buccal tube-feet; note that there are no ampullae to the other tube-feet, their protraction being probably brought about by contraction of the water-vascular canals and adjacent tube-feet.

10. Single *polian vesicle*, probably a reservoir for water-vascular fluid.

11. *Madreporic canal* and terminal body, usually very small and yellow-pink. The function of this limb of the water-vascular canal system is unknown, but is probably related to the maintenance of a satisfactory pressure in the system.

12. Circum-oral *water vascular ring*, from which the polian vesicle, madreporic canal and radial canals arise.

13. Longitudinal *muscle bands* of the body wall.

Examine the structure of a buccal tube-foot by extending the dorsal cut to split open the mouth. The animal collects food on these structures, bends them into the mouth and wipes them on the inside wall of the oesophagus, at the same time probably receiving a fresh covering of mucus from glands in the oesophagus wall.

Dissect out the *gonad*. Although it is single, it consists of two masses of tubules, one on each side of the midline. Remove the gonad. Pull aside the two respiratory trees and the mass of Cuvierian tubules. Note the arrow head at the distal end of each of these tubules. If living material is available, gently place a specimen of *Holothuria* in a glass dish of sea-water and prod the body until the white tubules are emitted from the anus. The propulsive force is hydraulic, the pressure being generated by the contraction of intrinsic muscles of the tubule wall so that the tubule is elongated.

Follow the alimentary canal, which should be separated from the mass of tubules filling the rest of the perivisceral coelom. The intestine ends in a cloaca, which has muscle strands radiating out from its wall to insert at the body wall. The activity of these muscles causes the cloaca to pulsate, thus forcing water into and out of the respiratory trees.

13.7.3. SPICULES

Place a small piece of body wall in neat sodium hypochlorite solution. Tiny particles, which are the spicules, should be seen as a sediment on the bottom of the tube. Pour off the supernatant. Either examine the ossicles in water, or, to make a permanent preparation, add 90% alcohol, then transfer to absolute alcohol, then xylene as usual, and mount in D.P.X; or from 90% alcohol directly to dimethyl hydrantoin formaldehyde (DMHF).

13.7.4. *Leptopentacta elongata*

This holothuroid, probably commoner than *Holothuria*, burrows in sandy-mud below low tide. If living specimens are available, and a little of the natural substrate from which they were taken, (or sandy-mud, if not) allow a few specimens to burrow. Note that with buccal tube-feet withdrawn the animal uses its proboscis to thrust the anterior part of the body into the mud, until it has disappeared. It breaks surface a short distance away, and extends its tube-feet for feeding (see Fig. 89). Note that of the ten tentacles, two are shorter than the rest and 'Y'-shaped (morphologically, this is the mid-ventral pair). In the feeding process, the other eight tube-feet are bent one by one towards the centre of the crown and thrust into the pharynx; on withdrawal, each is wiped between the limbs of the 'Y', so that the food-laden mucus is removed (Fish, 1967). Because respiratory exchange takes place across cloacal structures, the anal end of the burrow is also open. Defaecation takes place with some force, so that the water in the region of the cloaca is not polluted. Feeding is said to occur only in the summer months of May to September.

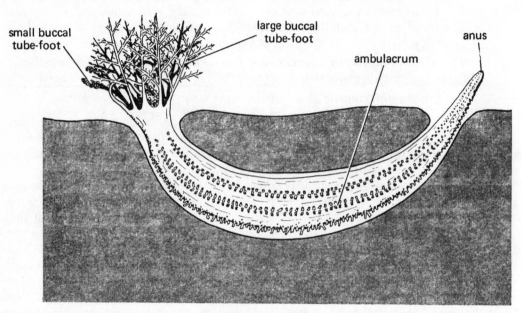

Figure 89

Diagram of the British dendrochirote sea-cucumber *Leptopentacta elongata* in its burrow in mud, though feeding. One of the 8 large tube-feet round the mouth is shown being grasped near the mouth by one of the 2 short tube-feet

14. CHAETOGNATHS

The chaetognaths are a small group of animals, many species of which form important elements of marine zooplankton. The most common genus is *Sagitta*. The transparency of the body makes it possible to see many features without staining, but this is necessary for complete examination. Specimens are usually preserved in sea water-formalin mixtures and should be stained in haematoxylin, or borax carmine and methylene blue, differentiating to obtain the correct degree of staining before clearing and mounting.

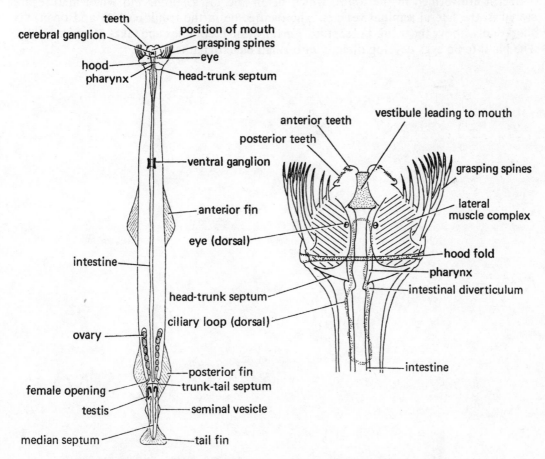

Figure 90 *Sagitta elegans*, A, dorsal view; B, head, ventral aspect

The following description is based upon *Sagitta elegans*. (Fig. 90). The head region is complex and bears feeding and sensory organs. When swimming the head is covered by the hood, a fold of body wall which can be retracted to allow the chitinous spines to be used in grasping prey. The spines are operated by a series of muscles attached to skeletal plates in the head. Two large ganglia are situated in the head, the dorsal cerebral and the ventral vestibular, and these are connected to each other and to the large ventral truck ganglion by commissures. In addition to the eyes (which have a characteristic arrangement of five fused ocelli) there are a number of sense receptors on the head and trunk. Between and posterior to the eyes is a ridge of ciliated epithelium, the corona ciliata, which is also thought to have a sensory function.

Food grasped by the jaws is pushed into the ventral mouth and passed back, *via* a short bulbous pharynx, into the intestine. This is a simple tube, supported by mesenteries, which gives rise to short anterior diverticula and opens posteriorly into a short rectum.

The coelomic cavities of the head, trunk and tail are separated from one another by transverse septa, the trunk coelom being incompletely divided dorso-ventrally by the intestinal mesenteries. The tail coelom is also divided by a median septum.

Sperm are formed in the testes, which lie in the tail coelom, and when mature are stored in the lateral seminal vesicles. The ovaries lie in the trunk coelom and open, via short ducts, above the trunk-tail septum. Fertilization is complex and takes place internally. The planktonic eggs develop directly into small chaetognaths.

15. CHELICERATES

The chelicerate arthropods, Xiphosura, Arachnida and related groups, form one of the major evolutionary lines of the phylum. They are characterized by the possession of preoral appendages in the form of *chelicerae*—never antennae—and by the presence of post-oral appendages (*pedipalps*) which are limb-like or sensory, never mandibulate. The body is typically divided into a *prosoma* or 'cephalothorax,' and an *opisthosoma* or 'abdomen.' Primitively the chelicerate body contained 20 segments, 7 prosomatic, of which usually 6 persist into the adult stage, and 13 opisthosomatic, which have been variously reduced in modern groups.

15.1. Xiphosurans

The Xiphosura form one Sub-class of the Class Merostomata. These are aquatic chelicerates, respiring by opisthosomatic appendages modified as book-gills. The Sub-class Eurypterida were Palaeozoic forms with free opisthosomatic segments.

The xiphosurans have the prosoma and opisthosoma covered by fused carapaces, hinged on one another. The pedipalps are used as walking limbs and appendages are present on the pregenital segment. There is a large caudal spine. Xiphosurans feed on particulate foods.

Only five species of Xiphosura are extant, of these the King Crab *Limulus polyphemus* is the most familiar. It occurs in large numbers on the east coast of the U.S.A.

Examination of dried or preserved specimens from the dorsal surface gives little indication of the segmental organization of the prosoma. The prosomatic carapace bears anterior median ocelli and larger lateral eyes of a compound nature, though differing from those of other arthropods. The opisthosomatic carapace bears lateral segmental spines and two median rows of depressions which indicate areas of segmental muscle attachment.

On the ventral surface (Fig. 91) it is possible to identify the six paired segmental appendages of the prosoma and the seven appendages of the opisthosoma. The most anterior appendages, the three-segmented chelicerae, are used to manipulate food.

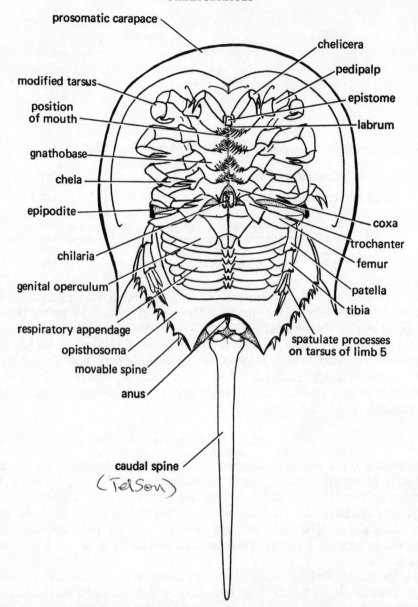

Figure 91 *Limulus polyphemus*, Male, ventral view

Behind them lies the mouth region, bordered anteriorly by the labrum, laterally by the pedipalps and first three walking limbs, and posteriorly by the chilaria, appendages of the opisthosomatic pregenital segment which are functionally incorporated into the prosoma. The coxae of the pedipalps and first three pairs of walking limbs carry coarse spines and form gnathobases, which are used to grind up the soft-bodied invertebrates on which the animal feeds. The last walking limb does not form a gnathobase, but bears a coxal process (epipodite) used in cleaning the gills. Unlike the anterior limbs the posterior limb has an additional seventh segment and is not chelate. The penultimate segment of the limb bears four spatulate processes used in locomotion and in digging.

The overlapping opisthosomatic appendages are broad, flattened plates, the two halves of each being united in the midline. The most anterior, the genital operculum, bears the genital openings on its posterior surface and partially covers the posterior gills. The five respiratory appendages carry numerous thin-walled lamellae on their posterior surfaces and it is across the surfaces of these that respiratory exchange with the blood takes place. In life these appendages are moved backwards and forwards, creating a respiratory and swimming current and at the same time pumping blood into and out of the internal spaces of the lamellae. (In fresh specimens it may be possible to find the commensal triclad turbellarian *Bdelloura* and its egg capsules on the lamellae.)

The large caudal spine is considered to be an extension of the non-segmental telson.

Where living *Limulus* are available for class study external examination should be supplemented by dissection, for which a goood account is given in Brown (1950).

15.2. Arachnids

15.2.1. INTRODUCTION

The arachnids are the largest and most successful class of the chelicerate arthropods and have had a long evolutionary history. It is possible to trace in the class a number of evolutionary trends, including a reduction in the length of the body axis, loss of distinct segmentation in the opisthosoma and an increasing adaptation to terrestrial life.

15.2.2. CLASSIFICATION

Class ARACHNIDA

Mainly terrestrial, respiring by book-lungs, tracheae or through body surface. Pedipalps rarely locomotory, often sensory or grasping. Pregenital segment usually absent or forming pedicel between pro- and opisthosoma. Feed on liquid foods.

The class contains 14 orders, four of which are wholly extinct and are not included here.

Order SCORPIONIDA

Opisthosoma divided into meso- and metasoma, tergites distinct. Chelicerae 3-segmented. Pedipalps large, chelate. Metasoma bears sting. Book-lungs.

Centurus, Buthus

Order PSEUDOSCORPIONIDA

Small forms. Opisthosoma not divided, tergites distinct. Chelicerae 2-segmented. Pedipalps chelate with poison glands. Tracheae.

Chelifer

Order SOLPUGIDA

Prosomatic carapace divided. Opisthosomatic tergites distinct. Chelicerae large, 2-segmented. Pedipalps adhesive. First walking limbs sensory. Fourth limbs with sensory racquets. Tracheae.

Galeodes

Order PALPIGRADI

Very small forms. Prosomatic carapace divided. Opisthosomatic tergites distinct. Long terminal flagellum. Chelicerae 3-segmented. Pedipalps locomotory. First walking limbs sensory. Respire through book-lungs or body surface.

Koenenia

Order UROPYGI

Opisthosomatic tergites distinct. Long terminal flagellum. Chelicerae 2-segmented. Pedipalps grasping. First walking limbs sensory. Book-lungs.

Mastigoproctus

Order AMBLYPYGI

Opisthosoma joined to prosoma by pedicel, tergites somewhat reduced. Chelicerae 2-segmented. Pedipalps grasping. First walking limbs antenniform, sensory. Book-lungs.

Charinus

Order ARANEAE

Opisthosoma joined to prosoma by pedicel, tergites reduced. Chelicerae 2-segmented with poison glands. Pedipalps sensory with copulatory function in male. Well developed silk glands. Book-lungs and/or tracheae.

Araneus, Tegenaria

Order RICINULEI

Opisthosomatic tergites reduced. Chelicerae 2-segmented. Pedipalps small, chelate. Third walking limbs with copulatory function in male. Tracheae.

Ricinoides

Order OPILIONES

Prosoma and opisthosoma not obviously separated. Chelicerae 3-segmented. Pedipalps sensory. Limbs very long and thin. Male with copulatory organ, female with ovipositor. Tracheae.

Phalangium

Order ACARINA

Prosoma and opisthosoma not separated. No external segmentation. Chelicerae and pedipalps often highly modified. First walking limbs may be sensory. Respire by tracheae or through body surface.

Ixodes

15.2.3. SCORPIONS

Scorpions are among the oldest arachnids and retain many primitive characters although they are highly specialized for their mode of life. A study of scorpions gives a good picture of basic arachnid organization (Fig. 92).

The body is long and retains 18 segments in the adult. The opisthosoma is clearly divided into a broad mesosoma, in which the segmental tergites and sternites are separated by flexible pleural membranes, and a narrow metasoma, in which the segmental exoskeleton forms complete rings. The telson is modified to form a terminal sting.

The prosomatic carapace bears median and lateral ocelli; no compound eyes are present. The ventral surface of the prosoma is made up largely of the coxal segments of the limbs; there is a small sternum.

The large chelate pedipalps are used to grasp prey, which is then torn apart by the chelicerae and digested by regurgitated enzymes. The mouth is situated at the end of a preoral cavity formed dorsally by the chelicerae and labrum, laterally by the pedipalpal coxae and ventrally by the coxal endites of the first and second walking limbs. Digested food is taken into the mouth via the preoral cavity by the sucking action of the pharynx.

The pregenital segment is lost in the adult and the genital (8th.) segment becomes the first mesosomatic segment, being represented ventrally by the small genital operculum, which covers the genital opening. The pectines, modified appendages of the second mesosomatic segment, are thought to have a sensory function, perhaps in substratum selection during mating and other activities.

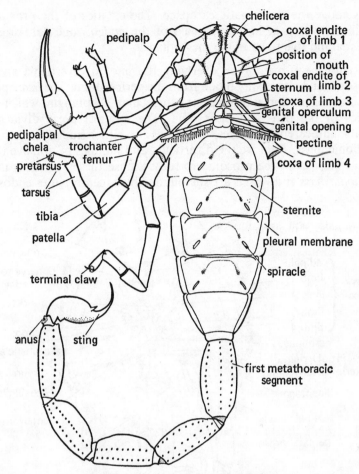

Figure 92 *Centurus* sp. Ventral view

Four of the mesosomatic sternites have lateral spiracles which open into the cavities of the book-lungs. These consist of internal lamellae of thin cuticle, across which respiratory exchange occurs. The metasomatic segments are articulated in such a way as to give greatest flexibility in the vertical plane, so that the sting may be brought over the prosoma to subdue prey held by the pedipalps.

Dissection

The following account is based upon the dissection of a male scorpionid species. Variations in certain features may be found if members of other families are used.

1. If the specimen is small, remove the limbs and embed the ventral surface in a wax-bottomed dissecting dish. Otherwise use pins to hold the body.

2. Beginning at the last mesosomatic segment, cut forwards through the lateral pleural membranes on each side of the body and then cut carefully through the tergites of the first and last mesosomatic segments. Remove the mesosomatic tergites without damaging the underlying heart.

3. Remove the prosomatic carapace. The cavities of the pro- and mesosoma are filled by the sac-like diverticula of the *midgut*, one pair originating in the prosoma and five pairs in the mesosoma.

4. Remove the membrane overlying the midgut diverticula and heart in the mesosoma and clear away blood from the pericardium. The *heart* passes forwards as the anterior aorta through the muscular diaphragm which separates the cavities of the pro- and mesosoma. The anterior *aorta* divides to form the cephalic and cerebral arteries, which pass on either side of the supraoesophageal ganglionic mass. Remove the diverticula and bundles of dorsoventral and longitudinal muscles in the prosoma to fully expose these vessels as in Fig. 93. The *ganglionic mass* and vessels are surrounded by a membranous endosternite, which

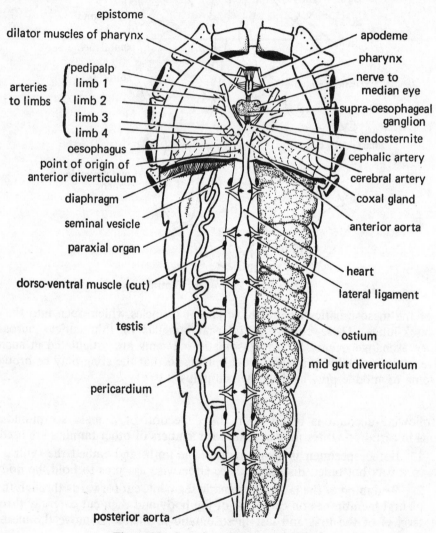

Figure 93 Scorpion. General dissection

connects with the epistomal apodemes and with the diaphragm, and which must be removed carefully in order to fully display all the organs. The nerves passing to the chelicerae will probably be removed in the process of displaying the pharynx and its musculature. Note the excretory coxal glands which open onto the coxae of the third limbs.

5. Remove the diverticula on one side of the body in the mesosoma, taking care not to damage the gonad, and study the relationship of the heart and its ventral ligaments to the book-lungs. The heart is suspended by intersegmental dorsal, lateral and ventral ligaments, the latter being attached to the roof of the ventral blood sinus. The movements of the heart are thereby transmitted to the sinus, pumping blood into and out of the book-lungs. Blood returns to the heart via sinuses and veins, entering through the ostia. Remove the dorsal region of the metasomatic sclerites and trace the posterior aorta to the end of the body.

6. Cut through and remove the arteries to the limbs, the anterior aorta, the heart and pericardium. Clear away muscle and the remaining diverticular tissue to expose the oesophagus and midgut, which passes through the diaphragm below the opening for the aorta and runs back through the mesosoma. Identify the six paired openings of the diverticula. At the posterior end of the mesosoma the midgut gives off two pairs of Malpighian tubules at its junction with the hindgut. Trace the hindgut back to the anus at the base of the sting.

7. Examine the paired *gonads* and accessory organs which will have been revealed by the removal of the diverticula. The gonads are a network of tubules and essentially similar in both sexes. The male possesses seminal vesicles and large paraxial organs which produce the spermatophore transferred to the female in mating (Fig. 93).

8. Remove the oesophagus, midgut, hindgut and gonads to completely display the suboesophageal ganglionic mass and ventral *nervous system* (Fig. 94). The ventral nerve cords are covered by the ventral artery which arises from the cerebral artery. Remove the ventral artery to display the segmental ganglia in the meso- and metasoma. The former has three free ganglia (belonging to segments 12, 13 and 14) and the latter four (15, 16, 17, 18+19). The anterior ganglia of the mesosoma are incorporated into the suboesophageal ganglionic mass and their segmental nerves pass posteriorly through the diaphragm. Carefully cut away the central area of the diaphragm to display the complete system.

15.2.4. SPIDERS

(a) *External features*

Spiders are among the more advanced arachnids and their bodies show many modifications from the primitive arachnid pattern. The most suitable and readily available specimens for class study are the larger species of web spinning spiders (*Araneus*), house spiders (*Tegenaria*) and wolf spiders (*Lycosidae*). These advanced araneomorph spiders should be compared with specimens of the primitive mygalomorph spiders.

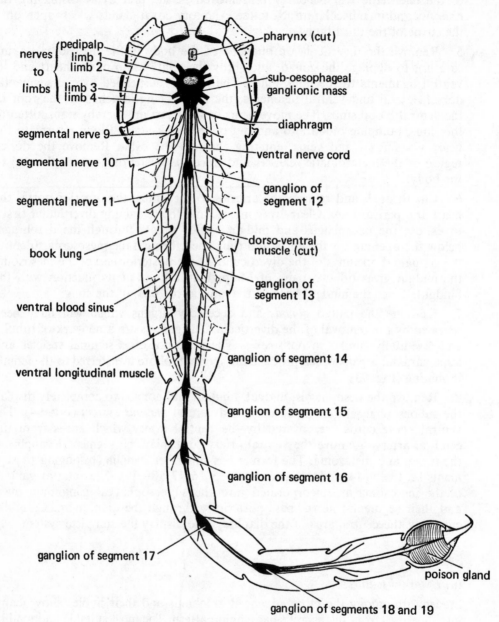

Figure 94 Scorpion. Dissection of nervous system

Specimens should be examined alive, and after preservation in 70% alcohol+ glycerine. Whole mount preparations can be made by boiling in 10% KOH to remove soft tissues, dehydrating in alcohols, clearing in xylol and mounting in canada balsam, D.P.X. or other resin.

The prosoma of spiders is separated by a narrow pedicel from the rounded opisthosoma, thus allowing considerable mobility of the latter (Fig. 95). On the anterior margin the prosomatic carapace bears eight eyes. The chelicerae each consist of a terminal unguis which articulates transversely upon the basal

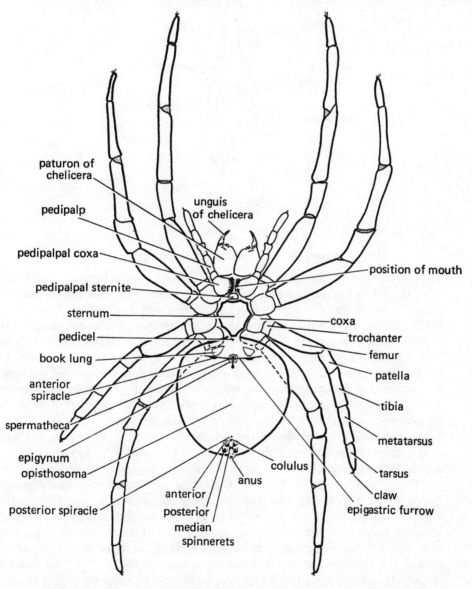

Figure 95 *Araneus diadematus*. Female, ventral view of prepared specimen

paturon. (In mygalomorph spiders the unguis articulates in the plane of the longitudinal axis of the body). The poison gland of each chelicera opens at the tip of the unguis. The pedipalps have a sensory function in females, but in mature males the terminal segments are enlarged and modified to form a copulatory organ; the sexes are easily distinguished by this character. The organ can be protruded by boiling the pedipalp in dilute KOH and may be mounted in the usual way or examined directly in glycerine (Fig. 96).

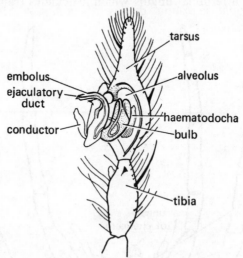

Figure 96 *Tegenaria* sp. Male pedipalp

The mouth lies at the base of a small preoral cavity formed largely by the sternite and coxae of the pedipalps. These bear thick clusters of hairs which project into the cavity and filter out particles from digested food material.

Little external evidence of segmentation is visible on the opisthosoma. On the ventral surface the region of the genital segment is indicated by the genital opening and the anterior spiracles which lie on the epigastric furrow. In the female there is a chitinous epigynal plate surrounding the genital opening and the openings of the spermathecae, which receive sperm from the male pedipalps during mating. The anterior spiracles open into the cavities of book-lungs, the posterior spiracle (on segment 9) opens into tracheae.

The spinnerets, through which the silk of the animal is extruded, are situated at the posterior end of the opisthosoma. They represent modified segmental appendages and consist of an anterior pair on segment 10 and a median and posterior pair on segment 11. Silk is spun through numerous fine openings at the tip of each spinneret and through larger tubes or spigots. Between the anterior spinnerets lies a small protruberence, the colulus, which probably represents vestigial anterior median spinnerets. In cribellate spiders, e.g. *Amaurobius*, the colulus is replaced by a small plate, the cribellum, through which special silk is spun and combed out by spines on the hind limbs. Behind the spinnerets, usually on a small tubercle, is the anal opening belonging to segment 18 of the body.

(b) *Internal anatomy: Araneus diadematus*

It is possible to study much of the internal anatomy of a spider by cutting the the animal longitudinally into equal halves. For this purpose specimens should have been hardened in alcohol for some time. The following description is based upon female specimens of *Araneus diadematus* (Fig. 97).

Remove the limbs and place the specimen with its ventral surface uppermost. Use a sharp razor blade to cut the animal in two along the mid-line. Both halves of the body should be examined to gain a complete picture. Identify the organs present.

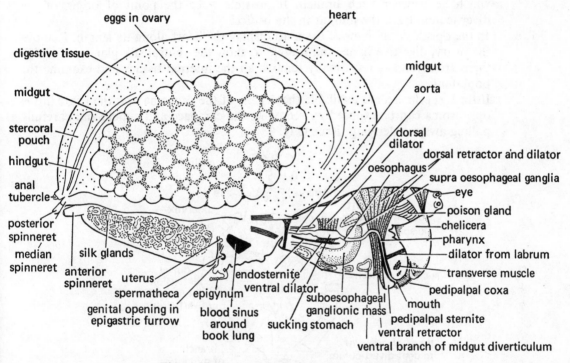

Figure 97 *Araneus diadematus*. Female, left half

The relationship of the mouth to the pedipalpal sternite and coxae can now be seen clearly. Notice the dorsal and ventral muscles which are attached to plates in the walls of the pharynx and 'stomach' and which, by their contraction, produce the suctorial action used in feeding. The 'stomach' opens into the midgut, which gives rise to a large diverticulum shortly before the pedicel. Numerous diverticula are given off in the opisthosoma, forming a mass of digestive tissue which obscures the midgut for most of its length. However, the terminal hindgut and dorsal stercoral pouch are often filled with dark waste material and therefore are easily seen. Malpighian tubules arise from the midgut-hindgut junction and pass laterally forward; they cannot be seen in the cut specimen.

When these organs have been examined tissues can be removed to display deeper-lying organs. Use fine forceps and pins and dissect in alcohol under a microscope.

Remove the muscles running from the prosomatic carapace to the pharynx and expose the large poison gland which passes into the chelicera of that side of the body. Note the spiral musculature of the gland. Below the gland lies an anterior extension of the midgut diverticulum. Use one half of the spider and carefully remove the prosomatic carapace and the underlying muscle and poison gland. Display the branches of the midgut diverticulum which pass into each walking limb by carefully pulling away the bundles of muscle fibres which lie between each branch. If possible trace the point of origin of the diverticulum from the midgut in the pedicel.

In the opisthosoma remove tissue to display the heart along its length. Remove the ovary, digestive tissue and muscle to show the ventral silk glands, of which there are several types. Finally remove all the internal tissues and examine the book-lung.

If fresh specimens are available the organs of the prosoma and the heart in the opisthsoma can be exposed by removing the carapace and cuticle and carefully pulling away underlying tissue (Fig. 98).

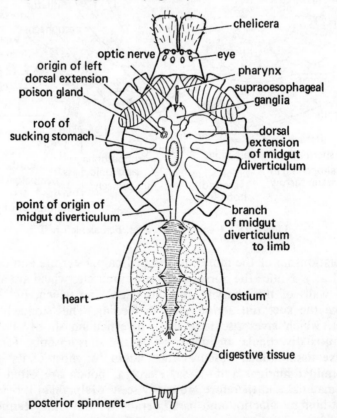

Figure 98 *Tegenaria* sp. Female, dissection to show dorsal organs

15.2.5. MITES AND TICKS: *Ixodes ricinus*

The mites and ticks or Acari constitute one of the largest orders of the Arachnida and occur in almost every terrestrial and aquatic habitat. Many are commensal or parasitic, some of the latter being of considerable economic importance. Representative free living species may be obtained from soil, leaf litter, stored farinaceous materials and fresh water habitats; commensal and parasitic species can be collected from many invertebrate and vertebrate hosts. The majority of species show the characteristic features of acarine organization, together with specializations of chelicerae and limbs for particular modes of life. Small species can be satisfactorily mounted in lactophenol with lignin pink, larger species should be boiled in 10% KOH to remove soft tissues before dehydration in alcohols, clearing in xylol and mounting in canada balsam, or other resin.

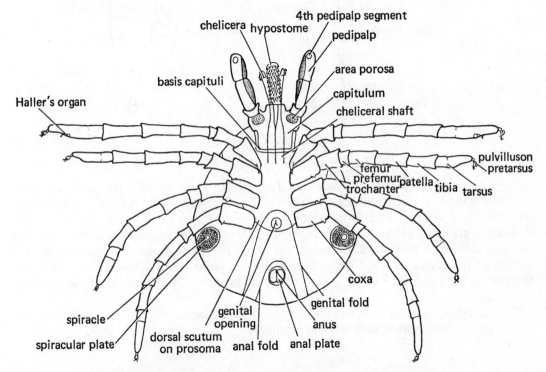

Figure 99 *Ixodes ricinus.* Female, ventral view of prepared specimen

Ixodes ricinus, the common sheep tick, is a familiar example of the order and may be used as a basis for description (Fig. 99). The body shows no external indication of segmentation or division into regions. Anteriorly the pedipalps and chelicerae, highly modified for piercing the skin of the host, are borne on the heavily sclerotized capitulum. The body behind the capitulum can be considered as an anterior leg-bearing prosoma and a posterior opisthosoma, although there is no obvious demarcation. In females the prosoma is covered by a dorsal scutum, in males the scutum covers the entire body. The cuticle of the opisthosomatic region is folded and flexible, allowing considerable expansion to take place when the animal is sucking blood from the host. Engorged females may be two or three times the size of unfed individuals.

The coxal segments of the limbs form flat plates on the ventral surface of the body, and are almost immovable. The first pair of limbs bears sensory structures, known as Haller's organs, on the tarsal segments; these organs play an important role in host detection.

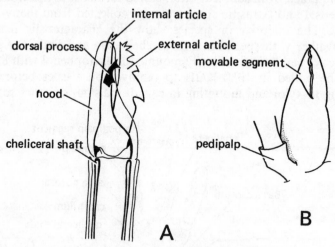

Figure 100 Chelicerae
A, *Ixodes ricinus* (female); B, flour mite (left chelicera)

The pedipalps, chelicerae and associated structures form an efficient mechanism for penetrating the host's skin and taking up blood. The four-segmented pedipalps are concave on their inner surfaces and form a protective cover for the chelicerae. In feeding the chelicerae are protruded by an increase in internal body pressure and the movable terminal articles (Fig. 100) cut through the skin. The large toothed hypostome, an extension of the capitulum (and derived from the pedipalpal coxae), is used to anchor the tick during the extended feeding period. Blood is taken up by the action of the suctorial pharynx.

Free living mites retain a chelate condition in the chelicerae (Fig. 100) and are therefore more directly comparable with other arachnids.

16.   CRUSTACEANS

16.1.   Introduction

Much of the structural diversity of the Crustacea is caused by variation in their appendages. For this reason the following account gives details of the external anatomy of a wide range of Crustacea, and describes the internal anatomy of fewer representative forms. The appendages are not described in great detail, but just sufficiently for identification, so that the student can investigate their structure and function. It is suggested that drawings should be made of those appendages which appear to be important in feeding and locomotion.

If time is not available to examine all the forms described in the following pages the following short list is suggested: *Artemia*, *Daphnia*, *Cyclops*, *Lepas*, *Mysis* and *Astacus*. If time is available for a more extended list then the work of McLaughlin (1980) would be useful.

16.2. Classification

Class CRUSTACEA

Two pairs of pre-oral antennae, at least three pairs of post-oral appendages act as jaws. Any or all of these appendages may be absent from adult parasitic forms.

Sub-class BRANCHIOPODA

Limbs usually phyllopodous, antennules simple and reduced, mandible without palp, maxillae reduced or absent.

Order ANOSTRACA

No carapace, eyes stalked, antennae prehensile in ♂, reduced in ♀. Trunk limbs 11 or 19 pairs. Eggs carried in median ventral sac at posterior end of thorax.

Artemia, Chirocephalus

Order NOTOSTRACA

Large horseshoe shaped carapace, eyes sessile, antennae vestigeal, numerous trunk limbs (up to 70 pairs).

Triops, Lepidurus

Order DIPLOSTRACA

Carapace large, bivalved, enclosing trunk. Antennae large, biramous, used for swimming, eyes sessile, closely apposed in the mid-line.

Sub-order CONCHOSTRACA

Trunk limbs 10–27 pairs. Carapace encloses head in adult.

Limnadia, Estheria

Sub-order CLADOCERA

Trunk limbs 4–6 pairs. Carapace does not enclose head.

Daphnia, Leptodora,
Polyphemus, Evadne

Sub-class CEPHALOCARIDA

Carapace small, covering first trunk segment. Trunk limbs with endopod of six podomeres and exopod of two. Egg carried under 10th trunk somite.

Hutchinsoniella, Lightiella

Sub-class MYSTACOCARIDA

No carapace, body narrow. Antennules and antennae well developed. Mandibles with large exopod. Thoracic limbs reduced to single podomeres with a few terminal setae. Caudal rami strong, claw-like, bearing setae.

Derocheilocaris

Sub-class COPEPODA

No carapace, antennules uniramous, mandibles may have palp. Typically 9 free trunk somites, and 6 pairs of trunk limbs (often fewer). Numerous parasitic forms provide exceptions to all definitions.

Cyclops, Calanus, Lernaea

Sub-class OSTRACODA

Carapace forms a bivalved shell. Not more than 5 pairs of limbs behind mandibles. Mandible with a palp.

Cythere, Eucypris,
Gigantocypris

Sub-class BRANCHIURA

Body flattened, large carapace fused with thorax. Paired sessile compound eyes. Mouthparts suctorial, forming a proboscis. First maxillae often form suckers. Abdomen short, bilobed. 4 pairs biramous swimming legs. Fish lice.

Argulus, Dolops

Sub-class CIRRIPEDIA

Sessile when adult. Carapace forms a mantle enclosing trunk and limbs, often with calcareous plates. Antennules vestigeal; antennae absent. Free living forms usually have 6 pairs of trunk limbs forming elongated feathered cirri. Often hermaphrodite. Larval stages include nauplius and cypris.

Order THORACICA

Typical barnacles, mantle distinct, 6 pairs of cirri.

Sub-order PEDUNCULATA
With distinct peduncle or stalk.

Lepas, Conchoderma

Sub-order OPERCULATA
Without peduncle. Outer plates form a turret-like wall. Scuta and terga form a movable valve.

Balanus, Elminius

Order ACROTHORACICA

Mantle distinct, fewer than 6 pairs of trunk appendages.

Alcippe

Order ASCOTHORACICA

Parasites on Echinoderms and Coelenterates. Mantle expanded, two lobed or sac like, having diverticula from gut and gonads. Trunk appendages reduced.

Laura, Dendrogaster

Order RHIZOCEPHALA

Parasites on Crustacea. With a system of absorptive roots. Mantle distinct. Adult has no appendages or alimentary canal.

Sacculina

Sub-class MALACOSTRACA

Trunk limbs differentiated into two tagmata: 8 thoracic and 6 abdominal. Eyes paired, stalked or sessile. Female genital aperture on 6th thoracic somite, male aperture on 8th thoracic somite.

Division PHYLLOCARIDA

Order LEPTOSTRACA

Abdomen with 7 somites, carapace large, covering thorax and part of abdomen. 8 thoracic limbs all similar. Eyes stalked. Telson bears caudal rami.

Nebalia

Division HOPLOCARIDA

Order STOMATOPODA

Carapace shallow, leaving 4 thoracic somites uncovered. Two movable somites in head, bearing stalked eyes and antennules. Second thoracic limbs form massive grasping organs. Abdomen large, with large pleopods bearing branchial filaments. Last abdominal appendages (uropods) form tail fan with telson.

Squilla

Division SYNCARIDA

No carapace. Mandible lacks lacinia mobilis. No oostegites on thoracic limbs.

Order ANASPIDACEA

First thoracic somite fused to head. Statocyst in antennule. 6 free abdominal segments. No trace of caudal rami.

Anaspides, Koonunga

Order STYGOCARIDACEA

First thoracic somite fused to head. Statocyst present in antennule. 6 free abdominal somites. Rudiments of caudal rami on telson.

Stygocaris

Order BATHYNELLACEA

First thoracic somite not fused to head. Pleopods reduced or absent. 5 free abdominal somites. Caudal rami well developed.

Bathynella

Division PERACARIDA

Females with oostegites on thoracic limbs, forming a brood pouch. Mandible with a lacinia mobilis. First thoracic segment fused to head.

Order MYSIDACEA

Carapace extends over most of thorax. Eyes stalked. Antennae with exopod forming a scale. Thoracic limbs with natatory exopodites. Uropods form a tail fan with telson.

Mysis, Hemimysis,
Gnathophausia

Order CUMACEA

Carapace fused to first three or four thoracic somites, and extends forwards to form a pseudorostrum. Eyes sessile, close together in mid-line. Abdomen long, narrow, giving a general 'tadpole' shape. Uropods styliform.

Diastylis, Bodotria

Order TANAIDACEA

Carapace small, fused dorsally to first two thoracic somites. Eyes if present on short immovable stalks. Second thoracic appendages large and chelate. Abdomen short, pleopods well developed, biramous.

Tanais, Apseudes

Order SPELAEOGRIPHACEA

Carapace very small, fused dorsally to first thoracic somite. First three pairs of walking legs bear exopods. Blind, cave dwelling. Single known species *Spelaeogriphus lepidops* from S. Africa.

Order THERMOSBAENACEA

Carapace fused with first thoracic somite, but extending over three more somites, and in the female forming a dorsal brood pouch unique among Malacostraca.

Thermosbaena, Monodella

Order AMPHIPODA

No carapace, eyes sessile, respiratory bracts attached at bases of thoracic limbs. Pleopods narrow. Antennal gland is functional excretory organ.

Gammarus, Cyamus

Order ISOPODA

No carapace, eyes sessile, thoracic limbs uniramous, lacking exopods. Pleopods flattened, respiratory in function. Maxillary gland is main excretory organ. Some adult parasites defy this definition.

Ligia, Idotea, Asellus,
Cryptoniscus

Division EUCARIDA

Carapace large, coalescing dorsally with all 8 thoracic somites. Eyes stalked.

Order EUPHAUSIACEA

No thoracic limbs modified as maxillipeds, gills in a single series attached to bases of thoracic limbs.

Euphausia, Meganyctiphanes

Order DECAPODA

First three pairs of thoracic limbs modified to form maxillipeds. Gills in several series attached to bases of thoracic limbs and to thorax wall.

Sub-order NATANTIA

Pleopods well developed and used for swimming. Antennal scale large and well developed.

Crangon, Palaemon

Sub-order REPTANTIA

Pleopods small, not used for swimming. Antennal scale usually small.

Section PALINURA

Rostrum small or absent, pincers, when present on front legs weakly developed, abdomen large, well armoured with large tail fan. Outer branch of uropod not subdivided.

Palinurus, Panulirus, Scyllarus

Section ASTACURA

Rostrum well developed, pincers on first legs large, second and third legs also chelate. Outer branch of uropod divided by a suture.

Nephrops, Homarus,
Astacus (=Potamobius)

Section ANOMURA

Abdomen flexed forwards beneath thorax, or soft and assymetrical with a reduction of the pleopods on the right side. Tail fan present. Last walking leg usually small and tucked under the thorax.

Galathea, Eupagurus,
Birgus, Lithodes

Section BRACHYURA

Carapace expanded laterally, abdomen small, tucked forwards under thorax. Tail fan reduced, usually absent. First legs always chelate.

Cancer, Carcinus, Maia, Uca

16.3. Branchiopods

16.3.1. *ARTEMIA*

(a) *Examination. Artemia* is best examined alive. Specimens may be narcotized with a few drops of chloroform in their water, but much can be seen in unnarcotized specimens held under a cover slip supported by plasticine. Specimens should be viewed both from the side and ventrally.

(b) *Form and segmentation of the body.* All members of the Anostraca lack a carapace. The head bears a small median or nauplius eye, and a pair of well developed stalked eyes. A large labrum projects posteriorly from the ventral side of the head, and overlaps the mandibles. The general form of the body is elongated and more or less cylindrical. In the female a brood pouch is formed on the ventral side behind the last pair of limbs. The posterior six segments of the trunk are limbless, and this region terminates in a caudal furca of two rami. The size of these rami varies with the salinity of the water in which the animal has grown. In low salinities the rami are long with numerous setae, but in high salinities they are short with few setae. For further details see Keunen (1939), Lockhead (1950), Gilchrist (1960) and Bowen (1962).

(c) *Appendages*

1. The antennules are small and simple with a few terminal setae.

2. The antennae show sexual dimorphism. In the female they are relatively small, but in the male they are enlarged and used as grasping organs for holding the female during copulation.

3.   The mandibles are relatively large, and lack a palp. In living specimens they can be seen making partial rotatory movements around their long axes. These movements serve to compact food collected by the filtering action of the trunk limbs.

4.   The maxillules and maxillae are rudimentary. The small projection representing the maxilla bears the opening of the maxillary gland, which can be seen coiled just behind the mandibles.

5.   The eleven trunk limbs are all built on the same plan (Fig. 101) with small variations in the proportions of each part. A food groove runs along the mid ventral side of the body between the two rows of limbs. Remove one of these limbs and examine the structure of the filter setae on the basal endite.

(d) *Internal anatomy*

1.   The *gut* forms a simple tube from mouth to anus; two small sac like diverticula are given off from the anterior part of the midgut.

2.   The *heart* is long, with ostia in each trunk segment except the first and last. The blood may be pink with haemoglobin if the animal has been living in poorly aerated conditions.

3.   The *nervous system* is difficult to see in living specimens. Parts of the cerebral ganglia and optic nerves can usually be seen, and if a specimen is mounted ventral side uppermost under a microscope with strong lighting some parts of the ventral nerve cord may be seen.

4.   The *reproductive system* of the male is simple, with tubular testes opening via vasa deferentia through small intromittent organs located just behind the last pair of legs.

In the female the ovaries are also tubular, but the ducts open into lateral brood pouches which communicate with a median brood pouch. The eggs first pass into the lateral pouches, and later into the median pouch. Shell glands supply the tough outer covering of the resting eggs. The ventral surface of the pouch bears a pair of small hooks which engage the antennae of the male during copulation. The external opening of the brood pouch faces posteriorly, and the release of the eggs is effected by muscles which pull on the lips of the opening.

(e) *Development*

Brine shrimp eggs remain viable when dry for many years and are obtainable from some commercial suppliers. Nauplii hatch in 24-48 hours.

16.3.2. *TRIOPS* AND *LEPIDURUS*

(a) *General remarks.* The differences between *Triops* and *Lepidurus* are relatively small, so that species of either genus may be used as typical representatives of the order Notostraca. In most laboratories only preserved material will be available. If living specimens are found they are well worth observing under a binocular dissecting microscope to see details of the feeding mechanism and movements of the limbs.

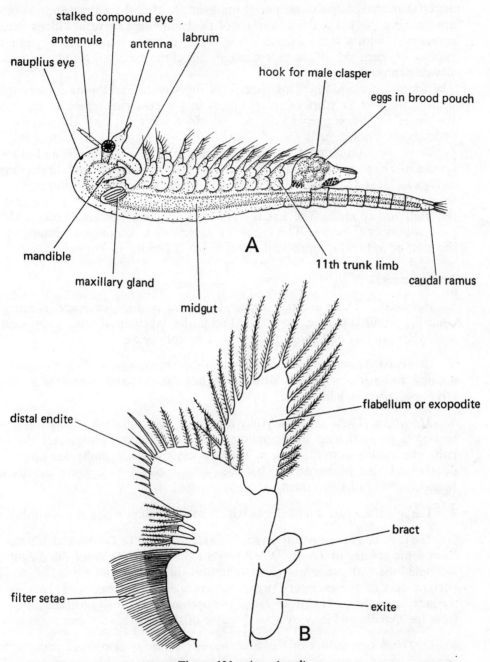

Figure 101 *Artemia salina*

A, lateral view of adult female; B, trunk limb

(b) *Form and segmentation of the body.* The large shield shaped carapace is attached to the posterior margin of the head, so that the thorax and abdomen are free (Fig. 102). The dorsal surface of the head bears a pair of kidney shaped sessile eyes, with a small median eye in front, and a median dorsal organ close behind. On each side of the carapace the coils of the maxillary gland can be seen showing through the integument.

The thorax and abdomen together form the trunk. The thorax is the region bearing the first 11 pairs of trunk limbs. In the posterior region of the trunk the limits of the segments become indistinct ventrally, and the number of appendages is not directly related to the number of segments. In some species of *Lepidurus* the total number of trunk segments may be as low as 26, and in some species of *Triops* as high as 44. The total number of pairs of trunk limbs ranges as high as 71, so that some of the posterior segments (discernable dorsally) carry several pairs of appendages. A few segments (about 7) at the posterior end of the trunk are apodous. The actual number of apodous segments varies slightly, even within one species. The telson of *Lepidurus* is prolonged posteriorly in the midline to form a supra-anal plate, which is lacking in *Triops*.

(c) *Appendages*

1. *Antennules.* These are very small. The basal portion, or scape, is narrow, while the terminal portion, corresponding to the flagellum of other Crustacea, is somewhat broader and ends in three short sensory setae.

2. *Antennae.* These are even smaller than the antennules, and each consists of a single rudimentary process situated slightly lateral to the antennules. Some large specimens lack antennae.

3. *Mandibles.* These are powerfully developed with about 8 strong teeth and lacking a palp. Behind the mandibles a bilobed structure separates the two pairs of maxillae from the mouth. The two lobes, or paragnaths, are often well developed in the Malacostraca, but the Notostraca are exceptional among the Branchiopoda in having them so well developed.

4. The *maxillules* are simple lobes with a row of spines along the ventral edge.

5. The *maxillae* are generally larger in *Lepidurus* than in *Triops*, and are absent from some species of *Triops*. When present the general form of the maxilla is a simple lobe with numerous short spines on the ventral and medial edge. The efferent duct of the excretory organ, or maxillary gland opens at the base of the maxilla on the outer side. In *Triops* the opening of the duct protrudes laterally from the maxilla and may overlap the base of the first thoracic appendage.

6. In *Triops* the endites of the first thoracic limb are elongated, and the 5th endite forms a long antenna-like structure. This presumably compensates for the minute size of the antennules and antennae. In *Lepidurus* the endites are not so elongated, and in *L. articus* they project only a short distance beyond the edge of the carapace.

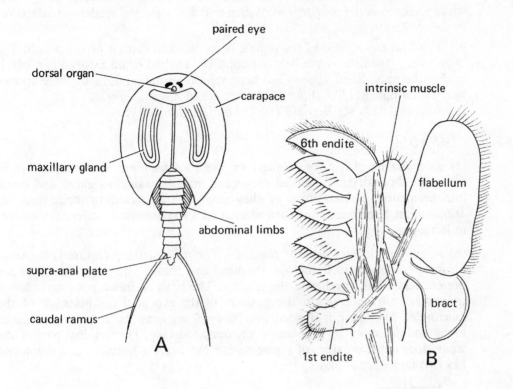

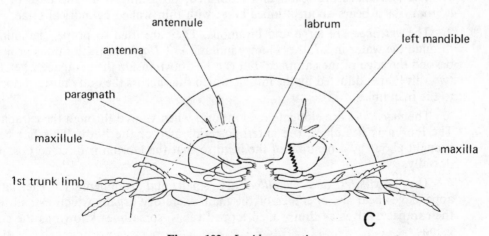

Figure 102 *Lepidurus arcticus*

A, dorsal view of adult; B, 7th trunk limb; C, ventral view of head

7. The remaining trunk limbs all have the same basic structure (Fig. 102), with variations in the proportions of each part and a general tendency to decrease in size posteriorly.

8. The 11th trunk limb of the female is modified to form a brood pouch. The main axis of the limb is swollen and cup-like. The flabellum forms a cap which retains the eggs. Some species are hermaphrodites. The adults of these species bear brood pouches like those of the females of bisexual species.
For further details see Bernard (1892) and Longhurst (1955).

16.3.3. *LIMNADIA*

(a) *Examination*. Living specimens of Conchostraca are rarely available in teaching laboratories. Preserved specimens may be stained, cleared and made into permanent preparations, or they may be transferred to lactic acid for examination. Specimens can be transferred back to preservative after examination in lactic acid.

(b) *Forms and Segmentation of the body*. The most striking feature is the large bivalved carapace which encloses the head and trunk. Two compound eyes are present very close together in the midline. The head is drawn out into a broad triangular rostrum between the position of the eyes and the insertion of the antennules. The trunk is divided into 22 to 27 segments, each bearing a pair of limbs and delimited by grooves on the dorsal surface. The terminal part of the trunk does not show signs of segmentation but bears a heavier exoskeleton and has two large termal claws.

(c) *Appendages*

1. The *antennules* are small and elongated, projecting from the base of the labrum. Each bears about 8 lobes beset with thin walled cylindrical setae.

2. The *antennae* are large and biramous. They are used to propel the animal through the water in a rather clumsy imitation of *Daphnia*. Each ramus projects beyond the edge of the carapace, but can be drawn inside the carapace when the two valves are adducted by the muscle which runs across the body postero-dorsal to the mandibles.

3. The *mandibles* are elongate oval in form when viewed through the carapace. The long axis lies along the anteroposterior axis of the body. This has been brought about by a bending of the head so that the labrum is directed postero-dorsally.

4. The *maxillules* and *maxillae* are small and rudimentary. The latter appendages bear the openings of the main excretory organs which extend into the carapace valves as thin walled looped tubes, sometimes known as the shell glands.

5. The first 15 *trunk limbs* are all similar in structure, but there is a gradual decrease in size from the anterior to the posterior limbs. Each limb is an elongated phyllopodium, basically similar to those of the Anostraca, but placed much closer together. The posterior 12 limbs are not typical phyllopodia; they

lack filtratory setae, and have enlarged gnathobases which assist in the break up of large food particles. In a sense *Limnadia* may be said to chew with its hind limbs. Legs 9, 10 and 11 bear long dorsally directed processes which project around the trunk and into the brood pouch. When eggs are laid they adhere to these elongated dorsal processes from the limbs.

(d) *Internal anatomy*. The following notes are restricted to the features that can be seen in a cleared specimen.

The gut forms a simple tube, without coils, but with a complex pair of diverticula from the anterior part of the midgut. The rectum is more obviously muscular than the midgut, and opens posteriorly between the abdominal claws.

The heart lies dorsal to the anterior part of the midgut, and extends from the region of the adductor muscle to the fourth trunk segment. There are four distinct pairs of ostia, and a posterior aperture through which blood enters the heart. There are thus nine incurrent apertures, but the blood leaves the heart via the single anterior opening.

It is difficult to see any details of the nervous system, but the cerebral ganglion can usually be seen in the space between the insertion of the antennules and the optic lobe, which connects via the optic nerve with the compound eyes. A triangular pigmented ocellus is connected to the anterior part of the cerebral ganglion via a short nerve.

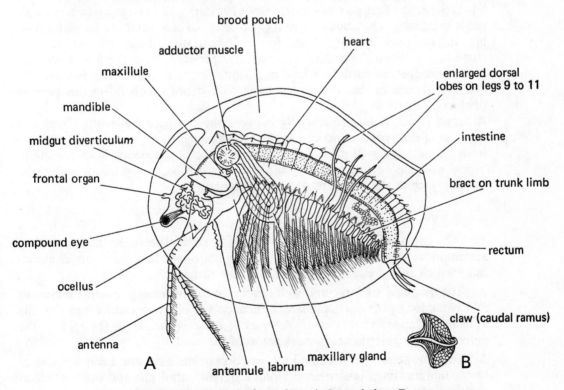

Figure 103 *Limnadia lenticularis*. A, Lateral view; B, egg

The ovaries are elongated and extend through most of the trunk in a somewhat spiral form. A short transparent oviduct opens at the base of the 11th leg on each side. When the eggs are laid into the brood pouch they adhere in a glutinous mass. The eggs are small, numerous and peculiar in shape (Fig. 103).

The male of *Limnadia lenticularis* is unknown, but in other Conchostraca the males are known to have modified first trunk limbs which are used to grasp the female during mating.

For further details, Sars '*Fauna Norvegiae*' Vol. 1 (1896) should be consulted.

16.3.4. *DAPHNIA*

(a) *Examination*. Living specimens are best examined in a drop of water under a cover slip supported by small pieces of 'Plasticine.' The cover slip can then be pressed gently against the plasticine until the animal is just held between the slide and cover slip.

(b) *Segmentation of the body*. External signs of segmentation have been lost (Fig. 104). The head has a single median compound eye formed by the fusion of two lateral eyes. (This process can be seen in developing embryos in the brood pouch.) The compound eye lies internal to the head skeleton and is moved by a series of muscles which cause it to oscillate at the end of the optic nerve. A large labrum projects backwards from the postero-ventral part of the head and lies between the first pair of trunk limbs. The carapace is large and encloses all the trunk limbs. The abdomen bears two terminal claws and can be withdrawn into the carapace by being flexed forwards ventrally between the two rows of trunk limbs. When the abdomen is extended forwards from the flexed position it can dislodge any particles which might interfere with the feeding mechanism. The dorsal edge of the abdomen is beset with spines which aid in this process of clearing the food groove.

A brood pouch is formed dorsally between the carapace and trunk. Eggs and developing embryos are retained in the brood pouch by means of a long process from the antero-dorsal part of the abdomen. The brood pouch is in communication with the outside medium, and the eggs can be moved about within the brood pouch as the female moves her abdomen.

(c) *Appendages*

1. The *antennules* are very small with a few terminal setae. In the female they are immovably fused to the underside of the rostrum. Males have longer antennules which are moveably articulated with the rostrum.

2. The *antennae* are large and well developed for swimming. Each is biramous and bears long plumose setae which increase the effective surface area as the antenna is flapped in the water. A series of muscles runs from the base of the antenna to the antero-dorsal exoskeleton of the head.

3. The *mandibles* lie just behind the junction of the head and carapace, dorsal to the labrum. Each is elongated along a dorso-ventral line and bears a milled surface on the distal median face. These milled edges are brought together by

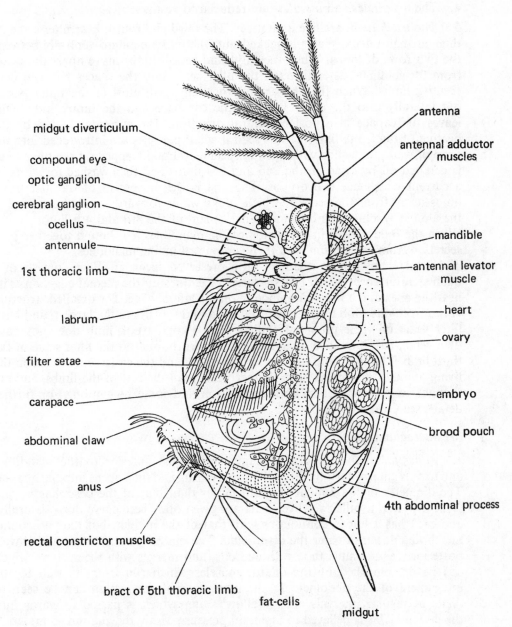

midgut diverticulum

compound eye

optic ganglion

cerebral ganglion

ocellus

antennule

1st thoracic limb

labrum

filter setae

carapace

abdominal claw

anus

rectal constrictor muscles

bract of 5th thoracic limb

fat-cells

midgut

antenna

antennal adductor
muscles

mandible

antennal levator
muscle

heart

ovary

embryo

brood pouch

4th abdominal process

Figure 104 *Daphnia magna*, lateral view of adult female

a rotatory action which compacts food into a small bolus and passes it up into the oesphagus.

4.   The *maxillules* and *maxillae* are reduced to vestiges.

5.   The *trunk limbs* are five in number. The third and fourth bear filter setae on their median edges. Water is sucked into the midline antero-ventrally between the two rows of limbs. When the third and fourth limbs move apart the water from the midline passes through the filter setae into the spaces between consecutive limbs. When the limbs move together again most of the water passes out laterally into the space between the carapace and the limbs and finally leaves the capace in a postero-ventral direction. The currents created by the beating of the trunk limbs can be seen if small particles are introduced into the drop of water on a slide with a living specimen. Food trapped on the filter setae is combed off by setae on the 2nd and 3rd limbs and then washed forwards by a current which has two origins. The first is the movement of the first trunk limbs away from the second, which sucks water forwards, and the second is the almost synchronous backwards movement of the 3rd and 4th limbs.

When the food arrives in the region of the mandibles it is bound together by a secretion from the labrum, and then compacted by the mandibles.

The trunk limbs can be dissected off a preserved specimen. Each consists of a main stem or corm which bears a series of endites on the medial edge, an exite near the apex and a bract or epipodite on the lateral edge. The detailed structure varies from one limb to the next. The third and fourth are the largest, and have filter setae on their proximal endites. The second trunk limb does not have filter setae, but does have some setae which brush food off the filter setae of the third limb. The fifth limb has the largest bract, and the easiest one to see in the living animal. The bracts have thinner walls than the rest of the limbs, and are thought to play a part in both respiration and ionic regulation. For further details see Cannon (1933).

(d) *Internal anatomy*

1.   *Alimentary canal.* The oesophagus opens just anterior to the mandibles and travels antero-dorsally through the nerve ring and then opens into the midgut. The diameter of the midgut is much larger than that of the oesophagus, and from the lateral walls two diverticula are given off. These curve dorso-laterally, and each has a lumen continuous with that of the midgut, but the peritrophic membrane does not enter the diverticula. The midgut curves round and travels posteriorly as a simple tube without coils, and merges with the rectum, which can be distinguished by the dilator muscles which run from its wall to the integument of the abdomen. In the living animal the rectum can be seen to dilate at regular intervals and 'swallow' water which is passed forwards into the midgut. This is believed to maintain pressure within the gut and so maintain the tonus of the gut muscles. It has also been suggested that the anal uptake of water may function like an enema, and aid in the evacuation of faeces.

2.   *Heart and circulation.* The heart is a simple thin walled muscular sac with a single pair of ostia and a very short aorta which is little more than an anterior

valve. In the living animal the course of circulation can be seen easily by observing the blood cells as they are carried along. If the specimen has been taken from a population living in poorly aerated water the blood may be red with haemoglobin dissolved in the plasma (not in the cells, which are small amoebocytes). Populations from well aerated water usually have colourless blood, although when algal food is abundant the blood may be green or blue. These colours are caused by carotenoid protein pigments, which also colour the ovary and embryos.

3. *Reproductive system.* The ovaries are paired, each lying lateral and slightly dorsal to the midgut. Each opens via a short oviduct, which is not normally visible. If several females with large ovaries and empty brood pouches are observed at intervals over the course of several hours it is often possible to observe the process of egg laying. The eggs pass through the narrow oviduct like toothpaste coming out of a tube. Once in the brood pouch they begin to swell and take up a rounded shape. The ovaries and eggs are normally coloured green or blue with a carotenoid-protein. Orange oil droplets may be present, and when the animal has been kept in poorly aerated water there may also be haemoglobin in the eggs.

In good conditions *Daphnia* is parthenogenetic. The numerous eggs in the brood pouch develop directly into females during the course of a few days. When conditions deteriorate some of the eggs develop into males, and some of the females produce eggs which require fertilization before they will develop. Normally a single egg of this type is produced by each ovary, and the two eggs become enclosed in a modified brood pouch which darkens and contracts around the eggs to form an ephippium. The ephippium is cast off when the female moults. The eggs within the ephippium are capable of withstanding freezing and desiccation, but will hatch and give rise to females in favourable conditions. For further details see Berg (1934), Scharfenberg (1910), Banta (1939).

4. *Fat body.* This system consists of loosely connected large cells which contain various inclusions, notably large oil or fat droplets. A major part of the fat body lies in the trunk on the ventro-lateral sides of the midgut. Strands from this region pass into the bases of the limbs. The fat body seems to function as a store of foods and as a centre of active metabolism. The number and size of the fat droplets shows a cycle which alternates with the cyclical growth of the ovary, and the cells often contain carotenoid-protein similar to that found in the ovary. There is also evidence that the fat cells are concerned in the storage of glycogen and in the metabolism of haemoglobin.

5. *Nervous system.* In the living animal the only parts of the nervous system which are clearly visible are the cerebral ganglia and the optic nerve. With careful observation the small nerve to the antennule can usually be seen. A small pigmented ocellus lies just in front of the cerebral ganglion and below the eye. From this region a fine nerve runs antero-ventrally to a couple of sensory cells which form the frontal organ. Another fine nerve runs dorsally to the nuchal or neck-organ, which is of unknown function. For further details see Claus (1876), Banta (1939).

16.4. Copepods: *CYCLOPS*

(a) *Examination*. *Cyclops* should be examined alive trapped in water between a cover slip and a glass slide. Movement can be restricted by adjusting the amount of water on the slide so that the specimen is held still by the weight of the cover slip. To avoid crushing the copepod the cover slip may be supported by very small pieces of 'Plasticine.' With practice the cover slip can be moved to roll the specimen so that both dorsal and ventral views can be examined. After the general structure has been determined by live examination it may be necessary to perform a simple dissection to see the detailed structure of the limbs. The simplest technique is to use very fine needles to divide the body in front of the fourth pair of legs. The posterior region separated in this way can then be arranged so that the fourth legs lie flat and point anteriorly. The second and third legs should then be separated from the rest of the anterior region. The first pair of swimming legs may be left attached to the head, but should be arranged so that they lie flat and point posteriorly. Details of the mouthparts can usually be discerned without further dissection if the anterior region is mounted ventral side uppermost (Fig. 105).

(b) *Segmentation of the body*. The body is composed of a cephalon bearing six pairs of appendages, a thorax of six segments, the first of which is united dorsally with the cephalon, and an abdomen of four segments. The fifth thoracic segment (4th free one when the animal is viewed from the dorsal side) is narrower anteriorly, and the junction between this segment and the one anterior to it forms a flexible joint between the fore and hind parts of the body. In the female the last thoracic (6th) segment and the first abdominal are joined together to form a genital segment. The last abdominal segment bears the caudal furca which consists of two rami bearing long terminal setae.

(c) *Appendages*

1.   *Antennules*. Different species of *Cyclops* differ in the number of podomeres in the antennule. Commonly the number is 17, but some species have only 11 or 12, and immature specimens may have as few as 4 podomeres in the antennule. The males have the antennules modified as prehensile organs for grasping females during copulation. Two main points of flexure occur: between podomeres 7 and 9, and podomeres 14 and 15.

2.   *Antennae*. These are simple uniramous appendages with four podomeres.

3.   *Mandibles*. Each mandible consists of a shaft drawn out into a distal masticatory process bearing numerous teeth. The shaft is twisted so that the teeth lie in a vertical plane in the entrance to the oesophagus. Two very long setae emerge from the ventral side of each mandible and extend backwards as far as the posterior border of the cephalon. A very small seta may also be seen emerging from the same pore.

4.   *Maxillules*. Each maxillule consists of a flat plate which becomes narrower towards the apex and bears three strong distal teeth. A short palp is present and bears numerous setae which may detect movements of living prey. The

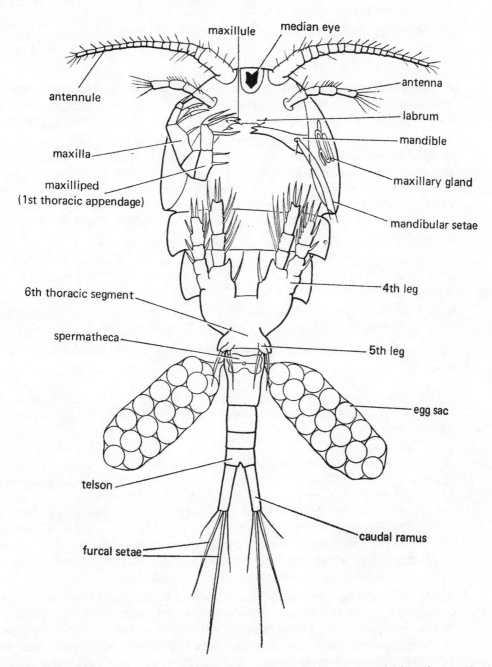

Figure 105 *Cyclops*, ventral view of adult female; swimming legs 1, 2, and 3 have been omitted

maxillules are used in seizing and cutting; a single snap of these appendages can cut a small oligochaete in half.

5. *Maxillae.* These are jointed, and bear stout spines at the apex. Internal muscles can flex the distal podomeres so that the maxillae can be used to hold prey against the maxillules.

6. *Maxillipeds.* Each maxilliped lies medial to the corresponding maxilla, and resembles that appendage in general structure, but is more lightly constructed with fewer, more delicate spines. These appendages are also used for grasping.

7. *Legs* 1—4 are similar in structure, each has a basal region, consisting of a basipodite and coxopodite. The two branches, exopod and endopod, each consist of three podomeres. Some species of *Cyclops* have a reduced number of podomeres on one or more pairs of legs, but the common large species have three podomeres in both endopod and exopod. There are many variations in the armament of spines and setae on these legs in different species of *Cyclops*.

8. *Leg* 5. This appendage is always reduced, but the degree of reduction varies in different species of *Cyclops*. In some there are two distinct podomeres and three setae, but in others the limb is reduced to a pair of setae springing from a small projection of the body wall.

9. *Leg* 6 is more reduced than Leg 5, and is not detectable in the female, but is often present as three small setae springing from a small plate in the male.

10. The abdomen does not bear appendages, but has the caudal ramus at its extremity.

(d) *Internal anatomy.* When a living specimen is examined from the dorsal side it is possible to see through the body wall and the following organs may be discernable.

1. In the *female* the single ovary is median in position, lying dorsally in the posterior part of the cephalon and first free segment. The two oviducts run from the ovary back to the junction of the last thoracic and first abdominal segments which are fused to form the genital segment. Each oviduct gives off several diverticula, or uterine processes, which when full of eggs are opaque and may give the female a banded appearance. In the genital segment there is a spermatheca which forms a double bag opening via a median copulatory pore through which the sperms enter. The sperms leave the spermatheca through two sperm ducts which open laterally into the ends of the oviducts.

2. In the *male* the single testis occupies a position similar to that of the ovary in the female. The paired vasa deferentia run backwards from the anterior end of the testis, then loop forwards again before finally running back into the last thoracic segment. Each vas deferens opens into a sac-like vesicula seminalis. Spermatophores are produced in the vasa deferentia and pass down into the vesiculae seminales. During copulation the spermatophores are attached to the copulatory pore of the female, and then the contents of the spermatophores swell up and drive the sperms into the spermatheca.

3. The *alimentary canal* has a relatively simple structure. The oesophagus passes upwards and forwards from the mouth, but halfway along its length it bends at a sharp angle so that the upper part passes obliquely backwards. The upper part of the oesophagus is circled by six constrictor muscles. These are not easy to see, but the two most posterior are sometimes visible if the copepod is examined from the side. The stomach, intestine and rectum form a simple tube which opens posteriorly at the anus. The anus is guarded by two anal valves and forms a longitudinal slit opening dorsally near the apex of the telson. Part of the telson extends dorsally over the anus, forming a supra anal plate. The walls of the stomach and intestine are often laden with various granules and vacuoles. There are no accessory glands or diverticula opening into the gut, but a pair of salivary glands lie near the outer edges of the labrum and open on the oral face of the labrum by a median salivary pore.

4. The maxillary gland is best seen with the animal lying on its side, when a coiled tube with colourless contents can be seen lying under the body wall above the maxillae and maxillipeds. This tube has a small end sac at its coelomic end, and opens on the posterior face of the maxilla. The opening is very difficult to see, but is sometimes visible in a copepod which has been slightly squashed.

5. It is not possible to see any details of the nervous system in a living specimen.

For further details see Hartog (1888); for feeding mechanism, Fryer (1957).

16.5. Ostracods: *EUCYPRIS*

(a) *Examination*. The general organization of an ostracod is best seen with one valve of the shell removed. For a detailed examination of the appendages the trunk and head should be dissected out of the shell using fine needles, and then be gently squashed under a cover slip. Lactic acid or polyvinyl lactophenol can be used as a mounting medium for the appendages, but not for the shell, because the calcium carbonate in the shell dissolves and liberates bubbles of carbon dioxide in acid media. Lignin pink and chlor-azol black can be used to stain the chitin of the appendages.

(b) *Form and segmentation of the body*. The head and trunk are enclosed in a bivalved carapace or shell (Fig. 106). The two halves of the shell can be drawn tightly together by adductor muscles, the insertions of which can be seen approximately in the middle of a lateral view of the shell. A single median eye is visible through the shell in an antero-dorsal position. The trunk is attached dorsally to the shell along the line of the hinge, and does not show any clear signs of segmentation. At the posterior end of the trunk there is a pair of elongated caudal rami which bear claws at their tips.

(c) *Appendages*. Two alternative systems of nomenclature have been used for the appendages:

System 1	System 2
antennule	antennule
antenna	antenna

System 1	System 2
mandible	mandible
maxillule	maxilla
maxilla	1st leg
1st leg	2nd leg
2nd leg	3rd leg

The second system is more used in taxonomic work, and has been adopted here.

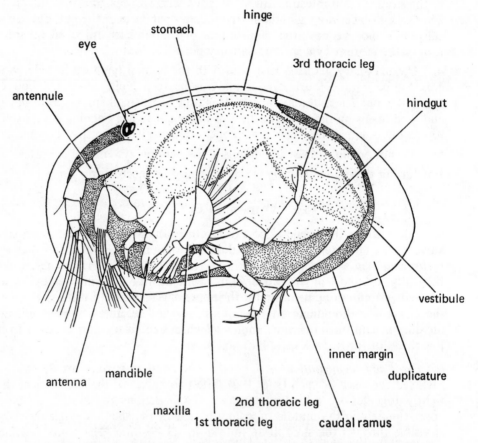

Figure 106 *Eucypris*, lateral view of female, with left valve of shell removed

1.   The *antennules* are well developed, with several podomeres and swimming setae.

2.   The *antennae* are also large, and bear swimming setae.

3.   The *mandibles* have strong elongated basal podomeres bearing teeth on their inner faces. The palp of the mandible has four podomeres; from the first of these there springs an exopod plate with numerous setae.

4. Each *maxilla* has four ventral projections with setae which aid in the feeding mechanism. A large delicate exopod plate, bearing setae along its ventral border, extends dorsally from the maxilla. In life this plate beats backwards and forwards beneath the shell and presumably aids the circulation of water past the body.

5. The *first thoracic legs* are relatively small, and closely concerned with the feeding mechanism. The setae on the anterior basal part help in passing food into the mouth. A small exopod plate with delicate setae is also present on this limb. In the male the palps of the first limb are enlarged and modified to form clasping organs used during copulation.

6. The *second thoracic limb* is leg like, with a strong apical claw.

7. The *third thoracic limb* is also leg like, but is normally carried reflexed inside the shell. This is often known as the cleaning limb, and bears three long setae near the apex which are used to sweep debris out of the carapace.

(d) *Internal anatomy*. The only really satisfactory way to study the internal anatomy of a small ostracod is by means of serial sections, but some features can be seen in specimens from which the shell valves have been removed.

The *gut* is divided into a stomach and hind gut, and when these contain food their form is visible through the body wall. The wall of the gut may be coloured, particularly when the ostracod has been feeding on blue-green algae, and granules of phycobilin are found in the gut cells. A diverticulum from the stomach extends into each shell valve, and is normally visible in an external lateral view of a whole specimen as a thick line running obliquely in the hind two thirds of the valve.

Diverticula from the *gonads* also extend into the shell valves, and are normally visible running parallel to and above the gut diverticula.

A squashed specimen of a male will reveal the paired Zenker's organs, which are striking cylindrical structures, each with a central duct and about 7 wreaths of chitinous spines at right angles to the duct. The ends and the sides of these spines are all interconnected by minute muscles so that the whole structure takes up a cylindrical form with a great intricacy of internal structure.

For further details see Klie (1938) and Kesling (1965).

16.6. Cirripedes

16.6.1. *LEPAS*

(a) *Examination*. *Lepas* is large enough to be dissected in a wax-bottomed dish. To orientate the animal in the same way as the drawings in this book lay the animal on its left side with the peduncle towards you and the mid-dorsal carinal plate to your right. Remember that the peduncle represents the anterior end.

(b) *Segmentation of the body*. The peduncle or stalk is formed from the anterior part of the head (Fig. 107). Traces of the antennules may be found where the stalk is attached to the substratum, but the antennae are completely lacking.

The mantle or carapace extends as two leathery folds from the posterior end of the peduncle. The two folds are joined along part of the mid-ventral line, and along the mid-dorsal line, where there is a single median calcareous plate, the

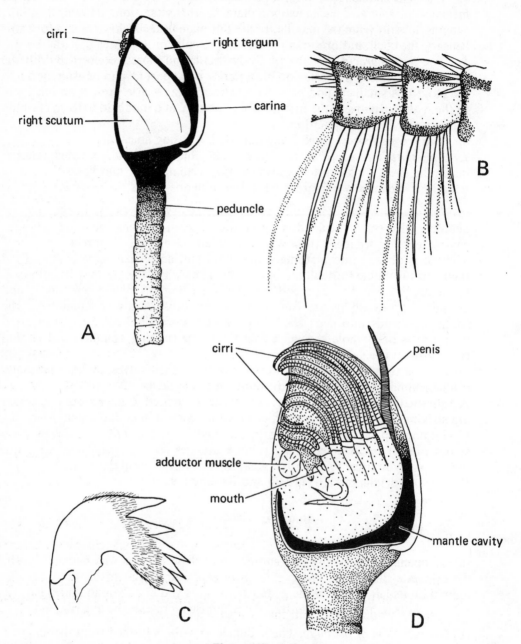

Figure 107 *Lepas*

A, lateral view from right side; B, detail of part of cirrus; C, mandible; D, lateral view, with right side of mantle removed

carina. Each side of the mantle has two calcareous plates: a large scutum and a smaller tergum.

Remove the right half of the mantle. Note the adductor muscle running across from one side of the mantle to the other.

The trunk and posterior part of the head are completely enclosed within the mantle. The body wall of the trunk is thin and translucent, without any distinct signs of segmentation.

(c) *Appendages*

1.   The *mandibles* are the first really distinct appendages; they are flattened and have five strong teeth.

2.   The *maxillules* lie in the same mound as the mandibles. They are smaller, but similar in general form. The teeth are not so large or sharply pointed, but at the anterior corner there is a very sharp spine followed by several setae.

3.   The *maxillae* form the posterior or lower lips. They are thicker than the mandibles and maxillae. Setae clothe their inner surfaces. The maxillae have been observed to move rapidly too and fro acting like brushes sweeping small particles from the cirri to the maxillules and mandibles. The outer edges of the maxillae bear small papillae on which open the excretory organs or maxillary glands.

4.   There are *six pairs of cirri* which can be protruded from the mantle cavity and swept through the water to catch small animals and plants. The first pair of cirri are smaller than the others and not so heavily beset with setae. Each cirrus has a protopodite of two podomeres, and two multiarticulate rami. At the bases of the first pair of cirri are two filamentous processes which are sometimes thought to have a respiratory function.

5.   The *caudal rami* are very small, but a large median penis extends forwards from the posterior end of the body, and normally lies between the two rows of cirri.

(d) *Internal anatomy*

Remove the integument from the right side of the peduncle and trunk (Fig. 108).

1.   *Cement Gland.* Inside the anterior end of the peduncle there is a cement gland opening on the antennules.

2.   *Reproductive system.* The *ovary* lies in the posterior part of the peduncle. The size of the ovary varies considerably with the breeding condition of *Lepas*. A narrow oviduct leads from the ovary to open near the base of the first pair of cirri.

The *testes* are the first organs encountered when the integument of the trunk is removed. Each is greatly subdivided with branching tubules opening into a vas deferens which expands to form a conspicuous white seminal vesicle. The two seminal vesicles unite at the base of the penis and form the ejaculatory duct which can be followed through the length of the penis.

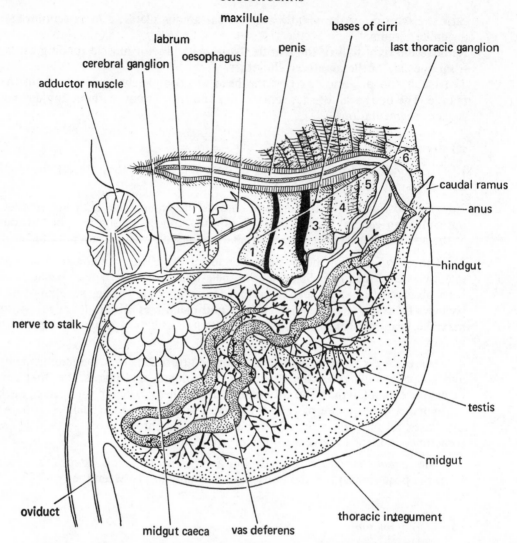

Figure 108 *Lepas*, lateral view of dissection from the right side

3. *Alimentary canal*. The mouth opens between the large labrum and the mandibles. From the mouth the oesophagus runs anteriorly and slightly ventrally then opens into a voluminous mid gut. At the anterior end of the mid gut a number of bulbous caeca are given off. At the posterior end the mid gut becomes narrower and joins the hind gut which is somewhat more muscular and leads straight to the anus.

4. *Nervous system*. This is best seen if the cirri of the right side are carefully removed. Cut as close as possible to the median edge of these cirri. The cerebral ganglion lies between the adductor muscle and the oesophagus at the base of the anterior edge of the labrum. A conspicuous nerve runs from the

cerebral ganglion to the peduncle. The circum oesophageal commissures run to the first thoracic ganglion, which is the largest ganglion in the body. The most conspicuous pair of nerves from this ganglion run to the viscera. Each thoracic ganglion gives off a pair of nerves to the corresponding cirri. The fifth or last thoracic ganglion gives off nerves to the fifth and sixth cirri as well as a large branch to the penis.

5.   The *circulatory system* and *excretory system* are not usually discernable in an ordinary dissection. There is no heart.

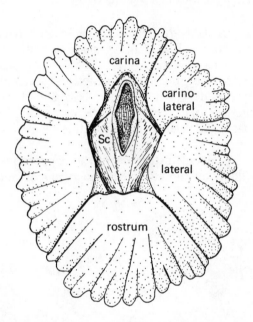

Figure 109   *Balanus balanoides*, viewed from above

16.6.2.   *BALANUS*

Examine a living *Balanus* attached to a stone in sea water. Identify the following parts of the mantle (cf. Fig. 109) rostrum, lateral, carino-lateral, carina, scutum, tergum.

Note the movements of the cirri. The rate of movement can be made to vary with variation in temperature. Place a small drop of milk in the sea water near the junction of the rostrum and lateral plates of a barnacle which is active. Note the way in which the milk flows into the mantle and is passed out again behind the cirri and in front of the carina.

The internal anatomy of *Balanus* is basically similar to that of *Lepas*. The major difference is that the ovary is housed in the trunk because the stalk is missing.

For further details on structure and biology of Cirripedes see Darwin (1851), Gruvel (1905), Batham (1944) and Cornwall (1953; nervous system).

16.7. Malacostracans

16.7.1. LEPTOSTRACANS: *NEBALIA*

(a) *Examination.* Small specimens can be examined alive in sea water on a glass slide under a cover slip supported by 'Plasticine.' Large preserved specimens can be examined in a dish under a binocular dissecting microscope. It is necessary to remove half the carapace in order to see the arrangement of the mouthparts and thoracic limbs. Specimens that have been cleared and mounted as permanent preparations can also be used.

(b) *Segmentation of the body.* The carapace extends backwards over the thorax and part of the abdomen (Fig. 110). A transverse adductor muscle, situated in the region of the maxilla can draw the two halves of the carapace more closely together. There are eight thoracic segments, and seven abdominal segments plus a telson bearing a caudal furca. At the anterior end of the body there is a distinct rostrum, articulated at its base with the carapace. The eyes are stalked and mobile; their bases are inserted close together at the front of the head.

(c) *Appendages.* The detailed structure of each appendage is seen at its best when dissected from the body and mounted in glycerine or lactic acid.

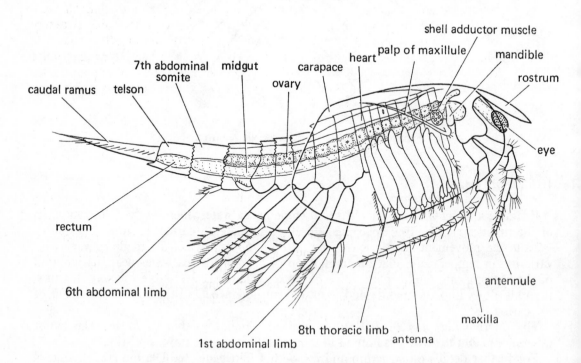

Figure 110 *Nebalia bipes*, lateral view of a cleared adult female

1. *Antennules.* The peduncle is stout and flattened, with four podomeres, while the flagellum is slender and multiarticulated. The outer ramus of the antennule is represented by an oval scale bearing fine setae.

2. *Antennae.* The peduncle consists of four podomeres, but the distal two are fused. In the male the flagellum is about as long as the body, but in the female the tip of the flagellum only reaches the posterior border of the carapace.

3. *Mandibles.* The molar process is large, and the incisor process small. The palp is large, with three podomeres, and normally projects forwards between the eyes and under the rostrum.

4. *Maxillules.* The palp is long and projects obliquely postero-dorsally underneath the carapace, where it is thought to act in a cleaning capacity. The base of the maxillule bears two endites beset with fine setae and short spines.

5. *Maxillae.* Each maxilla bears four endites, and distally an elongated endopodite of two podomeres, while the exopodite is somewhat shorter, consisting of a single podomere.

6. *Thoracic limbs.* All 8 limbs are similar in structure. Each forms a paddle-like structure. The endopod is narrow, with a series of setae modified for filter feeding. The exopod and epipod are flattened and expanded.

7. *Abdominal appendages.* The first four pairs of abdominal appendages are biramous and stoutly constructed; they are used in swimming. Each limb bears a small appendix interna which hooks on to the corresponding structure of the limb on the opposite side of the body so that the two limbs are used in unison when *Nebalia* swims. The last two pairs of abdominal appendages are rudimentary uniramous and immobile. The 7th abdominal somite does not bear any appendages, but the telson bears a caudal furca consisting of two rami which are slightly divergent.

For further details see Sars (1896), Cannon (1927; feeding mechanism), and Manton (1934; embryology).

16.7.2. MYSIDACEANS: *MYSIS*

(a) *Examination.* Mysids are best examined under a binocular dissecting microscope.

(b) *Segmentation of the body.* The carapace extends over all the thoracic segments, leaving a part of the last one uncovered (Fig. 111). The thoracic segments are free and not fused to the carapace. The eyes are stalked and movable.

The abdominal segments are all free, the sixth being formed during embryological development by the fusion of two segments. The telson is a well developed plate, which in some mysids has a split in the mid-line.

(c) *Appendages*

1. The *antennules* are biramous, with a long outer branch and a shorter inner branch.

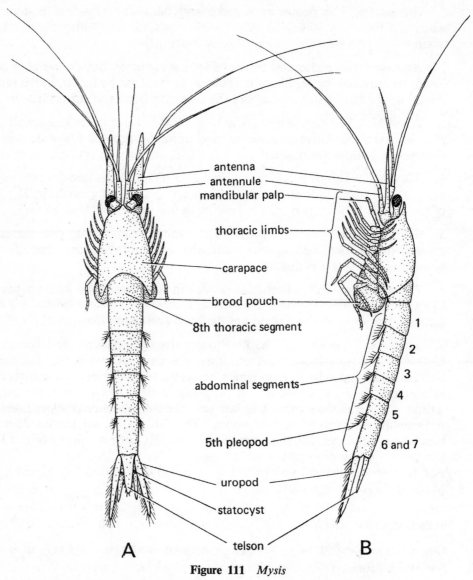

Figure 111 *Mysis*

A, dorsal view of adult female; B, lateral view of adult female

2. The *antennae* have the exopod modified to form an elongated scale which projects rigidly forwards. The flagellum is long, reaching at least the same length as the outer branch of the antennule.

3. The *mandibles* have a *lacinia mobilis*, and there are some differences in detail between the left and right mandibles. See if you can relate these differences to the way in which food is pushed into the oesophagus. Each mandible has a conspicuous palp projecting forwards between the bases of the antennae.

4.   The *maxillules* each consist of two lobes with setae and spines at their free ends.

5.   The *maxillae* form curved plates lying ventrally under the maxillules. There is a large proximal endite and a subdivided distal endite. Both bear dense inner fringes of plumose setae. The maxillary palp has two podomeres and is also densely set with plumose setae. An elongated exite forms a flap closing the gap between the outer edges of the maxilla and maxillule.

6.   The *first trunk limb* bears endites beset with strong setae which are directed towards the mouth. From the base of this limb an epipodite extends backwards beneath the carapace and its movements aid the circulation of water between the carapace and trunk. The remaining thoracic limbs lack endites, but have well developed endopods, which are basically walking limbs. All the thoracic limbs have exopods bearing plumose setae. In life these exopods project laterally from the body and are whirled around to produce currents for swimming and feeding. In the mature female some of the thoracic limbs bear oostegites at their bases.

7.   The first five pairs of abdominal appendages, or *pleopods* show sexual dimorphism. In the females each consists of a single podomere bearing a fringe of setae. In the male the first second and fifth pleopods resemble those of the female, but the third and fourth are larger and biramous.

8.   The last abdominal appendages, or *uropods* are large and biramous, forming a tail fan with the telson. At the base of each endopod there is a conspicuous statocyst.
For further details see Tattersall (1951), Cannon and Manton (1927; feeding mechanism) and Manton (1928; embryology).

16.7.3.   CUMACEANS: *DIASTYLIS*

(a) *Examination. Diastylis* is best examined from one side. Small specimens can be made into temporary or permanent mounts. Dissection of the mouth-parts is not easy, but it is worth attempting to dissect out the first maxillipeds.

(b) *Segmentation of the body*. The carapace coalesces with three thoracic segments, and extends forwards in front of the head as two lobes which meet in the midline to form a pseudorostrum (Fig. 112). Dorsolaterally on each side of the head and anterior thorax a branchial chamber is enclosed by the carapace. This chamber is closed posteriorly so that water can enter and leave only at the anterior end.
The eyes are located close together in the mid-dorsal line.
The free thoracic segments are five in number, but the first of these is often difficult to see. The last thoracic segment is drawn out into strong lateral spines (this does not apply to all species of *Diastylis*).
The abdomen is very much thinner than the thorax, giving the animal a tadpole-like appearance. There are six free abdominal segments plus an elongated telson.

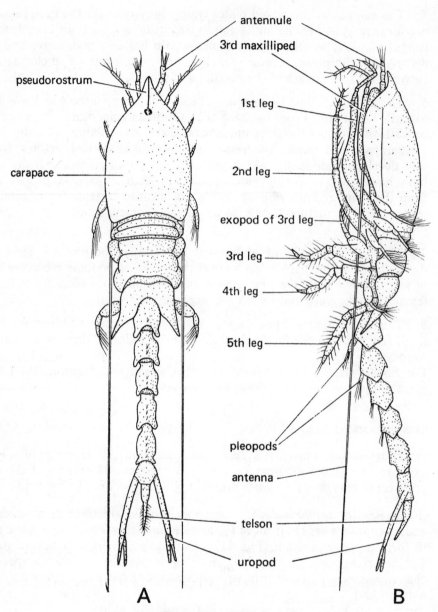

Figure 112 *Diastylis*

A, dorsal view of adult male; B, lateral view of adult male

(c) *Appendages*

1. The *antennules* are short and uniramous.

2. The *antennae* show sexual dimorphism. In the female they are vestigeal, but in the male they are extremely elongated and are normally carried projecting backwards alongside the body.

3. The *mandibles* have a *lacinia mobilis*, and both molar and incisor processes.

4. The *maxillule* consists of two small lobes bearing setae at their ends, and a small palp which bends backwards under the carapace.

5. The *maxilla* has a fairly large base and two small terminal lobes. The inner border bears fine filter setae which form part of the feeding mechanism.

6. The *first maxillipeds* are complex in structure. The basipodite bears a projection on the inner side that bears brush like setae which project upwards towards the mid-ventral line. From the base of the maxilliped a large epipodite extends backwards under the carapace. This epipodite bears a folded gill. A flattened exopodite extends forwards from the base of the maxilliped into the pseudorostrum. This flat exopodite, or rostral valve forms the lower side of a tube, with the pseudorostrum forming the upper part. Through this tube, or siphon, water flows out of the branchial chamber.

7. The *second maxillipeds* lack exopodites, but have elongated endopodites which project forwards under the other mouthparts. Water flows into the branchial chamber between the second and third maxillipeds.

8. The *third maxillipeds* have exopodites of several podomeres, and the elongated endopodites reach forwards under the head almost as far as the end of the pseudorostrum.

9. The first two pairs of *thoracic legs* are also directed forwards, and reach the anterior margin of the carapace. The last three pairs of thoracic limbs are shorter than the more anterior ones and seem to be used for digging. Well developed exopodites are present on all except the last pair of thoracic legs.

10. *Pleopods* are absent from the females, but are present on the anterior four segments of the abdomen of the males. Each consists of a simple plate bearing setae. The first abdominal segment has the best developed pleopod, and the size rapidly diminishes on the more posterior segments.

11. The *uropods* (on abdominal segment 6) are elongated and biramous, extending beyond the tip of the narrow telson.
For further details see Oelze (1931), Zimmer (1941), Fage (1951) and Dennell (1934; feeding mechanism).

16.7.4. AMPHIPODS: *TALITRUS*

(a) *Examination*. The lateral compression of most amphipods makes it necessary to examine the animal from one side. A binocular dissecting microscope is essential if the mouthparts are to be dissected out.

(b) *Segmentation of the body*. The first thoracic segment is incorporated in the head, and its appendages form the maxillipeds (Fig. 113). The remaining seven thoracic segments are free and distinct. The first three abdominal segments are also free, but the last three are fused together to form the urosome. The telson is a small plate at the end of the urosome.

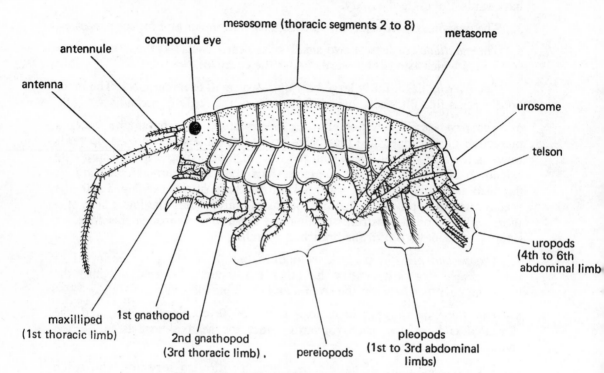

Figure 113 *Talitrus saltator*, lateral view of adult male

(c) *Appendages*

1.   The *antennules* are short and uniramous.

2.   The *antennae* are long, with a strong peduncle and a multiarticular flagellum.

3.   The *mandibles* have the three characteristic peracadrian processes: the incisor process, *lacinia mobilis* and the molar process. Between the *lacinia mobilis* and the molar process there are several densely plumose setae.

4.   The *maxillules* are well developed, with two lobes. The inner lobe is small, but the outer one is stouter and longer, with strong spines at the tip.

5.   The *maxillae* are small and relatively feeble, with two flattened lobes each with a brush of setae at the end.

6.   The *maxillipeds* are fused together in the midline at the base. On each side three inner lobes are given off from successive podomeres and the appendage terminates in a palp-like structure of two podomeres. The medial edges of all the podomeres and lobes are densely beset with setae.

7. The *first and second thoracic legs* (on thoracic segments 2 and 3) are often called gnathopods. The first is not much modified in *Talitrus*. but the second has a mitten-like ending. Other amphipods, such as *Gammarus*, often have subchelate gnathopods with large flattened subterminal podomeres and long reflexed claws.

8. The remaining five pairs of thoracic appendages form the *pereiopods*. These vary in length, but are otherwise all similar in structure, with stout terminal claws.

9. The abdominal appendages can be divided into two series. The first three pairs form the *pleopods*, which are attached to free segments. The pleopods are elongated, biramous and densely beset with plumose setae. The second series comprises three pairs of *uropods* attached to the urosome. The first two pairs of uropods are biramous, and the third is short and uniramous. These appendages are stoutly constructed and have short stout spines instead of plumose setae. When *Talitrus* jumps the urosome is flexed forwards ventrally and then suddenly extended so that the uropods strike the substratum and fling the animal into the air.

For further details see Cussans (1904: *Gammarus*), Dorsman (1935; *Orchestia*) and Reid 1947.

16.7.5. ISOPODS: *LIGIA*

(a) *Examination*. *Ligia* is large enough to enable one to see many details without resort to lens or microscope, but is best examined under a binocular dissecting microscope. The account which follows differs only in detail from other Oniscoidea, such as the woodlice *Porcellio* and *Oniscus*.

(b) *Segmentation of the body*. The first thoracic segment is incorporated in the head, and its appendages form maxillipeds (Fig. 114). The seven remaining thoracic segments bear walking legs. There are five distinct abdominal segments; the sixth is fused with the telson.

(c) *Appendages*

1. The *antennules* are minute, lying between the bases of the antennae.

2. The *antennae* are well developed with a 12 jointed flagellum.

3. The *mandibles* are stoutly constructed with a distal incisor process and a proximal molar process, between these two processes is a small movable tooth or *lacinia mobilis*.

4. The *maxillules* have two lobes, the inner lobe bears subterminal setae which project towards the midline. The outer lobe bears short terminal setae which merely project distally.

5. Each *maxilla* is formed of a simple lobe with a few fine setae.

6. The *maxillipeds* forms a lower lip, comparable to the labium of the insects. The two appendages are not however fused. Each has an inner undivided lobe and an outer palp composed of five short podomeres.

7.   The walking legs or *pereiopods* are all similar in structure, but in the mature female the first five pairs bear oostegites which overlap in the midline and form the brood pouch.

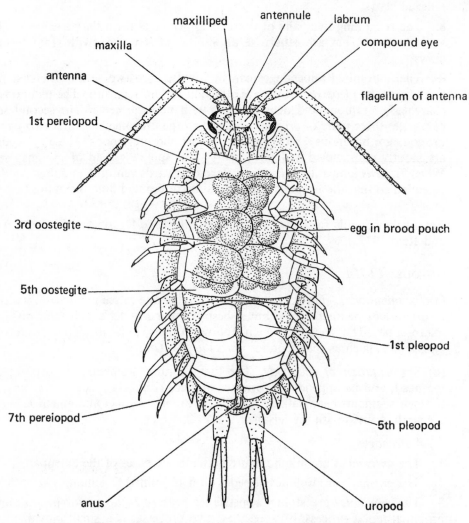

Figure 114   *Ligia oceanica*, ventral view of adult female

8.   The abdominal appendages, or *pleopods*, are flattened and overlap in sequence. There are five pairs of these flattened plates, which serve as respiratory organs. The last (6th) pair of abdominal appendages form a single pair of *uropods*.

For further details see Hewitt (1907), and Gruner (1965–66). See also 16.8.2. and 16.8.3., below.

16.7.6. DECAPODS: *ASTACUS* AND *CANCER*

In general terms the following description applies to any member of the Section Astacura (Figs. 115, 116, 117, 118).

(a) *Form and segmentation of the body*

1.   The head bears a rostrum, two pairs of antennae and stalked eyes.

2.   The carapace extends backwards and is fused dorsally with all 8 thoracic segments. The branchiocardiac grooves on either side of the dorsal surface indicate where the carapace extends laterally to form the branchiostegites which cover the gills, forming a branchial chamber. The gills are attached to the bases of the thoracic limbs, and to the lateral walls of the thorax.

3.   The abdomen has six segments and ends in a tail fan formed by the uropods and telson.

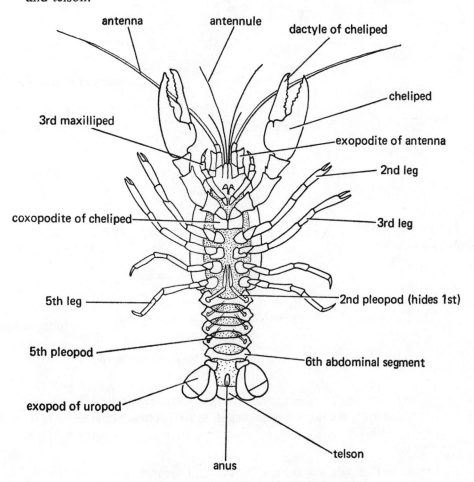

Figure 115 *Astacus*, ventral view of adult male

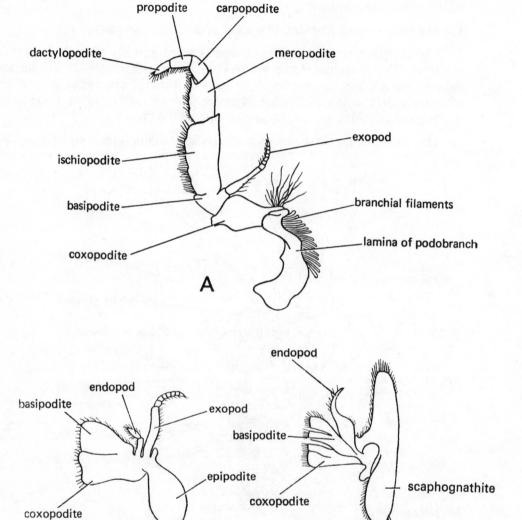

Figure 116 *Astacus*, A, 3rd maxilliped; B, 1st maxilliped; C, maxilla

(b) *Appendages*

1.   The *antennules* are biramous, and each has a peduncle of 3 podomeres, the most proximal one containing a statocyst.

2.   The *antennae* are longer than the antennules, and have scale like exopods. The antennal gland (green gland, or excretory organ) opens on the ventral side of the basal podomere (coxopodite).

3.   The *mandibles* are massively constructed, and transversely elongated with a semicircular masticatory surface bearing teeth on the medial border. A stout palp of 3 podomeres represents the endopod.

4.   The *maxillules* each consist of two small flattened plates (the coxopodite and basipodite) and a simple unjointed endopodite.

5.   The *maxilla* (Fig. 116, C) also has an undivided endopod, but has a large outer lobe, or scaphognathite, which may represent both exopodite and epipodite. The scaphognathite acts as a pump maintaining a flow of water through the branchial cavity.

6.   The three pairs of *maxillipeds* form a series in which there is a change in the relative proportions as one passes backwards from the first pair. Three main trends are discernable.

  (i) The protopodite (coxopodite+basipodite) decreases in size from the first to the third maxilliped.

 (ii) The endopod increases in size, so that in the third maxilliped (Fig. 116, A) it is much larger than the exopod.

(iii) The relative size of the exopod decreases.

The third maxilliped shows the most complete development of its parts, and has better developed branchial filaments than the other maxillipeds.

7.   The first three pairs of legs are chelate. The first pair (*chelipeds*) are equipped with very powerful pincers. The last two pairs of legs are not chelate. All the *thoracic limbs* bear gills at their bases. Those gills actually inserted on the bases of the limbs are termed podobranchs, those inserted on the articular membranes are arthrobranchs, and those inserted on the thorax walls (pleura) are called pleurobranchs.

8.   The five pairs of *pleopods* are all similar in the female, but in the male the first two pairs are modified for sperm transfer. In the normal male the first pair of pleopods is hidden by the forwardly directed second pair.

9.   The *uropods* are broad and flat, forming a tail fan with the telson. The exopodite is subdivided by a transverse suture.

(c) *Dissection*

Cut through the posterior margin of the carapace on each side just medial to the branchiocardiac groove. Continue the cut forwards to the level of the eyes. Lift the flap of carapace carefully and cut away the tissues as close as possible to the inner surface, taking care not to damage the heart. Continue the cut backwards through the abdominal terga to the middle of the telson. Remove the terga carefully, avoiding damage to the dorsal abdominal artery which lies on top of the hind gut (Figs. 117, 118).

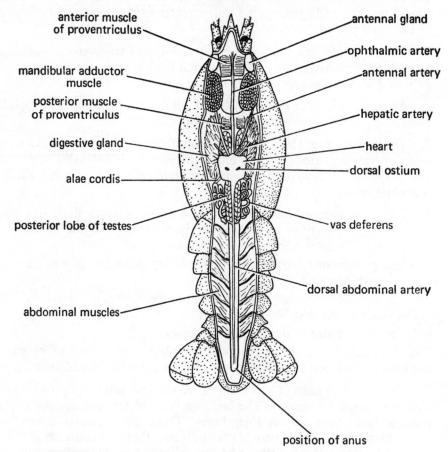

anterior muscle of proventriculus

antennal gland

ophthalmic artery

mandibular adductor muscle

antennal artery

posterior muscle of proventriculus

hepatic artery

digestive gland

heart

dorsal ostium

alae cordis

posterior lobe of testes

vas deferens

dorsal abdominal artery

abdominal muscles

position of anus

Figure 117   *Astacus*, general dissection of adult male from dorsal surface

(d) *Internal anatomy*

1.   The *heart* lies in a pericardial sinus, and has six ostia: two dorsal, two lateral and two ventral. Three pairs of fibrous bands (alae cordis) connect the heart with the walls of the pericardial sinus.

The *arteries* leading from the heart are best seen if carmine is injected with a fine pipette through one of the dorsal ostia. Anteriorly there are five arteries leaving the heart: the median opthalmic artery, paired antennary arteries and paired hepatic arteries. Posteriorly there is a single median artery. Just after leaving the heart this artery gives off a branch the sternal artery, which passes downwards, either to the left or right of the intestine, and passes through the nerve cord between the fourth and fifth thoracic ganglia. The sternal artery then divides and passes forwards and backwards beneath the nerve cord.

2. The most easily discernable part of the *alimentary canal* is the hind gut, running along the midline of the abdomen. Anteriorly it is possible to see the proventriculus, with ossicles in its wall. The most conspicuous ossicle is the cardiac ossicle which lies transversely in the dorsal wall of the proventriculus.

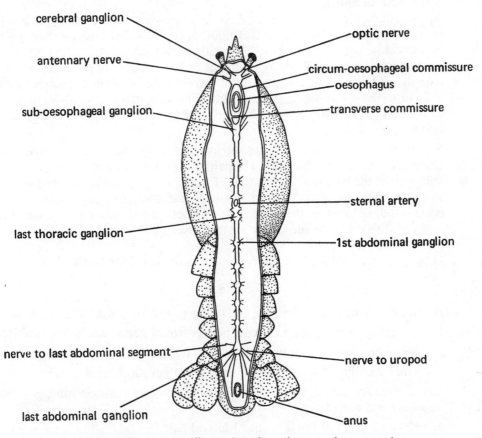

cerebral ganglion

optic nerve

antennary nerve

circum-oesophageal commissure

oesophagus

sub-oesophageal ganglion

transverse commissure

sternal artery

last thoracic ganglion

1st abdominal ganglion

nerve to last abdominal segment

nerve to uropod

last abdominal ganglion

anus

Figure 118 *Astacus*, dissected to show the ventral nerve cord

On either side of the proventriculus are two large muscles which attach to the inner wall of the carapace and adduct the mandibles.

The anterior and posterior proventricular muscles run from the wall of the proventriculus to the exoskeleton. Their contractions move the ossicles in relation to each other so that food in the proventriculus is triturated. Posterior to the proventriculus the short mid gut gives off two large lateral digestive glands which fill much of the body cavity on either side of the gut. The mid gut also gives off a small mid-dorsal caecum.

3. The *antennal glands* lie antero-lateral to the proventriculus, and at a some-what lower level. Each has a thin walled bladder opening via a pore on the base of the antenna. This bladder overlies a tubular and a spongy gland-like structure.

4. In order to trace the *nervous system* it is necessary to cut away the endo-phragmal skeleton which projects inwards from the sternal plates. The last abdominal ganglion is the easiest to find. Once this ganglion has been located cut away the abdominal muscles and the endophragmal skeleton to trace the nerve cord forwards.

At the anterior end of the cord there is a complex sub-oesophageal ganglion, giving off numerous nerves to the mouthparts. The first free thoracic ganglion lies very close behind the sub-oesophageal ganglion. A pair of circum oesophageal connectives run around the oesophagus to the cerebral ganglia. A short post oesophageal commissure connects the two circum oesophageal connectives just in front of the sub-oesophageal ganglion. The three main nerves from the cerebral gangli run to the eyes, antennules and antennae. For further details see Keim (1915).

5. In the male the *testis* is bilobed in front of the heart, and forms a single elongated lobe behind the heart. The paired vasa deferentia are highly convoluted, and open at the bases of the last pair of walking legs (8th thoracic segment).

In the female the *ovary* is also trilobed, but the lobes are shorter than the corresponding lobes in the male. The oviducts are simple and not convoluted, running directly from the ovary to the bases of the 2nd pair of walking legs (6th thoracic segment).

Huxley's classic (1880) will still be found most useful for further details.

CANCER

A typical brachyuran shows the following points of comparison with *Astacus*.

1. Widening of carapace and reduction in size of abdomen, lack of tail fan.

2. Reduction in size of antennules and antennae.

3. Third maxilliped forms 'doors' over the other mouthparts.

4. Alimentary canal: two long coiled mid-gut caeca, single hind gut caecum also long and coiled.

5. Heart: 2 pairs of dorsal ostia, 1 lateral pair.

6. Arteries: large lateral artery (=antennal artery) gives off branches to gonads, fore gut, hind gut, digestive gland. Hepatic artery plunges from heart into digestive gland—care needed to find it.

7. Gills: number on each side reduced to 9.

8. Nervous system: all the thoracic and abdominal ganglia are concentrated into a single mass pierced in the middle by the sternal artery descending from the heart.

9. Testes join in mid-line, but do not form a posterior lobe. Ovaries have a posterior lobe, and are connected directly with spermathecae which open to outside via short oviducts. For further details see Pearson (1908).

16.8. Experiments

16.8.1. SIMPLE EXPERIMENTS WITH *Daphnia*

1.   Heart rate. Specimens mounted on a slide under a cover slip can be irrigated by means of a fine tube leading to one side of the coverslip and a soakaway of filter paper on the other. Observations can be made on changes in heart rate with temperature, light, oxygen content of the water or with various drugs. At high temperatures the rate of beating may be too fast for normal counting, but dotting on paper with a pencil in rhythm with the heart can produce reasonable results. More elaborate methods can involve the use of a stroboscope, or illumination with a stroboflash (see Tonolli, 1947).

2.   Reactions to light. Free swimming *Daphnia* or other small Crustacea can be subjected to unidirectional light and show varying responses according to age, temperature, pH, and the colour of the light (cf. Baylor and Smith, 1957; Harris and Mason, 1956).

3.   Haemoglobin synthesis. Pond dwelling species of *Daphnia* kept in poorly aerated water become bright red with haemoglobin. Experiments comparing *Daphnia* kept in poorly and well aerated water yield clear results in a week (cf. Fox, Hardcastle and Dresel, 1949). See also Fox (1956).

16.8.2. WATER LOSS AND HUMIDITY REACTIONS OF ISOPODS

Humidity is an important factor influencing the distribution of land isopods. Humidity preferences of different species (of genera such as *Porcellio*, *Armadillidium* and *Ligia*) can be compared using choice-chambers. These can be made from plastic boxes, preferably blackened on the outside and with perforated zinc or plastic floors supported above the bottom. A stoppered hole is provided in the lid through which animals may be dropped onto the perforated floor, beneath which water or a saturated salt solution may be placed in shallow dishes. When different solutions are provided at each end, a humidity gradient is set up after 30 minutes. The animals are then introduced and allowed to distribute themselves. For details of such experiments see Gunn and Kennedy (1936). For solutions for maintaining different humidities, see O'Brien (1948). Simple acid dilutions are not appropriate because of toxicity. Convenient ones are: K_2SO_4, 98% R.H.; NaCl, 76%; $Ca(NO_3)_2 4H_2O$, 56%; CH_3COOK, 20%; $ZnCl_2 \times H_2O$, 10% R.H., at 20°C. Some of these salts are extremely soluble and to produce a saturated solution add water to the salt, not *vice versa*. O'Brien gives many other salts and lists the relative humidities their saturated solutions will maintain at different temperatures. Survival times of different species at low humidities and different temperatures can also be determined. For details, see Edney (1957).

16.8.3. COLOUR CHANGE

Many crustaceans change colour according to the background. The large chromatophores of many shrimps and prawns are easily seen. *Ligia* is particularly good for class demonstration. In this isopod chromatophore control is partly through a direct light response and partly by hormones produced in the optic lobe. Adapt animals to black and to white backgrounds. Paint over the eyes with a cellulose paint, completely, or over the lower or upper part alone. Place the treated animals on a contrasting background. Decapitate dark-adapted animals. Grind up the heads. Add a little seawater, mix and centrifuge. Inject a small quantity of the fluid into other animals.

The chromatophores may be examined under a binocular microscope if the animal be trapped between microscope slides kept apart by 'Plasticine' and lightly bound together.

In many Malacostraca with stalked eyes, the sinus gland-X organ complex may be clearly visible (see the figures in Carlisle and Knowles, 1959). *Hippolyte*, *Crangon*, *Leander*, *Palaemonetes* or *Uca* can all be used for demonstration of chromatophore response as available. *Uca*, like most brachyurans, shows a diurnal response rather than a background response shown by macrurans or *Ligia*. The effect of the sinus gland may be demonstrated by tying a ligature tightly round the base of the eyestalk, rather than by actual eyestalk removal. Eyestalk extract may also be injected. For details of such experiments, see Welsh and Smith (1949). For further details see also Kleinholz (1937), Smith (1938) and Carlisle and Knowles (1939).

16.8.4. Osmoregulation in Decapods

Carcinus maenas is common in British estuaries and can withstand considerable lowering of salinity. It regulatory ability in different salinities is best studied using freezing-point depression as a measure of fluid concentration in crabs adapted to (say) 100%, 50% and 25% seawater overnight. Body fluid can be collected by snapping off the legs. Roughly dry the crab with a cloth, kill and snap off the legs and allow the fluid to drain from the sockets. Chloride determinations can also be done (see section 8.10.2., above). For details see Webb (1940). If a flame photometer is available, Na^+ and K^+ may be determined on small volumes extracted with a hypodermic syringe passed through one of the basal arthrodial membranes of a leg. For details of such experiments see Clark (1966). For general information, see Potts and Parry (1964).

16.8.5. Moulting in Decapods

Eyestalk removal in crayfishes (see 16.8.3., above) hastens moulting, and is readilly demonstrated, provided 4-5 weeks are available to observe the effect. Ligature the eyestalks at the base before removal to avoid loss of blood. Use immature crayfishes, leaving animals of similar age, of course, as controls. See Smith (1940), Welsh and Smith (1949) and Carlisle and Knowles (1959) for details.

17. ONYCHOPHORANS

The Onychophora is a small phylum containing less than 70 species but is of considerable interest in that both annelidan and arthropodan characteristics are apparent in the organization of the body. Although considered primitive in some respects Onychophora are specialized for life in confined spaces. Their bodies are highly deformable, enabling the animals to squeeze through narrow cracks into places where they are safe from predators and in which the atmosphere is humid. This ability to deform the body is reflected in many aspects of onychophoran morphology and biology.

If living specimens are available their locomotion and general behaviour should be observed. When walking waves of limb movements pass along the body. This movement is produced by the action of extrinsic limb muscles and there is little undulation of the body. Discharge of the adhesive secretion from the oral papillae (a defensive reaction) may be elicited by disturbing the animal.

(a) *External features*

The body is covered by a continuous cuticle of thin chitin, overlying the dermo-muscular body wall. This comprises a single layered epidermis, a collagenous dermis and circular, diagonal and longitudinal layers of unstriated muscle. The arrangement of fibres in the outer two layers can be clearly seen by stripping away the cuticle and epidermis. The outer body surface is raised into numerous scale-covered tubercles, the larger of which bear sensory bristles. Anteriorly (Fig. 119) there is a pair of annulated antennae, each of which bears a small ocellus at the base. On the ventral surface a pre-oral cavity, surrounded by raised lips, leads to the mouth. Within the cavity are paired lateral mandibles and a median tongue which bears sensory spines, as do the lips. The mandibles have long cutting blades and are operated by a number of muscles, the largest of which, the retractors are attached to the long mandibular apodemes. On either side of the mouth region lies an oral papilla bearing the opening of a slime gland. The antennae, mandibles and papillae represent the modified first, second and third pairs of segmental appendages. The remaining appendages form walking limbs attached latero-ventrally to the body and having distal pads and terminal claws. The cutting blades of the mandibles are derived from the terminal claws of the modified appendages. At the posterior end of the body, on the ventral surface, are the terminal anus and subterminal genital opening.

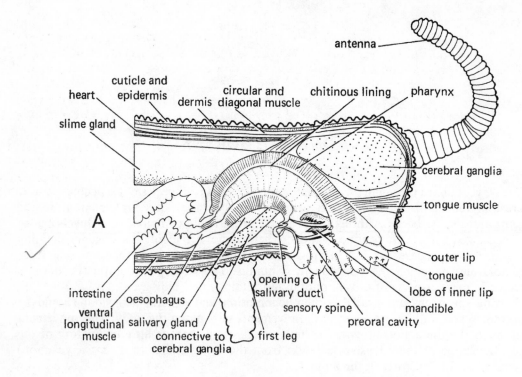

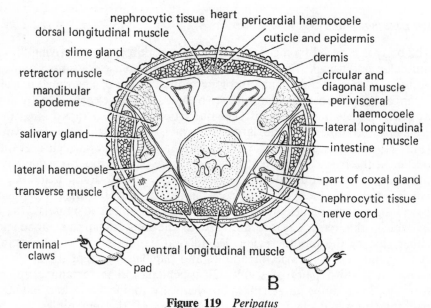

Figure 119 *Peripatus*

A, animal cut sagittally through anterior region (some tissue removed to show connective and salivary gland); B, posterior view of animal cut transversely between legs 1 and 2

(b) *Internal anatomy*

The internal anatomy can be studied in dissected specimens (opened dorso-laterally) and in histological preparations. Alternatively animals which have been preserved in alcohol may be cut, using a sharp razor blade, in vertical and transverse planes through various regions of the body.

The haemocoelic body cavity is partitioned into a number of interconnecting sinuses (Fig. 119, B) and forms the hydrostatic skeleton against which the muscle layers act. In the anterior region of the body the large perivisceral haemocoele contains the slime glands, the mandibular retractor muscles and the alimentary tract. The lateral haemocoeles contain the salivary glands, which open into the pre-oral cavity, the segmental coxal glands and the ventral nerve cords. In the posterior region the perivisceral haemocoele contains the alimentary tract and the reproductive organs.

The pre-oral cavity leads into a muscular pharynx, which is lined by chitin and has a characteristic almost tetraradiate lumen. A short oesophagus opens into the long tubular intestine running posteriorly to the rectum and anus. There are no Malpighian tubules, nitrogenous excretion is carried out largely by cells of the intestine.

Dorsal to the pharynx are the cerebral ganglia, which innervate the antennae, eyes and mandibles. The ganglia are joined by circum-pharyngeal connectives to the ventral nerve cords, which are widely separated but linked by numerous commissures.

The majority of body segments contain paired coxal glands or 'nephridia' situated latero-ventrally and opening at the bases of the limbs. Each organ is derived from the segmental coelomoduct and consists of a small coelomic end sac, a tubule and a terminal bladder. The most anterior and posterior coelomoducts form the salivary glands and the genital ducts respectively. Crural glands may be found ventral to the coxal glands and these also open at the bases of the limbs.

Respiration is carried out by diffusion through numerous spiracles into tufts of fine tracheae. The spiracles cannot be closed and thus evaporation of water occurs freely through them.

The reproductive system of mature female onychophorans consists of fused ovaries and paired ducts which open via a common duct at the genital pore. The paired ducts are differentiated into seminal receptacles and uteri and in the majority of species the uteri will be found to contain developing embryos. These should be removed and examined. Male animals have paired testes and genital ducts, the latter joining to form a single duct in which spermatophores are formed. Accessory organs may be present.

18. MYRIAPODOUS ARTHROPODS

18.1. Introduction

The Myriapoda are a group of mandibulate arthropods characterized by the possession of a segmented body divided into an anterior head and a long posterior region bearing many walking limbs. The diverse morphology of the myriapodan classes has led to the view that the grouping is one of convenience only, with little systematic value. Manton (1964) however, considers that the Myriapoda represent a distinct line of arthropod evolution, their diversity resulting from adaptations to widely varying modes of life.

The functional morphology of the Myriapoda has been analysed very completely by Manton in a series of papers between 1952 and 1966. A useful precis of this information, together with detailed references, is given by Manton in Gray (1968).

The two Classes Pauropoda and Symphyla are not treated in detail here.

18.2. Chilopods

The 'centipedes' are carnivores with one pair of legs per body segment, the first pair modified as poison claws. The mouthparts include mandibles and two pairs of maxillae. The reproductive ducts open posteriorly.

18.2.1. CLASSIFICATION

The chilopods may be divided into a number of orders.

Class CHILOPODA

    Order SCUTIGEROMORPHA
    Fast moving surface forms. Spiracles dorsal. 15 pairs of legs. Tergites reduced in number.
                                              Scutigera

    Order LITHOBIOMORPHA
    Fast moving crevice and soil forms. Spiracles lateral. 15 pairs of legs. Tergites alternately long and short.
                                              Lithobius

Order SCOLOPENDROMORPHA

Slower crevice and soil forms. Spiracles lateral. 21–23 pairs of legs. Tergites equal in size.

Scolopendra

Order GEOPHILOMORPHA

Slow, burrowing soil forms. Spiracles lateral. More than 30 legs. Body very flexible, segments jointed.

Geophilus

Scutigeromorph centipedes do not occur in this country, but representatives of the other orders are common and easily obtainable.

18.2.2. *Lithobius*

Lithobius forficatus is the largest British species of its order and is described below. The animal is capable both of fast surface running and of penetrating into narrow spaces by twisting and bending its body and morphological modifications for both habits are therefore present. Specimens should be examined live, in order to observe the characteristic locomotory patterns, and after killing in 70% alcohol. Preparations of mouth parts can be made by boiling the head in 10% KOH, dehydrating in alcohols, clearing in xylol and mounting in canada balsam.

In *dorsal view* (Fig. 120, A) note the flattened head, the unequal lengths of the legs and the alternating short and long tergites which lie anterior and posterior to the long tergites 8 and 9. The latter features are modifications for fast locomotion, reducing interference with leg movement and restricting lateral undulations of the body in running. Behind the last leg-bearing segment is the pregenital segment which bears small gonopods. The body terminates in a telson.

In *ventral view* the mouthparts and large poison claws can be seen (Fig. 120, B). The latter contain posion glands which open near the tips. At the posterior end of the male (Fig. 121, A) the vestigial gonopods and the penis lie on and behind the 17th sternite; in the female the gonopods and genital opening occupy a comparable position. The anus is terminal.

The *legs* are attached laterally to the flexible pleural membranes of each segment, the coxa forming a flattened ring against the side of the segment. (Fig. 121, B). The greatest amount of movement in the limb occurs at the articulation between the coxa and the trochanter, but during locomotion the coxa and katopleure rotate, thus allowing a large stepping movement to be made. Above each limb base in segments 4, 6, 9, 11, 13 and 15 (i.e. segments with long tergites) lies a spiracle opening into the tracheal respiratory system.

In *reproduction* the male produces a spermatophore, which is picked up by the female and manipulated by the gonopods into the genital opening. The eggs hatch as small forms with seven pairs of limbs, the remainder being added over a long period of development (anamorphic development in contrast to the epimorphic development seen in scolopendromorph and geophilomorph centipedes, where the newly hatched young have a complete set of segments).

Large fresh specimens of *Lithobius* may be dissected to show the internal body organs. The body should be opened by a lateral incision followed by careful removal of the tergites.

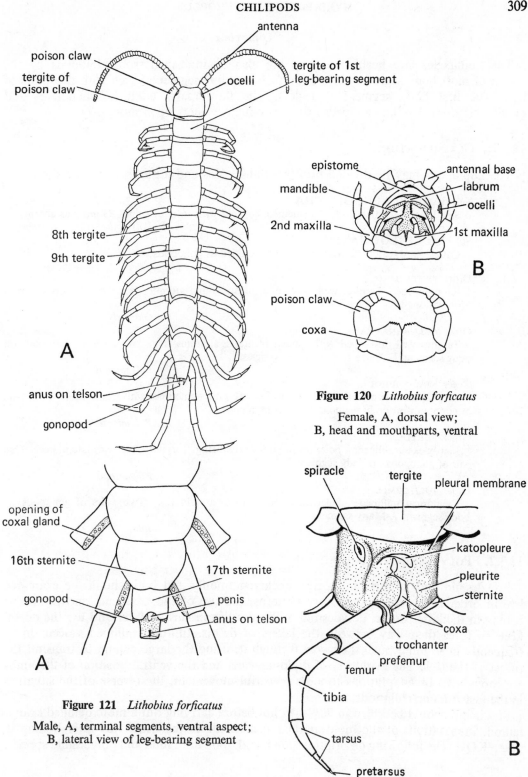

Figure 120 *Lithobius forficatus*

Female, A, dorsal view;
B, head and mouthparts, ventral

Figure 121 *Lithobius forficatus*

Male, A, terminal segments, ventral aspect;
B, lateral view of leg-bearing segment

18.3. Diplopods

The 'millipedes' are herbivores, and are also distinguished from chilopods by the fusion of most body segments into pairs to form 'diplosegments' each with two pairs of legs. The first body segment is legless. The mouthparts comprise mandibles and gnathochilarium. Unlike centipedes the reproductive ducts open anteriorly.

18.3.1. CLASSIFICATION

The Diplopods may be divided into two Sub-classes and a number of orders.

Sub-class PSELAPHOGNATHA
Small forms. Cuticle soft, not calcified, covered with tufts of spines. Gonopods absent.
Polyxenus

Sub-class CHILOGNATHA
Cuticle hard, calcified. Male has limbs modified as gonopods.

Order PENTAZONIA
Pill-millipedes. Tergites well developed, pleurites and sternites free. Last 3 pairs of limbs form gonopods.
Glomeris

Order COLOBGNATHA
Tergites well developed with hinged pleurites, sternites free. Two pairs of gonopods on body segment 7. Coxae bear eversible vesicles.
Polyzonium

Order NEMATOPHORA
Tergites and pleurites fused, sternites free. Tergites sometimes with lateral keels. Two pairs of gonopods on body segment 7. Many segments.
Polymicrodon

Order POLYDESMIDA
Flat-backed millipedes. Sclerites fused into rigid ring. Tergites often with lateral keels. One pair of gonopods on body segment 7.
Polydesmus

Order IULIFORMIA
Cylindrical millipedes. Sclerites fused to form a rigid ring. Two pairs of gonopods on body segment 7. Many segments.
Iulus

18.3.2. POLYDESMIDS AND IULIDS

Polydesmid and iuliform millipedes occur commonly and both should be examined live in order to observe locomotory patterns, the use of the antennae, and protective coiling in the latter order. Polydesmid millipedes live in surface litter and use the dorsal keels to force their way between the layers of debris. Iuliform millipedes occur more frequently in soil and push their way through it, using the large collum to transmit the thrust of the legs. The evolution of diplosegments and the ventral position of the limbs are thought to be adaptations to slow powerful movement, the reverse of the situation in the fast-running chilopods.

Specimens should be killed in 70% alcohol before carrying out a more detailed examination. Preparations of the mouthparts can be made in the usual way after boiling in dilute KOH. The following descriptions are based upon both iulid and polydesmid species.

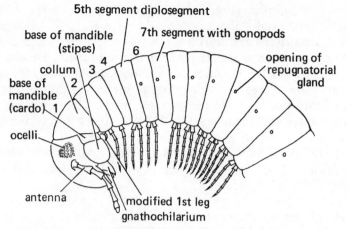

Figure 122   Iuliform millipede. Lateral view of anterior region

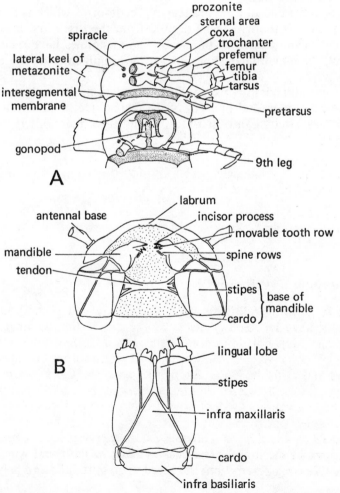

Figure 123   *Polydesmus* sp. A, male gonopods; B, head and mouthparts, ventral aspect

The head of a iulid millipede (Fig. 122, A) is small and bears the short antennae and lateral ocelli. Behind the head is the large dorsal tergite of the legless first body segment, the collum. The second, third and fourth segments carry only one pair of legs each and are probably not diplosegments. The first pair of legs are modified in the males of certain species. The genital opening lies on the third sternite. Segment five is the first of the diplosegments with two pairs of legs and succeeding segments are similar, except for segment seven, where the limbs are modified to form gonopods. The gonopods of polydesmids (one pair only) are more easily studied (Fig. 123 A,). At the end of the body there are a number of legless diplosegments and a terminal telson bearing the anus. The legbearing diplosegments carry lateral openings of the repugnatorial glands used for defensive purposes. Spiracles are placed ventrally, anterior to the limb bases, and open into hollow apodemes from which tracheae arise.

A region of the body should be boiled in dilute KOH to remove soft tissues in order to study the nature of the articulation between successive diplosegments. In coiling of iuliform species the ventral regions of the diplosegments telescope, thus protecting the soft intersegmental membranes.

The mouth lies at the end of a preoral cavity, the floor of which is formed by the large gnathochilarium. This is thought to be derived from the fused first maxillae, no second maxillary segment being present in the head. The mandibles have a characteristic, segmented structure, the mandibular stipes appearing prominently on the sides of the head (Fig. 123, B).

Reproduction in millipedes involves indirect copulation, the male using the gonopods to convey sperm into the genital opening of the female. Development is anamorphic (cf. Chilopoda), the eggs hatching into small forms with three pairs of legs.

19. INSECTS

19.1. Introduction

The anatomy of the larger insects may be studied by dissection, and some of the best for general study such as the cockroach and locust, are easily cultured in the laboratory. Other large species may be dissected following the same methods (see p. 317). Smaller species may be examined as whole preparations and in section. Polyvinyl lactophenol and Berlésé's fluid are good mountants. Living specimens may be placed directly in these fluids. Preserved specimens should be passed through 10% acetic acid before being placed in Berlésé's fluid, or 70% alcohol before placing in polyvinyl lactophenol. For special techniques see Oldroyd (1958). For general histology and fine structure, fixation in Duboscq-Brasil is probably best, followed by colouring the sections with Mallory or Azan. Because of the hardness of chitin, hard waxes or double-embedding is necessary. If the cuticle has been recently moulted or if the chitin is not especially hard or thick, sections may be cut in 58° paraffin wax containing up to 5% ceresin. Alternatively ester wax may be used. Vacuum-embedding is to be recommended whichever embedding compound is used. If paraffin-ceresin or ester wax prove inadequate, then the material may be embedded first in methyl benzoate celloidin for some weeks before embedding in paraffin wax. Sledge microtomes usually produce the best results. See Eltringham (1930), Gatenby and Beams (1950) or other manuals of histological techniques for details. Two exercises are specially useful in a general study: (1) preparation of sections of compound eye; (2) preparation of sections of cuticle. Oothecae of cockroach or larvae of *Drosophila* may be used for the latter. Compound eyes of *Drosophila*, housefly, or honey bee are good for sectioning. For class work the material will need to have been double-embedded before-hand, to give really good results. See Imms or Chapman (1971) for details and references.

19.2. Classification

The Class Insecta includes more species than any other. It is divided into a large number of orders which are, nevertheless, easily distinguished; the beginner should not be deterred by their number. The Class is divided into two sub-classes; Apterygota and Pterygota, the former including the primitively wingless (apterous) insects, the latter the winged insects. The winged insects are further subdivided into the Exopterygota (Hemimetabola) in which the wings develop gradually from instar to instar, and the Endopterygota (Holometabola) in which development proceeds through a series of larval instars termi-nating in a pupal phase in which metamorphosis to the adult form takes place. A rather different classification will be found in Grassé (Vol. 9, 10) but this is not generally accepted.

For the general student Imms (1964) is perhaps the best and most concise source of information and includes a useful 'appendix on literature'. For further details, see Imms or Chapman (1971) and Wigglesworth (1965). The sequence of orders given here follows that of Imms (1964). The most important orders are in heavy type.

Minor orders are placed in brackets.

Class INSECTA

Arthropods in which the body is divisible into 3 distinct tagmata: head, thorax and abdomen; thorax consisting of 3 segments each of which bears a pair of walking legs, and typically the posterior 2 of which bear wings. Head with a single pair of antennae and compound eyes. Gonoducts opening terminally at the tip of the abdomen. Tracheae.

   Sub-class APTERYGOTA

   Without wings (primitive condition). In practice, unlikely to be confused with secondarily apterous pterygotes because of various other characteristic features (see under orders, below).

   Order 1. THYSANURA (BRISTLE-TAILS, SILVERFISH)

   Small. Antennae long, setaceous. Mouthparts biting. Abdomen with long cerci; 11th segment forming a long median tail-filament. Body often scaly.

   Order 2. DIPLURA

   Small. Mostly unpigmented insects. Antennae moniliform. Mouthparts biting. Abdomen with cerci or forceps (no tail filament as in Thysanura).

   Order 3. PROTURA

   Minute (less than 1 mm). No antennae. Mouthparts piercing (concealed). Abdomen clearly with 11 segments plus telson. No cerci.

   Order 4. COLLEMBOLA (SPRINGTAILS)

   Small. Antennae 4-jointed. Abdomen only 6-segmented, 1st segment with a ventral tube; 4th with a forked spring reflexed under body.

Sub-class EXOPTERYGOTA

Normally winged: wings developing gradually from instar to instar, though in more specialized forms greater change in the final larval instar.

Order 5. EPHEMEROPTERA (MAYFLIES)

2 pairs membranous wings; hind pair smaller. Antennae very small. 2 long cerci with a median caudal filament. Nymphs aquatic, with 3 caudal gills.

Order 6. ODONATA (DRAGONFLIES, DAMSELFLIES)

2 pairs of generally similar, membranous wings with a prominent dark spot on the leading edge towards the tip. Antennae very short. Nymphs aquatic with a prehensile labial 'mask'.

Order 7. PLECOPTERA (STONEFLIES)

2 pairs of membranous wings, the hindwings with an anal lobe folded fanwise. Mouthparts biting. Antennae long, thread-like. Nymphs aquatic, with tufted gills along the sides.

Order 8. (GRYLLOBLATODEA)

Few living representatives of ancient group of very restricted distribution. See Imms (1964).

Order 9. ORTHOPTERA (CRICKETS, LOCUSTS)

Fore wings thickened (tegmina) protecting hind wings which often have a large anal area folded fanwise. Cerci short. Antennae very variable. Females with well-developed ovipositor. Mouthparts biting. Tarsi 3–4 jointed.

Order 1. PHASMIDA (STICK INSECTS)

Often apterous, very elongate stick or leaf-like insects. Mouthparts biting. Cerci short.

Order 11. DERMAPTERA (EARWIGS)

Forewings forming short covers (tegmina) covering the hind wings which are large and membranous and folded in a complex manner. Mouthparts biting. Abdomen with terminal forceps.

Order 12. (EMBIOPTERA) (WEB SPINNERS)

Males with 2 pairs of similar wings with few veins; females apterous. Mouthparts biting. Insects constructing silken webs. Tropical and sub-tropical.

Order 13. DICTYOPTERA (MANTIDS, COCKROACHES)

Wings forming thickened tegmina. Antennae long, usually filiform. Cerci jointed. Ovipositor short or concealed. Tarsi typically 5-jointed.

Order 14. ISOPTERA (TERMITES)

Winged forms with 2 pairs of similar wings. Mostly apterous, social; polymorphic; living in wood or forming earth-mounds.

Order 15. (ZORAPTERA)

Minute pscocid-like insects found in tropics or sub-tropics. See Imms (1964).

Order 16. PSOCOPTERA (BOOKLICE)

Small, often colourless insects; mostly apterous. Mouthparts biting. Cerci absent. Tarsi 2–3 jointed.

Order 17. MALLOPHAGA (BIRDLICE)

Small, apterous ectoparasites of birds (rarely, mammals). Mouthparts biting (specialized). Tarsi 1 or 2-jointed. Meso- and metathoracic segments partly fused. Cerci absent.

Order 18. SIPHUNCULATA (LICE)

Apterous ectoparasites of mammals. Mouthparts piercing and sucking (highly specialized), retractable. Pro-, meso- and metathorax fused. Tarsi 1-jointed with a single claw. Cerci absent.

Order 19.   HEMIPTERA (PLANT LICE, APHIDS, LEAFHOPPERS, SCALE INSECTS, CICADAS)

Mouthparts for piercing and sucking, characteristic: labial rostrum bearing piercing mandibles and maxillae; palps absent. Wings all membranous, or forewings hardened; often apterous.

> **Sub-order 1.   HOMOPTERA** (APHIS, CICADAS, FROGHOPPERS)
> Forewings membranous or leathery, but condition uniform. Wings folded roofwise.

> **Sub-order 2.   HETEROPTERA** (CAPSIDS, BED-BUGS, WATER BOATMEN, ETC.)
> Forewings not uniformly hardened. Wings folded flat.

Order 20.   THYSANOPTERA (THRIPS)

Minute, usually black, slender insects with narrow, feathery wings. Mouthparts highly specialized, rasping.

Sub-class ENDOPTERYGOTA

Larval instars terminating in a pupal phase with metamorphosis to adult form, wings developing internally; larvae often specialized.

Order 21.   NEUROPTERA (LACEWINGS)

Wings membranous, without anal areas. Antennae long. Mouthparts biting. Cerci absent. Larvae campodeiform.

> **Sub-order 1.   MEGALOPTERA** (ALDERFLIES)
> Venation relatively primitive with few cross veins. Larvae aquatic. For other details see Imms (1957).

> **Sub-order 2.   PLANIPENNIA** (LACEWINGS)
> Venation somewhat more specialized, cross veins more numerous and more branching towards wing margins. Larvae aquatic or terrestrial.

Order 22.   MECOPTERA (SCORPIONFLIES)

Wings similar, venation relatively primitive. Head prolonged into a rostrum. Mouthparts biting. Larvae oligopod or polypod (terrestrial).

Order 23.   LEPIDOPTFRA (BUTTERFLIES AND MOTHS)

Wings scaly. Mouthparts modified for sucking; mandibles usually absent. Larvae polypod (caterpillars); pupae partly free or obtect, often protected by cocoons.

Order 24.   TRICHOPTERA (CADDISFLIES)

Wings hairy (rather than scaly). Mouthparts reduced. Larvae aquatic, commonly in cases made from plant debris. Pupae also aquatic.

Order 25.   DIPTERA ('TRUE' FLIES)

Forewings only, membranous; hind wings modified as halteres. Mouthparts for piercing and lapping, piercing and sucking or simply sucking. Larvae apodous 'maggots'; pupae obtect or exarate and in puparium (not a cocoon).

> **Sub-order 1.   NEMATOCERA** (CRANEFLIES, MOSQUITOS ETC.)
> Antennae with many joints.

> **Sub-order 2.   BRACHYCERA** (HORSEFLIES ETC.)
> Antennae short, usually 3-jointed.

> **Sub-order 3.   CYCLORRHAPHA** (HOUSEFLIES ETC.)
> Antennae short, 3-jointed; terminal joint with a dorsal arista.

Order 26.   SIPHONAPTERA (FLEAS)

Apterous, laterally compressed ectoparasites of birds and mammals. Mouthparts specialized, piercing and sucking. Larvae maggot-like. Pupae exarate, in cocoons.

Order 27. HYMENOPTERA (ANTS, BEES AND WASPS)

Hind wings smaller than fore wings to which they are linked. Venation reduced, obviously specialized. Mouthparts biting and lapping. 1st abdominal segment fused to thorax. Female with a more or less prominent ovipositor (or sting). Larvae usually polypod or apodous; pupae exarate, usually cocooned.

> **Sub-order 1. SYMPHYTA** (SAWFLIES)
> Body with a broad connection between 'thorax' and 'abdomen', (i.e. 1st abdominal segment not petiolated (not forming a 'waist').

> **Sub-order 2. APOCRITA** (ANTS, BEES, WASPS)
> Body with 1st abdominal segment petiolate.
> *Aculeata* (ants, bees, wasps)
> 8th sternite retracted: ovipositor (characteristically forming a sting) thus appearing to arise from the tip of the abdomen.
> *Parasitica* (ichneumons, etc.)
> 8th sternite not retracted (normal), so that ovipositor appears to arise sub-terminally Larvae parasitic.

Order 28. COLEOPTERA (BEETLES)

Forewings forming hard elytra, without venation; hindwings normally membranous and folded beneath. Mouthparts biting. Larvae variable.

(Order 29. STREPSIPTERA) (STYLOPS)

Minute. Female endoparasite of other insects (including bees); male with club-shaped forewing; hind wing membranous and folded fanwise.

19.3. Basic Structure

If dissection is to be limited to a single species, then the locust *Locusta migratoria* is to be recommended. The American cockroach, *Periplaneta americana* is also relatively easy to dissect, and indeed is often studied in elementary classes. The biology and physiology of both are well known so that laboratory study may be related to other work. Both are readily cultured in the laboratory. So are stick insects such as *Carausius morosus*. Some other, specially large, species of stick insect are maintained in various laboratories and are quite good for general dissection. The large *Dytiscus* species may also be used. Dissection of *Dytiscus* is described by Balfour-Browne (1932). The blowfly or housefly larva can also be dissected under binocular microscopes (see p. 336). The honeybee is also well documented. For anatomy of the bee, see Snodgrass (1925). It is fairly difficult to dissect, like the housefly, and is more an exercise for the specialist. Though old, Hewitt (1912) for the anatomy of the housefly, and Lowne (1890, 1895) for the anatomy of the blowfly, are best for details.

Insects are best killed for dissection with chloroform vapour. A small desiccator makes an excellent killing chamber; the chloroform can be poured onto absorbent cotton beneath a perforated zinc platform on to which the insects can be tumbled.

19.3.1. THE COCKROACH

The American cockroach *Periplaneta americana* is readily available, but other large species are as good for dissection and do not differ significantly in general features. The main external head features of *P. americana* are shown in Fig. 124. For details of structure and biology see Guthrie and Tindall (1968) Cornwell (1968).

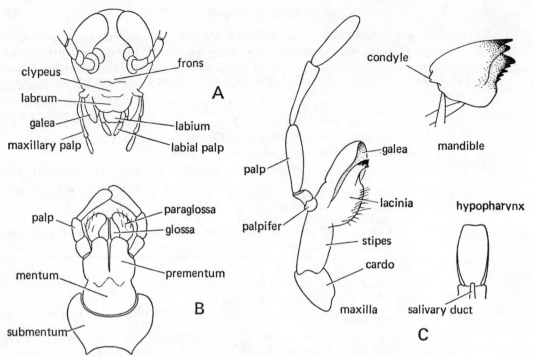

Figure 124 *Periplaneta americana*, head and mouthparts
A, frontal view; B, labium; C, left maxilla; mandible and hypopharynx

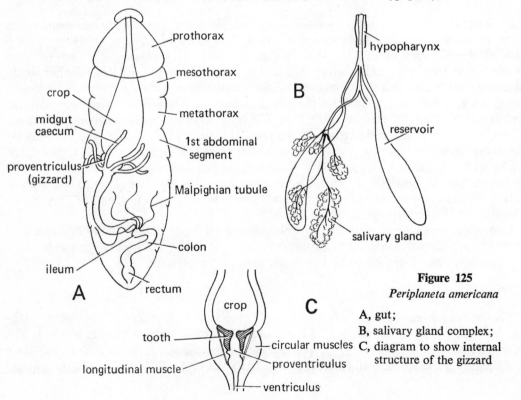

Figure 125
Periplaneta americana

A, gut;
B, salivary gland complex;
C, diagram to show internal structure of the gizzard

Cockroaches are most easily kept in tall glass tanks. These need no cover providing there is a band of petroleum jelly 50 mm wide smeared round the inside rim. Some shelters (inverted plastic boxes with notches cut in them for entry and exit) on sawdust should be provided, and dry food, and water. Guthrie and Tindall recommend rat cake or Purina dog chow. Water should be provided in shallow dishes in which the insects cannot drown. A jar full of water inverted on a Petri dish containing cotton wool is quite satisfactory.

(a) *Dissection*
1.   First remove the wings. To open the body the terga must be removed. This may be done either with the insect pinned down in a wax dish, or the initial cuts

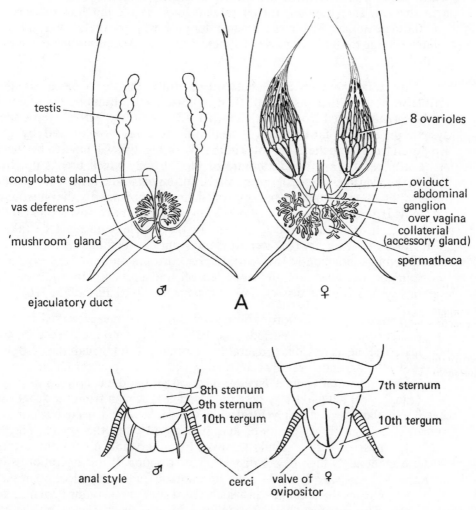

Figure 126 *Periplaneta americana*
A, reproductive systems, male (left) and female (right); B, external sexual differences: male (left) and female (right)

may be made with the insect held in the hand, the terga being actually removed after it is pinned down. Two cuts, one on each side and from one end to the other of the abdomen should be made. Use fine spring (bow) scissors if available, holding the blades almost horizontally. Pin the specimen down through the femora and through posterior tip of the abdomen, using fine entomological pins. Alternatively, the wax of the dish may be partly melted and the insect firmly pressed into the wax. The wax may be easily melted with a flame directed onto the top surface. Do not melt from underneath.

2.   After securing the specimen remove the terga one at a time, working backwards. Lift with forceps and clean from underlying muscle and other tissue with a fine scalpel. Then continue the lateral cuts forward into the thorax and remove the thoracic terga. The mesothoracic tergites will prove fairly resistant to removal owing to the flight muscles attached to them. The pronotum comes away easily.

3.   If the terga have all been removed carefully, the dorsal heart should be visible without further dissection. (If it is not obvious, examine the undersides of the terga removed from the abdomen.) To expose the *gut* and reproductive system dissect the fat-body away with fine forceps, needles and by gentle irrigation from a pipette. The water in the dissecting dish will have to be changed frequently because of the oily nature of the fat-body which breaks up. In the *female* the ovarioles are easily recognized. In the *male*, the testes are sometimes more difficult to recognize because of their resemblance to the surrounding fat-body. If a male is being dissected, the fat-body must be removed very carefully in the posterior third of the abdomen. In the female, care must be taken to distinguish the accessory (collaterial) gland from the surrounding fat-body. The last abdominal ganglion of the ventral nerve cord lies on top of the vagina, and the ganglion must be carefully dissected off and displaced to one side if the nervous system is to be dissected on the same specimen Figs. 125, 126).

4.   The *nervous system* is more easily dissected in a specimen preserved in alcohol. If a fresh cockroach is to be used, pour 95% alcohol onto the dissection, cover, and leave as long as possible (preferably overnight). Cut through the oesophagus in the prothorax and carefully remove the whole gut and reproductive system. Flush out with water using a pipette, pour off and add alcohol. To display the sub-oesophageal ganglion it is best to dissect a specimen from the front of the head. Pin the insect down ventral side up, pressing the head back into the (normal) hypognathous position, and pinning through the genae. The head schlerites must be very carefully removed and the whole dissection performed under a moderately high power dissecting microscope. If this operation is to be attempted in this way it is best done on another (preferably alcohol-preserved) specimen, not on the specimen in which the rest of the nervous system is to be displayed. The sub-oesophageal ganglion can also be displayed by dissection from the dorsal side, and related to the rest of the nerve cord and other ganglia. In such a dissection, the vertex of the head should be carefully removed to reveal the brain. It is simplest to follow the nerve cord forward from the abdominal

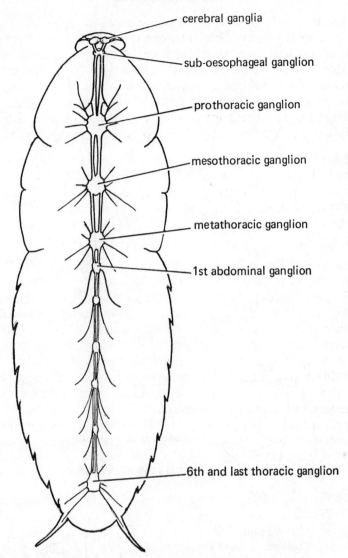

Figure 127 *Periplaneta americana.* Nervous system

ganglia which are obvious after removal of the gut and reproductive organs and treatment with alochol (Fig. 127).

(b) *Function and structure of the gut*

The structure of the proventriculus (gizzard) is best appreciated by longitudinal section; this reveals the teeth projecting into the lumen acting as a valve (Fig. 125, C). If one half is opened and pinned out the other parts of the intima will then be seen. The gut should also be opened (following dissection) at the junction of the proventriculus and mid-gut. Here the peritrophic membrane is secreted. Rate of movement of food through the gut may be studied with carmine mixed up

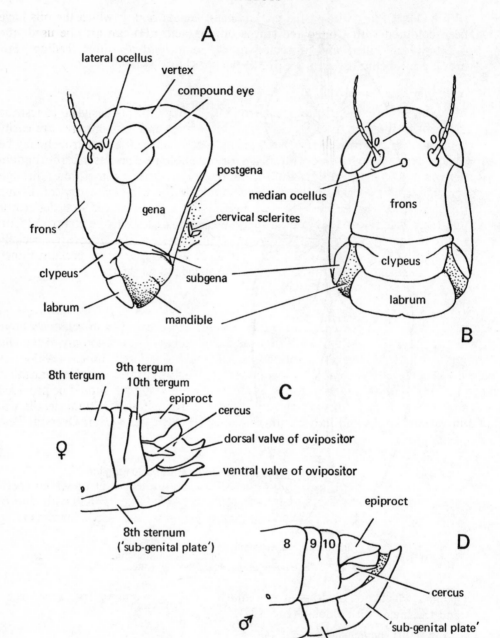

Figure 128 *Locusta migratoria*. External features

A, head from the left side; B, front; C, D, tip of the abdomen of female (C), and male (D) to illustrate the external sexual differences

with banana. Start with fasted cockroaches. Other food in which the oils have been coloured with Congo red (turns blue in acid pH) can also be used; the insects being killed and dissected at different intervals after feeding. For histological methods see under p. 313.

(c) *Water loss*

Many insects are adapted to dry surroundings. Experiments designed to test the survival of different species at different humidities and temperatures are easily conducted in the laboratory. The epicuticular greases in the cockroach may be removed or partly removed with solvent or by rubbing the insects in carborundum dust on a glass plate, (lightly chloroform or etherise first). Loss of water through spiracles can be checked by painting these over with quick-drying cement. Living (respiring) and just killed (not respiring) insects may be placed in a desiccator and their weight loss followed at intervals. Cockroaches may be compared with other insects such as mealworm adults and larvae. The latter are specially good for experiments of this kind. For ideas and details see especially Edney (1957), Beament (1958, 1961), Wigglesworth (1965).

19.3.2. THE LOCUST

The locust is easily kept under laboratory conditions and because of its relatively large size provides perhaps the best insect for dissection. For details of the anatomy of the Old World migratory locust *Locusta migratoria*, see Albrecht (1953) and Thomas (1963).

Cages should be easily cleanable. A low power light-bulb provides sufficient warmth if the cages are kept in ordinary laboratories. A plentiful supply of grass should be provided as food, together with twigs for resting, and sand for oviposition. Cultural details will be found in various books and in a pamphlet published by the Centre for Overseas Pest Research, London.

(a) *Dissection*

The external features (Figs. 128, 129, 130) can be studied from preserved specimens, but dissections should be made of freshly killed insects. Kill with chloroform (see p. 317). Ethyl acetate vapour may also be used. Specimens so

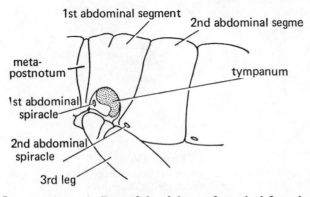

Figure 129 *Locusta migratoria.* Base of the abdomen from the left to show tympanum

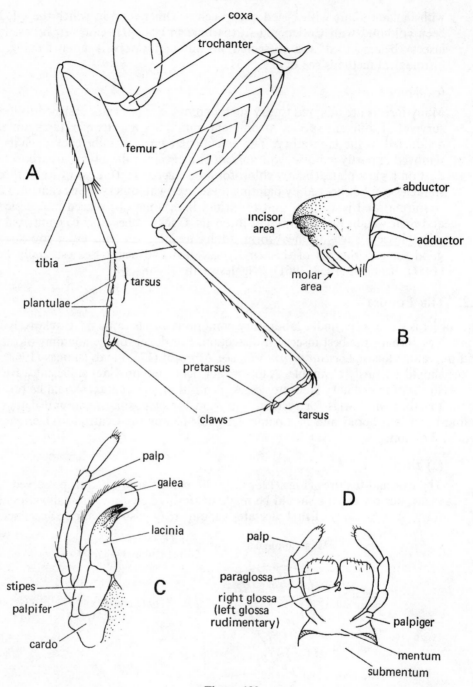

Figure 130

A, hind legs of *Periplaneta* (left) and *Locusta* (right); B, C, D, mouthparts of *Locusta migratoria*; B,
mandible; C, maxilla; D, labium

killed keep well in a deep-freezer. The most important features to note externally are: structure of the head and mouth parts; the general form of the wings and legs, the tympanum, and the secondary sexual differences.

1. Pin out the insect with fine pins through the legs and tip of the abdomen. Cut along each side of the abdominal terga from behind forward and then peel the terga off one at a time. Continue the cuts forward to the front of the pronotum. The meso and metathoracic terga are more difficult to remove than those of the abdominal segments owing to the flight muscles attached to them. The *heart* should be visible lying mid-dorsally (Fig. 131).

2. To display the *heart* satisfactorily a good method is to inject a living locust with aqueous ammonia-carmine by means of a hypodermic syringe through the side of the abdomen. Kill 1 hour after injection. Open as before. Again, if the locust is very lightly chloroformed the heart may be seen to beat, if the insect is opened up carefully under invertebrate saline. Another method to display the heart is to cut the terga on one side only and pin the specimen down in the dish with the cut side uppermost. The diaphragm and alary muscles may then be seen by continuing the dissection under 70% alcohol.

3. The *tracheal system* is complex and a fairly complete picture can only be obtained by a number of dissections. Use freshly killed insects whenever possible. Irrigation with methylene blue during dissection helps in that the tracheae do not dye and stand out clearly by virtue of the air within them. Wigglesworth's injection method (1950) works well, but a slightly simpler method which can display the major parts of the system quite well is to immerse the insect in a 1:1, kerosene: olive oil mixture coloured with any oil-soluble dye (Sudans, Oil red O, Fettrot etc). Place a live insect in this mixture and connect the vessel (through a trap) to a vacuum pump and evacuate for several hours. Then release the pressure and leave (preferably) overnight. Great care must be taken not to damage the tracheae on subsequent dissection or the fluid leaks out. Wigglesworth's (1950) method avoids this, and reference should be made to his paper for details of evacuation. Wigglesworth used hydrogen after each evacuation which is done repeatedly with the insect thrown into cobalt naphthanate dissolved in 1-2 volumes of light petroleum. After injection excess is washed off with light petroleum and then the specimen is transferred to fresh light petroleum and H_2S bubbled through for 5-60 minutes (according to size of specimen: locusts would need at least 1 hour), preferably after puncturing the cuticle in places. The tracheae are revealed by the precipitated sulphide. These are not methods for class use, but are useful for demonstration.

4. If the heart has been examined by one of the methods outlined under (2) above and another specimen is available for dissection then this may be more simply opened by a single mid-dorsal incision from the tip of the abdomen to the anterior edge of the pronotum. Then pin out laterally. The main parts of the tracheal system will be obvious by virtue of the air within them. The gut and reproductive organs may be displayed by separation of the parts with fine forceps and needles and removal of the fat-body with gentle irrigation with a pipette.

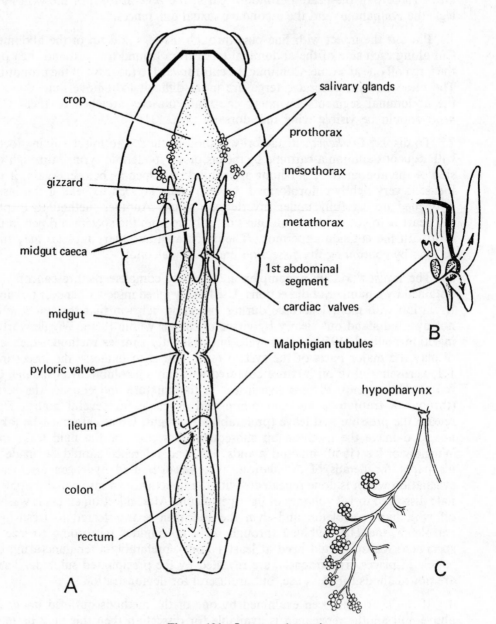

Figure 131 *Locusta migratoria*

A, gut; B, internal view to show relationship of mid-gut caeca and pyloric valves; C, salivary glands

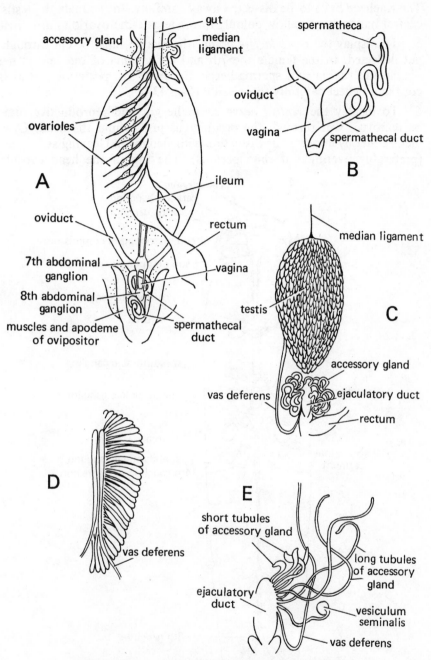

Figure 132 *Locusta migratoria.* Reproductive systems

A, female, rectum cut and gut deflected to the left; B, spermathecal duct unravelled after removal of the 7th and 8th abdominal ganglia and connectives; C, male, gut deflected to the left; D, testis, ventral aspect; E, ejaculatory duct, accessory glands unravelled

The tracheae have to be dissected away carefully. In the male the testis forms a central mass with a yellow colour. In the female the ovarioles are obvious.

5.   To display the *reproductive organs* the rectum must be cut through and the gut deflected. In the female the 7th and 8th ganglia of the ventral nerve cord overlie the vagina and spermathecae. Separate this posterior end of the nerve cord and deflect carefully to one side (Fig. 132).

6.   To display the *ventral nerve cord*, the gut and reproductive organs must be removed. Cut through the oesophagus just behind the head. Dissection is made easier by flooding the dissection with alcohol and leaving as long as possible (preferably overnight if time permits). The top of the head capsule is best

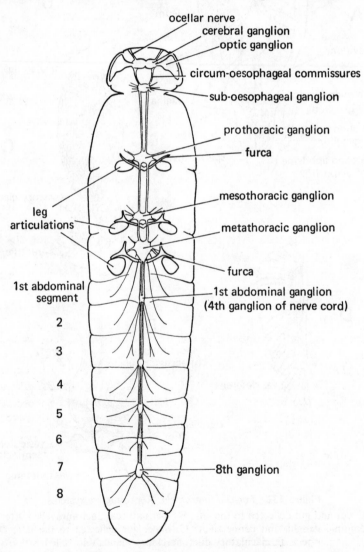

Figure 133 *Locusta migratoria*. Nervous system

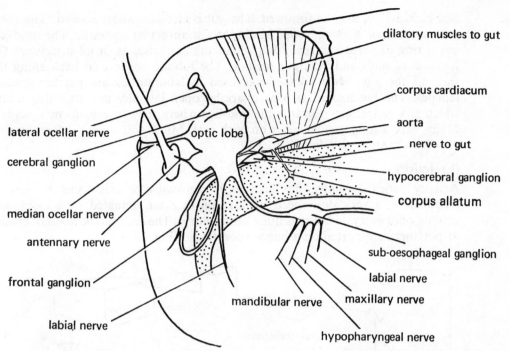

Figure 134 *Locusta migratoria.* Nervous system of the head, dissected from the left side

removed with fine scissors or a sharp scalpel in order to trace the circum-oesophageal commissures (connectives) to the brain. Dissection of the head can be done from the side. After removal of the top of the head, dissect down one side, pinning the head down firmly through the mouthparts. The brain is concealed from above by air-sacs and muscles. If this dissection is done under a dissecting microscope it is possible to reveal the corpus cardiacum of that side (Figs. 133, 134).

(b) *Flight*

If a locust is lightly chloroformed a metal or glass rod may be stuck to the top of the prothorax with 'tacky wax' or some cement, and the insect suspended. When the insect has recovered, flying may be induced by directing a stream of air onto it. Various simple experiments may be performed with insects treated in this way. The movements of the wings can be seen by suitable illumination with a stroboscope. The flying response may be induced by directing a stream of air onto the sensory hairs on the head. Peripheral responses may be induced by bringing a file card up under the legs. For ideas and details, see Weis-Fogh (1956) and Pringle (1957).

19.3.3. The Stick Insect

(a) *Dissection*

Open by a single dorsal cut from the tip of the abdomen forward, in the mid-line. With large specimens it may be possible to see the dorsal vessel if this cut is made slightly to one side of the mid-dorsal line. The general method for locust (see p. 323-329), may be followed. The gut is elongate and uncoiled. The fore-gut consists of a long crop leading to a muscular gizzard. The mid-gut has a ring of slender caeca. Malpighian tubules arise, as in all insects, at the junction of mid- and hind-gut regions. The tubules are looped back along the sides of the gut. Most stick-insects used in laboratories are parthenogenetic females. The bunches of slender ovarioles open laterally into paired oviducts which pass posteriorly towards the mid-line where they fuse to form a vagina. Males have a long slender testis on each side. Much coiled vasa deferentia join to form a median ejaculatory duct.

(b) *Activity*

Activity varies diurnally. An 'actograph' can easily be made and the activity recorded on a very slow-speed kymograph (a drum actuated by a clock and turning once every 24 hours is quite satisfactory). The cage can be made of muslin or polythene supported on a balsa-wood frame (Fig. 135).

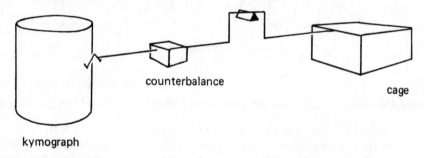

Figure 135   Actograph for recording diurnal activity of stick insects (after Kalmus, 1948)

19.4.   Demonstrations of variety and range of form

Examples and preparations designed to illustrate the classification of insects and their great diversity of form will clearly depend on specimens available. Three structural components: *mouthparts, wings* and *limbs* are specially important, and good class demonstrations can usually be arranged with common insects. Another demonstration of basic importance is one illustrating the differences in development, and the form of different larvae and pupae.

19.4.1.   MODIFICATIONS OF MOUTHPARTS

Examples chosen will be determined by the species available, but there must be few regions where at least some representatives of each type are not readily available at some time of year. Either cockroach or locust (somewhat more specialized) may be used for demonstration of the basic pattern of biting mouthparts. The most important modifications for lapping or sucking are: Hemiptera Homoptera (cicada, aphis, *Psylla* (apple sucker)) or Hemiptera Heteroptera (*Notonecta* (long rostrum), *Corixa* (short rostrum), Lepidopera, Diptera Nematocera (mosquito), Diptera Brachycera (horsefly), Diptera Cyclorrhapha (housefly), and Hymenoptera (hivebee) Figs. 136, 137, 138).

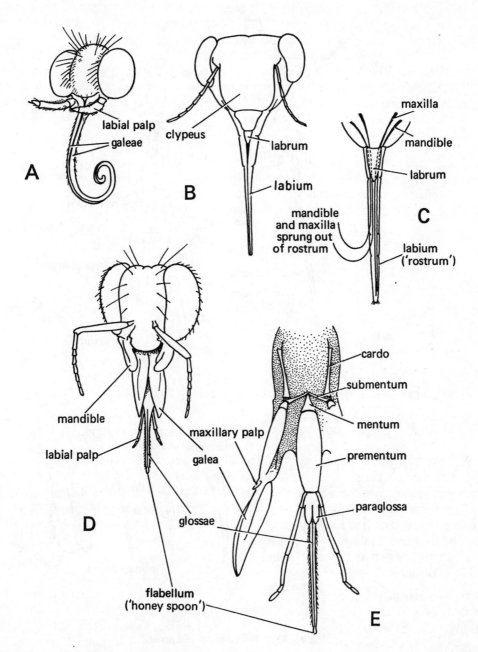

Figure 136   Mouthparts of A, Lepidoptera; B, C, Hemiptera, and D, E, Hymenoptera (honeybee)

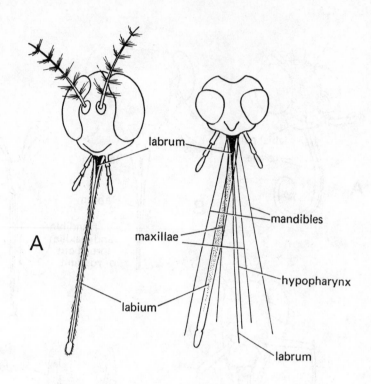

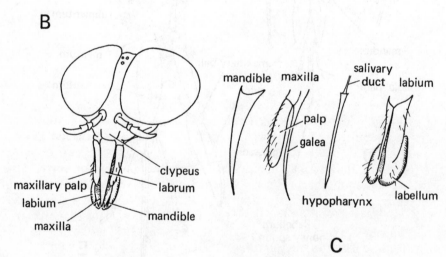

Figure 137 Mouthparts of Diptera

A, mosquito, frontal view of the head (left), view from above with stylets sprung from the labial rostrum (right); B, horsefly (*Tabanus*) frontal view; C, mouthparts of *Tabanus* separated

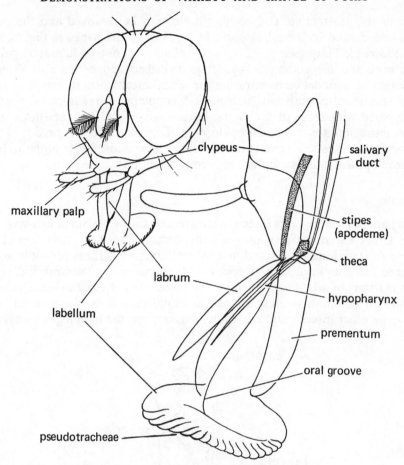

Figure 138 Mouthparts of housefly

Small insects such as aphides or the apple sucker (*Psylla*) are best mounted whole in lactic acid or in polyvinyl-lactophenol (with or without lignin pink) if more permanent preparations are required. Larger specimens should be decapitated and the head and mouthparts boiled in 5% KOH. It is best to do this in a beaker with an inverted test-tube to prevent 'bumping'. Boil until all muscles and soft tissues have disappeared. Then wash well with water. The mouthparts can be dissected off at this stage if desired. Permanent preparations may be made by washing in 70%, 95%, and absolute alcohol, clearing in benzyl alcohol or cedarwood oil and mounting in DPX or other resin. Coverglasses can be supported on small columns of 'Plasticine' if necessary. It is instructive to mount the mouthparts of houseflies frontally and on the side.

19.4.2. Wings

If it is desired to study venation, wings (all left or all right) should be removed from the base and may be mounted in lactic acid or polyvinyl lactophenol. Separation of a mesothoracic segment of a locust to demonstrate the indirect flight muscles and wing articulation is also instructive. Fat body, gut etc. can be removed and the preparation dehydrated and cleared in benzyl alcohol. Movement of the wings in flight can be seen using a stroboscopic lamp (see p. 329), or by placing a spot of luminous paint on the tip of each wing and suspending the insect as described above. See also Pringle (1957) for construction of a model demonstrating the 'click' mechanism in flies.

Primitive venational pattern is best seen in Mecoptera or Trichoptera. Such patterns may be compared with those of dragonfly, horsefly or bee. Other variations suitable for general demonstration are those of Lepidoptera, Diptera (hind wing reduced to haltere-tipulid craneflies are best); Hymenoptera (coupling mechanism); Hemiptera Heteroptera (tegminal development); Thysanoptera (extreme reduction).

19.4.3. LIMBS

The basic pattern of limb joints is seen in cockroach; the hind limb of locust is specialized for leaping. Other variations providing a good demonstration of the range of form are: *Notonecta* or *Corixa*, hind limb (modified for swimming); various phytophagous insects (tarsi modified for gripping), such as coccinellid, curculionid or cerambycid beetles, and more extreme forms of adaptation such as the 'copulatory' front tarsi of *Dytiscus*. Other specializations of common interest are the antennal combs on the fore limbs of bees, wasps and some other insects, and the 'pollen baskets' on the hind legs of worker bees.

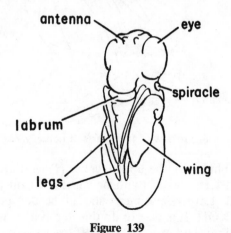

Figure 139

Pupa of *Drosophila* removed from puparium

19.4.4. DEVELOPMENT

There are two basic types of insect development: *hemimetabolous* and *holometabolous*, and these distinguish the two main subclasses of insects (see p. 314). There are many examples to choose from for demonstration, and choice will depend on species available.

Two dissections are specially worthwhile: that of the anisopterous dragonfly nymph, and of a Dipterous larva such as that of the blowfly. *Drosophila* larvae have essentially the same structure, and if *Drosophila* is maintained in the laboratory these can provide as good material for study. For this purpose, cultures of *Drosophila* kept at 15°C (rather

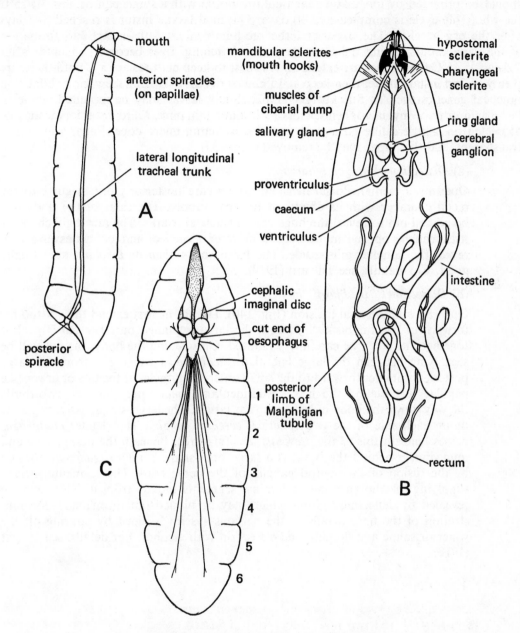

Figure 140   Dipterous larvae (Cyclorrhapha)

A, View from left side showing spiracles and main tracheal trunks; B, gut; C, nervous system

than 25°C) during later larval life will provide larger larvae. Optimal conditions for growth should be provided by inoculating each culture bottle with a single pair of flies. At 25°C the whole life-cycle is completed in 10 days. The final larvae instar is reached 3-4 days after the egg hatches. The puparium, the pre-pupal moult (4th instar) and formation of the pupa occurs during the 5-6th days after hatching. Eyes become pigmented after 7 days. At 15°C these times are extended. It is best to keep cultures at 25°C until there are plenty of 2nd instar larvae, then keep at 15°C. For details of rearing see various books on practical genetics such as Strickberger (1962). Small larvae may be mounted whole in lactic acid or polyvinyl lactophenol with or without lignin pink. Third instar larvae may be 'dissected' in a little saline on a slide under a dissecting microscope. Puparia are semi-transparent, and the pupae may be removed (Fig. 139).

(a) *Dissection of dragonfly larva*

Open by a median dorsal incision, taking care posteriorly not to cut into the rectal chamber with its abundant tracheal supply. The sides of the body can be pinned out to reveal the huge lateral tracheal trunks. The rectal gill chamber itself can be opened mid-dorsally, or later removed and cut transversely to reveal the lamellar gills inside. The larvae of *Aeshna* or *Libellula* are equally good. For details, see Tillyard (1917).

(b) *Dissection of the blowfly larva*

Open by a mid-dorsal incision (Fig. 140). Do not at first extend the cut too far forward or too far backwards until the larva is pinned out. The *gut* is packed together and must be carefully separated. On opening the body, the gut will be seen lying between the large lateral *tracheal trunks* which run from the large posterior spiracles forward to the small anterior spiracles at the tips of finger-like papillae just behind the head. Rudimentary *gonads* may be seen two-thirds the way down the body in last the instar larvae. The large, simple salivary glands lie on each side of the two spherical *cerebral ganglia*. The slender *oesophagus* passes beneath this to the *proventriculus* (gizzard). Beneath the oesophagus and immediately behind the brain is a large boat-shaped '*ventral ganglion*' formed by the fusion of the ventral ganglia of the nerve cord. The segmental nerves supplying the thoracic and abdominal segments radiate from it. They may be revealed by deflecting the gut which may be pinned out to one side. Demonstration of the finer details of the nervous system is aided by pouring off the water or saline and flooding the dissection with alcohol. For details, see Hewitt (1912).

20. GENERAL METHODS

Most of the methods recommended are described under the appropriate sections and may be found through the Index. A few methods have been referred to in several sections and the details of these are given here. Pantin (1946), Hale (1958) and Humason (1972) will be found useful as sources for further information.

20.1. Invertebrate salines

For physiological experiments salines must be made up accurately (see Hale, 1958), Dissections of most marine species can be done in seawater. If this is not available in sufficient quantity, the following is adequate for most purposes:

NaCl	KCl	$CaCl_2$	$NaHCO_3$	$MgCl_2$	
25	—	5	—	7	g/l distilled water

For dissection of leech, Hedon-Fleig's solution is adequate:

NaCl	KCl	$CaCl_2$	$NaHCO_3$	$MgSO_4$	
7	0.3	0.1	1.5	0.3	g/l distilled water

20.2. Narcotizing agents

Menthol
> Scatter crystals on surface of the water and cover the dish. Slow.

Magnesium chloride
> Saturated solution in distilled water. For marine species dilute with an equal volume of seawater.

MS 222 (*Sandoz*)
> The amount required varies greatly according to the animal. For marine annelids 0.75% w/v in 75% seawater is best. Rapid.

10% *alcohol*
> For freshwater animals

CO_2
> Good for many freshwater animals. Add from a soda-siphon

20.3. Fixatives

Bouin's Fixative

Saturated aqueous picric acid . . . ·	75 ml
Formalin	25 ml
Acetic acid	5 ml

Add the acetic acid just before use. Material may be left in Bouin almost indefinitely.

Duboscq-Brasil's Fixative

Picric acid	1 g
Glacial acetic acid	15 ml
Formalin	60 ml
80% alcohol	150 ml

Fix for 2–3 hours, then transfer to 90% alcohol. Specimens become very hard and brittle if the procedure is prolonged.

Susa's Fixative

Mercuric chloride	45 g
Sodium chloride	5 g
Trichloracetic acid	20 g
Glacial acetic acid	40 ml
Formalin	200 ml
Distilled water	800 ml

Material is best fixed overnight. Transfer to 95% alcohol containing a little iodine.

Schaudinn's Fixative

Saturated aqueous mercuric chloride . .	100 ml
Absolute alcohol	50 ml
Glacial acetic acid	5 ml

Add the acetic acid just before use. For Protozoa.

Helly and Zenker's Fixatives

The stock is the same: to make Helly's Fixative add 5 ml formalin
to make Zenker's Fixative add 5 ml glacial acetic acid just before use.

Stock:

Mercuric chloride	5 g
Potassium dichromate	2.5 g
Sodium sulphate (Na_2SO_4, H_2O) . . .	1 g
Distilled water	100 ml

Material should be left in the fixative for several hours. Wash in running water overnight and then with alcohol containing iodine (to avoid Hg-ppt).

20.4. Staining Mixtures

Azan

Stain 'A'
0.1% Azocarmine G	100 ml
Glacial acetic acid	1 ml

Stain 'B'
Aniline blue WS	0.5 g
Orange G	2.0 g
Distilled water	100 ml
Glacial acetic acid	8 ml

Staining is done in a paraffin embedding oven at 60° *To make* 'A' Dissolve the azocarmine by boiling. Filter when cold. Add the acid. (Any precipitate will redissolve at 60°). 'B' Mix the components and boil. Filter when cold. *For use, dilute with twice the volume of distilled water.*

Method

1. Slides to water. Stains 'A' and 'B' should be at 60°.
2. Stain 'A', 45 minutes.
3. Rinse off in distilled water.
4. Differentiate in aniline alcohol (0.1% v/v 90% alcohol) watch under a microscope.
5. Into 1% acetic-alcohol (1% glacial acetic acid v/v 95% alcohol).
6. Mordant in phosphototungstic acid 2–3 hours (5% in distilled water freshly made).
7. Rinse distilled water.
8. Stain in 'B' 1.5 hours.
9. Flick off excess stain into 95% alcohol.
10. Dehydrate, clear, mount.

Masson's Trichrome Mixture

(Baker's and Pantin's modifications).

Solutions:

'A' Hanson's Iron trioxyhaematin

'B' Xylidine Ponceau
Xylidine Ponceau SS	0.25 g
1% aqueous acetic acid	100 ml

'C' Light green
Light green SS	2 g
2% aqueous acetic acid . . .	100 ml

To make 'A'

1. Ammonium ferric sulphate ('iron alum')   10 g
Ammonium sulphate	1.4 g
Distilled water	150 ml

2. Haematoxylin . . . .   1.6 g
Distilled water	75 ml

Dissolve '1' by heating *gently*. '2' may be heated more strongly. Cool. Pour '2' into a large porcelain dish and add '1' to it *gradually*, with stirring. Bring to boil, then cool rapidly by floating the dish in the sink filled with cold water. When cold, filter into a bottle which should be filled to prevent oxidation. The mixture can be used repeatedly for 6 months provided the bottle is full.

Method

1. Stain in 'A', 3 minutes.
2. Wash in running water 10–20 minutes.
3. Stain in 'B'.
   (stain 'progressively' watching under a microscope; may take 5 minutes).
4. Rinse distilled water.
5. Differentiate in 1% phosphomolybdic acid, 4 minutes.
6. Rinse distilled water.
7. Stain in 'C'.
   as under 3, above.
8. Dehydrate as rapidly as possible, clear, mount.

Mallory

Stains:
'A' Acid fuchsin 1% w/v in distilled water.
'B' Aniline blue-Orange G.

Aniline blue WS	0.5 g
Orange G	2.0 g
Oxalic acid	2.0 g
Distilled water	100 ml

(the dyes are very soluble and may be dissolved cold).

Method

1. Slides to water.
2. Mordant in 5% aqueous acetic acid in saturated mercuric chloride.
3. Rinse in distilled water.
4. Stain in 'A', 15–30 seconds.
5. Differentiate by washing off with distilled water. Examine under microscope.
6. 1% phosphomolybdic acid, 1 minute.
7. Rinse distilled water.
8. Stain in 'B', 1.5 minutes.
9. Drain off excess stain and rinse in distilled water.
10. 90% alcohol.
    Watch under microscope till aniline blue sufficiently differentiated.
11. Dehydrate rapidly, clear, mount.

Note: Staining times and differentiation required should be adjusted to suit material; then standardize (without subsequent examination under microscope in steps 5, 10).

Haematoxylin and eosin

This simple, well-tried combination is remarkably good for many purposes.

A. *Ehrlich's acid haematoxylin*

Haematoxylin	2 g
Glacial acetic acid	10 ml
Glycerol	100 ml
Absolute alcohol	100 ml
Distilled water	100 ml
Potassium iodate	0.2 g

B. *Eosin*
1% in absolute alcohol

To make 'A' dissolve the haematoxylin in the alcohol. Then add the acetic acid, glycerol and the iodate solution. Allow to stand overnight with excess alum (potassium aluminium sulphate). Then filter. The potassium iodate (10% by weight of the haematoxylin) will oxidize half the haematoxylin. This gives a working strength which will be maintained by the gradual natural oxidation of the remainder.

Method

1. Bring slides to 70% alcohol.
2. Stain in 'A' until dark.
3. Wash in 70% alcohol.
4. Differentiate in 3% acetic acid in 70% alcohol watching under the microscope
5. 'Blue' in 70% alcohol made alkaline with a drop of ammonia.
6. Pass through alcohols into 'B'.
7. Rinse off in fresh absolute alcohol. Examine under a microscope; when the desired effect is reached, clear and mount.

Note: Many people 'blue' with tap-water. It can be quite good if the water is alkaline. See variant on p. 10.

0.1% Aqueous Toluidine Blue

This simple dye solution can give an effect as good as Azan with some annelid material, owing to its metachromasia. Bring the slide to water and place in the dye overnight. Quickly rinse off excess stain, dehydrate, clear and mount.

20.5. Mounting Media

Berlésé's fluid

Gum arabic	15 g
Saturated aqueous glucose	10 ml
Distilled water	20 ml
Glacial acetic acid	5 ml

To make: Dissolve the gum arabic in the water, then add the glucose, chloral hydrate and acetic acid in that order. Berlésé's original mixture also contained 160 g chloral hydrate. This is a dangerous substance and was included only in order that living insects could be placed directly in the mixture.

D.M.H.F.

(Dimethyl hydrantoin formaldehyde). Useful in that material may pass to it directly from 90% alcohol.

D.P.X.

(Distrene-plasticiser-xylene). Now generally used in place of Canada balsam.

Lactophenol

Phenol	1 g
Lactic acid	1 ml
Glycerol	2 ml
Distilled water	1 ml

Polyvinyl lactophenol

This may be bought made up. *To make*: Stir in polyvinyl alcohol to the lactophenol as made above. Polyvinyl alcohol is solid and does not dissolve readily. Add and stir in daily or from time to time until no more is taken up.

20.6.   Benzidine method for blood vessels

This method is that of Pickworth (1935) and Knox (1954) and is useful for studying the relationships of the smaller vessels of annelids or other small invertebrates with haem pigments in the blood. For fixation see p. 129. The fixative must be well washed out (running water) since benzidine forms an insoluble sulphate. Material can be embedded in gum phenol and thick slices cut on a freezing microlone, if desired.

gum arabic (coarse powder) . . .	500 g
phenol	20 g
distilled water	2 l

1. Transfer material or section to sodium nitroprusside-benzine mixture 1 h
    (sod. nitroprusside . . . . 0.1 g
    distilled water . . . . 75 ml
    add 25 ml. 0.5% benzidine in 2% acetate acid)*
2. Rinse quickly in water.
3. Put in weak hydrogen peroxide (H_2O_2 '20 vols.' 2ml + 400 ml distilled water).
4. Wash in water, dehydrate, clear and mount.

*Benzidine is a carcinogen and should be handled with care with protective gloves in a fume cupboard.

BIBLIOGRAPHY

AARONSON, S., and BAKER, H. (1959) A comparative study of two species of *Ochromonas*. *J.Protozool.* **6**, 282.

ABDEL-MALEK, E. T. (1951) Menthol relaxation of helminths before fixation. *J.Parasit.* **37**, 321.

ALBRECHT, F. O. (1953) The anatomy of the migratory locust London (Athlone Press).

ALEXANDER, R. McN. (1962) Visco-elastic properties of the bodywall of sea anemones. *J.exp.Biol.* **39**, 373.

ANDERSON, D. T. (1966) The comparative embryology of the Polychaeta. *Acta.zool.Stockh.* **47**, 1.

ARNDT, W. (1934–5) Porifera. *Tierwelt N.U.Ostsee*, Teil 3, a1, 1.

ARTHUR, D. R. (1965) Form and function in the interpretation of feeding in lumbricid worms. *Viewpoints in Biology*, Vol. 4, 204. London.

ATKINS, D. (1932) The ciliary feeding mechanism of the Entoproct Polyzoa, and a comparison with that of the Ectoproct Polyzoa. *Q.Jl.microsc.Sci.* **75**, 393.

ATKINS, D. (1936–38) On the ciliary mechanisms and interrelationships of lamellibranchs. *Q.Jl.microsc.Sci.* **79–80**.

BALFOUR-BROWNE, F. (1932) *A textbook of Practical Entomology*. London (Arnold).

BANTA, A. M. (1939) Studies on the physiology, genetics and evolution of some Cladocera. *Pap.Dep.Genet. Carneg.Instn.* No. 39.

BARNES, G. E. (1955) The behaviour of *Anodonta cygnea* L. and its neurophysiological bases *J.exp.Biol.* **32**, 158.

BARNES, R. D. (1980) *Invertebrate Zoology*. Philadelphia. (Saunders College) 4th. ed.

BARRINGTON, E. J. W. (1967) *Invertebrate structure and function*. London (Nelson).

BATHAM, E. J. (1944) *Pollicipes spinosus* Quoy and Gaimard I: Notes on the biology and anatomy of the adult barnacle. *Trans.R.Soc. N.Z.* **74**, 359.

BATHAM, E. J., and PANTIN, C. F. A. (1951) The organisation of the muscular system of *Metridium senile*. *Q.Jl.Microsc.Sci.* **92**, 27.

BATHAM, E. J., and PANTIN, C. F. A. (1950) Muscular and hydrostatic action in the sea anemone *Metridium senile* (L.). *J.exp.Biol.* **27**, 264.

BATHAM, E. J., and PANTIN, C. F. A., and ROBSON, E. A. (1960) The nerve net of the sea anemone *Metridium senile*: the mesenteries and the column. *Q.Jl.microsc.Sci.* **101**, 487.

BAYLOR, E. R. and SMITH, F. E. (1957) Diurnal migration of plankton crustaceans. In: *Recent advances in Invertebrate Physiology*. Ed. B. T. Scheer. Univ. Oregon Press.

BEADLE, L. C. (1937) Adaptation to changes of salinity in the polychaetes. I Control of body volume and of body fluid concentration in *Nereis diversicolor*. *J.exp.Biol.* **14**, 56.

BEADLE, L. C., and BOOTH, F. A. (1938) The reorganization of tissue masses of *Cordylophora lacustris* and the effect of oral cone grafts, with supplementary observations on *Obelia gelatinosa*. *J.exp.Biol.* **15**, 303.

DE BEAUCHAMP, P. (1965) Classe des Rotiferes. In: *Traité de Zoologie*. Ed. P. Grassé. Paris (Masson).

BEALE, G. H. (1954) *The genetics of Paramecium aurelia*. Cambridge (University Press).

BEAMENT, J. W. L. (1958) The effect of temperature on the waterproofing mechanism of an insect. *J.exp.Biol.* **35**, 495.

BEAMENT, J. W. L. (1961) *Biol.Rev.* **36**, 281.

BERG, K. (1934) Cyclic reproduction, sex determination and depression in the Cladocera. *Biol.Rev.* **9**, 139.

BERNARD, H. M. (1892) The Apodidae, a morphological study. London (MacMillan).

BHATIA, M. L. (1941) *Hirudinaria*. Indian Zool. Memoir No. 3. Lucknow.

BIDDER, A. M. (1950) The digestive organs of *Loligo* and *Alloteuthis*. *Q.Jl.microsc.Sci.* **91**, 1.

BLUM, J. J., SOMMER, J. R. and KAHN, V. (1965) Some biochemical, cytological and morphogenetic comparisons between *Astasia longa* and a bleached *Euglena gracilis*. *J.Protozool.* **12**, 209.

BOOLOOTIAN, R. A., and GIESE, A. C. (1958) Coelomic corpuscles of echinoderms. *Biol.Bull.mar.biol.Lab.*, *Woods Hole.* **115**, 53.

BOOLOOTIAN, R. A. (1966) *Physiology of Echinodermata*. New York (Wiley), 822.

BOWEN, S. T. (1962) The genetics of Artemia salina. 1. The reproductive cycle. *biol.Bull.mar.Biol.Lab.*, *Woods Hole.* **122**, 25–32.

BRINKHURST, R. O. (1963) A guide for the identification of British aquatic Oligochaeta. Freshwater Biol.Ass.Sci. Publ. No. 22.

BROWN, F. A. (Ed.) (1950) *Selected Invertebrate Types*. New York (Wiley).

BUCK, J. B. (1943) Quietening *Paramecium* for class study. *Science*. 97, 494.

BULLOCK, T. A. and HORRIDGE, G. A. (1965) *The Nervous Systems of the Invertebrates*. San Francisco.

BUNTING, M. (1926) Studies on the life-cycle of *Tetramitus rostratus*. *J.Morphol.Physiol*. 42, 23.

BURTON, M. (1967) Sponges. in-*Larousse encyclopedia of animal life*. London (Paul Hamlyn).

CALKINS, G. N. (1933) *The biology of the protozoa*. 2nd ed. Lea and Febiger, Philadelphia.

CAMPBELL, A. C. (1976) Observations on the activity of echinoid pedicellariae: III, Jaw responses of globiferous pedicellariae and their significance. *Mar.Behav.Physiol*. 4, 25.

CANNON, H. G. (1927) On the feeding mechanism of *Nebalia bipes*. *Trans.R.Soc.Edinb*. 55, 355.

CANNON, H. G. (1933) On the feeding mechanism of the Branchiopoda. *Phil.Trans.R.Soc.B*. 222, 267.

CANNON, H. G., and MANTON, S. M. (1927) On the feeding mechanism of a mysid crustacean, *Hemimysis lamornae*. *Trans.R.Soc.Edinb*. 55, 219.

CARLISLE, D. B., and KNOWLES, F. G. W. (1959) *Endocrine Control in Crustaceans*. Cambridge (University Press).

CARRIKER, M. R. (1946) The digestive system of *Lymnaea stagnalis adpressa*. *Biol.Bull.mar.biol.Lab*., Woods Hole. 91, 88

CHEN, Y. T. (1950) Investigations of the biology of *Peranema trichophorum* (Euglenineae).*Q. Jl.microsc.Sci*. 91, 279.

CHAPMAN, G. (1953) Studies of the mesoglea of coelenterates I Histology and chemical properties. *Q.Jl.microsc. Sci*. 94, 155.

CHAPMAN, R. F. (1971) *The insects, structure and function*. London (English Univ. Press).

CHITWOOD, B. G., and CHITWOOD, M. B. (1950) *An Introduction to Nematology*. Baltimore (Monumental Printing Co.).

CHUANG, S. H. (1956) The ciliary feeding mechanisms of *Lingula unguis* (L.). (Brachiopoda) *Proc.zool.Soc.Lond*. 127, 167.

CHUBB, J. C. (1962) Acetic acid as a diluent and dehydrant in the preparation of whole, stained helminths. *Stain Tech*. 37, 178.

CLARK, A. M. (1962) *Starfishes and their relations*. London, British Museum (Natural History).

CLARK, R. B. (1961) The origin and formation of the heteronereis. *Biol.Rev*. 36, 199.

CLARK, R. B. (1962) On the structure and functions of polychaete septa. *Proc.zool.Soc.Lond*. 138, 543.

CLARK, R. B. (1966) *A practical course in experimental zoology*. London (Wiley).

CLARK, R. B., and CLARK, M.E. (1960) The ligamentary system and the segmental musculature of *Nephtys*. *Q.Jl.microsc.Sci*. 101, 149.

CLAUS, C. (1876) Zur Kenntniss der Organisation und des feineren Baues der Daphniden und verwandter Cladoceren. *Z.wiss.Zool*. 27, 362.

COLEMAN, A. W. (1962) Sexuality. In LEWIN, R. A. (ed.). *Physiology and biochemistry of algae*. New York (Academic Press).

CORLISS, J. O. (1961) *The ciliated protozoa: characterization, classification, and guide to the literature*. Oxford (Pergamon).

CORNWALL, I. E. (1953) The central nervous system of barnacles (Cirripedia). *J.Fish.Res.Bd.Canada*. 10, 76.

CORNWELL, P. B. (1968) *The Cockroach*. Vol. 1. London (Hutchinson).

COX, F. E. G. (1967) A quantitative method for measuring the uptake of carbon particles by *Tetrahymena pyriformis*. *Trans.Am.microsc.Soc*. 86, 261.

CROFTS, D. R. (1929) *Haliotis*, L.M.B.C. Memoir 29. Liverpool (University Press).

CURTIS, A. S. G. (1962) Pattern and mechanism in the reaggregation of sponges. *Nature*, 196, 245.

CUSSANS, M. (1904) *Gammarus*. L.M.B.C. Memoir 12. Liverpool (University Press).

DAKIN, W. J. (1912) *Buccinum*, L.M.B.C. Memoir 20. Liverpool (University Press).

DALES, R. P. (1955) Feeding and digestion in terebellid polychaetes. *J.mar.biol.Ass.U.K*. 34, 55.

DALES, R. P. (1962) The polychaete stomodeum and the interrelationships of the families of Polychaeta. *Proc.zool.Soc.Lond*. 139, 389.

DARWIN, C. (1851) A Monograph on the sub-class Cirripedia. The Lepadidae or pedunculated cirripedes. London (Ray Soc.).

DAWES, B. (1946) *The Trematoda*. Cambridge (University Press).

DAY, J. H. (1967) *Polychaeta of Southern Africa*. Part 1 *Errantia*, Part 2 *Sedentaria*. London (British Museum).

DEFRETIN, R. (1949) Recherches sur la musculature des Néréidiens au cours de l'épitoquie, sur les glandes parapodiales, et sur le spermiogenèse. *Annls.Inst.Océanogr.Monaco*. 24, 117.

DENNELL, R. (1934) The feeding mechanism of the cumacean *Diastylis bradyi*. *Trans.R.Soc.Edinb*. 58, 125.

DIXON, G. C. (1915) *Tubifex*. L.M.B.C. Memoir 23. Liverpool (University Press).

DIXON, M. (1943) *Manometric Methods*. Cambridge (University Press).

DOFLEIN, F. and REICHENOW, E. (1927–9) *Lehrbuch der Protozoenkunde*. 5th ed. Jena (G. Fischer).

DOFLEIN, F. and REICHENOW, E. (1949–53) *Lehrbuch der Protozoenkunde*. 6th ed. Jena (G. Fischer).

DORSETT, D. A. (1967) Giant neurons and axon pathways in the brain of *Tritonia*. *J.exp.Biol.* **46**, 137.

DORSMAN, B. A. (1935) Notes on the life-history of *Orchestia bottae* Milne Edwards. Thesis, Leiden.

DREW, G. A. (1919) Sexual activities of the squid *Loligo pealic* (Les.) *J.Morph.* **32**, p. 379–418.

EALES, N. B. (1921). *Aplysia*, L.M.B.C. Memoir 24. Liverpool (University Press).

EDNEY, E. B. (1957) *The water relations of terrestrial anthropods*. Cambridge (University Press).

EDMONDSON, W. T. (1959) *Fresh-water Biology* 2nd ed. (Ward and Whipple). New York (Wiley).

EDWARDS, O. A. and LOFTY, J. R. (1977) *Biology of earthworms*, 2nd ed. London (Chapman & Hall).

ELTRINGHAM, H. (1930) *Histological and illustrative methods for Entomologists*. Oxford (Clarendon Press).

EWER, D. W. (1941) The blood systems of *Sabella* and *Spirographis*. *Q.Jl.microsc.Sci.* **82**, 587.

FAGE, L. (1951) Cumacés. Faune de France **54**.

FAUCHALD, K. (1977) *The polychaete worms*. Los Angeles (Nat. Hist. Mus. Los Angeles County/Allan Hancock Foundation).

FELL, H. B. (1963) The phylogeny of sea stars. *Phil.Trans.R.Soc.B.* **246**, 381.

FISH, J. D. (1967) The biology of *Cucumaria elongata* (Echinodermata: Holothuroidea). *J.mar.biol.Ass.U.K.* **47**, 129.

FONTAINE, A. R. (1965) The feeding mechanism of the ophiuroid *Ophiocomina nigra*. *J.mar.biol.Ass.U.K.* **45**, 373.

FOX, H. M. (1956) L'hémoglobine de la Daphnie et les problemes qu'elle soulève. *Bull.Soc.zool.Fr.* **80**, 288.

FOX, H. M., HARDCASTLE, S. M. and DRESEL, E. I. B. (1949) Fluctuations in the haemoglobin content of *Daphnia*. *Proc.R.Soc.B.* **136**, 388.

FOX, H. M. and WINGFIELD, C. A. (1938) A portable apparatus for the determination of oxygen dissolved in a small volume of water. *J.exp.Biol.* **15**, 437.

FRETTER, V. (1937) Structure and Functions of the Digestive System in Chitons (Polyplacophora). *Trans.R.Soc. Edinb.* **59**, 119.

FRETTER, V. and GRAHAM, A. (1962) *British Prosobranch Molluscs*. London (Ray Soc.).

FRETTER, V. and GRAHAM, A. (1976) *A functional anatomy of invertebrates*. London (Academic Press).

FRYER, G. (1957) The feeding mechanism of some freshwater cyclopoid copepods. *Proc.zool.Soc.Lond.* **129**, 1.

GARNHAM, P. C. C. (1966) *Malaria parasites and other haemosporidia*. Oxford (Blackwell).

GILCHRIST, B. M. (1960) Growth and form of the brine shrimp *Artemia salina* (L.). *Proc.zool.Soc.Lond.* **134**, 221.

GRAHAM, A. (1949) The molluscan stomach. *Trans.R.Soc.Edinb.* **41**, 737.

Grassé, P. P. (Editor) *Traité de Zoologie*. Paris (Masson).

    1 (1) (1952) Phylogénie. Protozoaires Généralités, Flagellés.
    1 (2) (1953) Protozoaires: Rhizopodes, Actinopodes, Sporozoaires, Cnidosporidies.
    4 (1) (1961) Platyhelminthes, Mésozoaires, Acanthocéphales, Némertiens.
    4 (2) (1965) Nemathelminthes (Nématodes).
    4 (3) (1965) Nemathelminthes, Nématodes, Gordiaces, Rotifères, Gastrotriches, Kinorhynches.
    5 (1) (1959) Annélides, Myzostomides, Sipunculiens, Priapuliens, Endoproctes, Phoronidies.
    5 (2) (1960) Bryozoaires, Brachiopodes, Chétognathes, Pogonophores, Mollusques (généralités).
    6   (1949) Onychophores, Tardigrades, Arthropodes (généralités) Trilobitomorphes, Chélicérates.
    9   (1949) Insectes, Inférieurs et Coléoptères.
    10 (1) (1951) Insectes Supérieurs, Néuropteroides, Mécopteroides, Hyménopteroides (1).
    10 (2) (1951) Insectes Supérieurs Hyménopteroides (2), Pscopteroides, Hémipteroides, Thysanopteroides.
    11   (1948) Echinodermes, Stomocordés, Procordés.

GRAY, J. (1968) *Animal Locomotion*. London (Weidenfeld and Nicolson).

GRAY, J., LISSMANN, H. W., and PUMPHREY, R. J. (1938) The mechanism of locomotion in the leech (*Hirudo medicinalis* Ray). *J.exp.Biol.* **15**, 408.

GRUNER, H. E. (1965–6) Isopoda. *Tierwelt Deutschlands*. **51** and **53**.

GRUVEL, A. (1905) *Monographie des Cirrhipedes ou Thecostraces*. Paris (Masson).

GUNN, D. L., and KENNEDY, J. S. (1936) Apparatus for investigating the reactions of land arthropods to humidity. *J.exp.Biol.* **13**, 450.

GUTHRIE, D. M., and TINDALL, A. R. (1968) The biology of the cockroach. London (Arnold).

HALL, R. P. (1953) *Protozoology*. New York (Prentice-Hall).

HAMBURGER, V. (1942) *Manual of Experimental Embryology*. Illinois (University of Chicago Press).

HANSON, J. (1950) The blood system of the serpulimorpha (Annelida, Polychaeta). I The anatomy of the blood system in the Serpulidae. *Q.Jl.Microsc.Sci.* **91**, 111.

HARRIS, J. E. and MASON, P. (1956) Vertical migration in eyeless *Daphnia*. *Proc.R.Soc.B.* **145**, 280.

HARTOG, M. M. (1888) The morphology of Cyclops and the relations of the Copepoda. *Trans.Linn.Soc.Lond.* (*Zool.*). **5**, 1.

HEDLEY, R. H. (1958) Tube formation by *Pomatoceros triqueter* (Polychaeta). *J.mar.biol.Ass.U.K.* **37**, 315.

HERLANT-MEEWIS, H. (1961) Régénération du système nerveux chez *Eisenia foetida* (Sav.). *Bull.biol.Fr.Belg.* **95**, 695.

HEWITT, C. G. (1907) *Ligia*. L.M.B.C. Memoir 14. Liverpool (University Press).

HEWITT, C. G. (1912) *The House-fly*. Cambridge (University Press).

HOARE, C. A. (1949) *Handbook of medical protozoology*. London (Baillière, Tindall and Cox).

HOARE, C. A. (1966) The classification of mammalian trypanosomes. *Ergebn.Mikrobiol.Immunforsche.Exp.Ther.* **39**, 43.

HOLMES, W. (1949) Colour change in *Sepia*. *Proc.zool.Soc.Lond.* **110**, 17.

HOWELLS, H. H. (1942) The structure and function of the alimentary canal of *Aplysia punctata*. *Q.Jl.microsc.Sci.* **83**, 357.

HORRIDGE, G. A. (1959) Analysis of the rapid response of *Nereis* and *Harmothoë* (Annelida). *Proc.R.Soc. B*, **150**, 245.

HUGHES, G. M. and TAREC, L. (1965) An electrophysiological study of the anatomical relations of two giant nerve cells in *Aplysia depilans*. *J.exp.Biol.* **45**, 469.

HUMASON, G. L. (1972) *Animal tissue techniques*, 3rd ed. San Francisco (Freeman).

HUXLEY T. H. (1880) The Crayfish. London (Kegan Paul).

HYMAN, L. H. *The Invertebrates.* New York (McGraw Hill).
    Vol. I   (1940)   Protozoa through Ctenophora.
          II   (1951)   Platyhelminthes and Rhynchocoela—the acoelomate Bilateria.
         III   (1951)   Acanthocephala, Aschelminthes and Entoprocta—the pseudocoelomate Bilateria.
         IV   (1955)   Echinodermata; the coelomate Bilateria.
         V   (1959)   Smaller coelomate groups: Chaetognatha, Hemichordata, Pogonophora, Phoronida, Ectoprocta, Brachiopoda, Sipunculida—the coelomate Bilateria.
         VI   (1967)   Mollusca I: Aplacophora, Polyplacophora, Monoplacophora, Gastropoda—the coelomate Bilateria.

IMMS, A. D. (1957) *A General Textbook of Entomology.* 9th ed. revised by O. W. Richards and R. G. Davies. London (Methuen).

IMMS, A. D. (1964) *Outlines of Entomology.* 5th ed. revised by O. W. Richards and R. G. Davies. London (Methuen).

ISGROVE, A. (1909) *Eledone.* L.M.B.C. Memoir 18, Liverpool (University Press).

JAHN, T. L. and JAHN, F. F. (1949) *How to know the protozoa.* Iowa (W. C. Brown).

JENNINGS, H. S. (1931) *Behaviour of lower organisms.* New York (Columbia University Press).

JENNINGS, J. B. (1956) A technique for the detection of *Polystoma integerrinum* in the common frog (*Rana temporaria*). *J.Helminth.* **30**, 119.

JEPPS, M. W. (1946) Vernalization of sponge gemmules. *Nature.* **158**, 485.

JEPPS, M. W. (1947) Contribution to the study of the Sponges. *Proc.R.Soc.B.* **134**, 408.

JOHNSTONE, J. (1899) *Cardium.* L.M.B.C. Memoir 2. Liverpool (University Press).

JOHRI, L. N., and SMYTH, J. D. (1956) A histochemical approach to the study of helminth morphology. *Parasitology*, **6**, 306.

JONES, W. C. (1954a) The orientation of the optic axis of spicules of *Leucoselenia complicata*. *Q.Jl.microsc.Sci.* **95**, 33.

JONES, W. C. (1954b) Spicule form in *Leucosolenia complicata*. *Q.Jl.microsc.Sci.* **95**, 191.

JØRGENSON, C. B. (1955) Quantitative aspects of filter feeding in invertebrates. *Biol.Rev.* **30**, 391.

JØRGENSEN, C. B. (1966) *Biology of Suspension Feeding.* Oxford (Pergamon).

KALMUS, H. (1948) *Simple experiments with Insects.* London (Heinemann).

KEIM, W. (1915) Das Nervensystem von *Astacus fluviatilis* (*Potamobius astacus* L.). Ein Beitrag zur Morphologie der Dekapoden. *Z.wiss.Zool.* **113**, 485.

KERKUT, G. A. (Ed.) (1968) *Experiments in Physiology and Biochemistry.* Vol. 1. London and New York (Acad. P.).

KESLING, R. V. (1965) Anatomy and dimorphism of adult *Candona surburbana* Hoff. National Science Foundation Project GB–26. Report No. 1, 1.

KITCHING, J. A. (1951) The physiology of contractile vacuoles. VII. Osmotic relations in a suctorian with special reference to the mechanism of control of vacuolar output. *J.Exp.Biol.* **28**, 203.

KLEINHOLZ, L. H. (1937) Studies in the pigmentary system of Crustacea I. Colour changes and diurnal rhythm in *Ligia baudiniana*. *Biol.Bull.mar.biol.Lab.*, *Woods Hole.* **72**, 24.

KLIE, W. (1938) Ostracoda. *Tierwelt Deutschlands* **34**, 1.

KNOX, G. A. (1954) The benzidine staining method for blood vessels. *Stain Tech.* **29**, 139.

KRASNE, F. B. (1965) Escape from recurring tactile stimulation in *Branchiomma vesiculosum*. *J.exp.Biol.* **42**, 307.

KUDO, R. R. (1966) *Protozoology.* 6th ed. Springfield, Illinois (C. T. Thomas).

KUENEN, D. J. (1939) Systematical and physiological notes on the brine shrimp, *Artemia*. *Arch.néerl.Zool.* **3**, 365.

LAVERACK, M. S. (1963) *The Physiology of Earthworms.* Oxford (Pergamon).

LAVERACK, M. S. and DANDO, J. (1979) *Lecture notes on invertebrate zoology*, 2nd ed. Oxford (Blackwell).

LISSMANN, H. W. (1945) The mechanism of locomotion in gastropod molluscs. I Kinematics, II Kinematics· *J.exp.Biol.* **21**, 58 and **22**, 37.

LOCHEAD, J. H. (1950) *Artemia.* In: *Selected Invertebrate types.* Ed. F. A. Brown. New York (Wiley).

LONGHURST, A. R. (1955) A review of the Notostraca. *Bull.Brit.Mus.Nat.Hist.Zool.* **3**, (1), 1.

LOWNE, B. T. (1890, 1895) The Anatomy, Physiology, Morphology and Development of the Blowfly (*Calliphora erythrocephala*) Vols. 1, 2. London.

LLEWELLYN, J. (1965) The evolution of parasitic Platyhelminthes. In: Taylor, A. E. R. *Evolution of Parasites.* Oxford (Blackwell).

MACKIE, G. O. (1966) Growth of the hydroid *Tubularia* in culture. In: Rees, W. J., *The Cnidaria and their Evolution.* London (Acad. Press).

MACKINNON, D. L. and HAWES, R. S. J. (1961) *An introduction to the study of protozoa.* Oxford (Clarendon Press).

MACGINITIE, G. E., and MACGINITIE, N. (1949) *Natural History of Marine Animals.* New York (McGraw Hill).

MANN, K. H. (1954a) The anatomy of the horse leech, *Haemopis sanguisuga* (L.). *Proc.zool.Soc.Lond.* **124**, 69.

MANN, K. H. (1954b) A key to the British freshwater leeches. Freshwater *Biol.Ass.Sci.Publ.* **14**

MANN, K. H. (1962) *Leeches (Hirudinea).* Oxford (Pergamon).

MANTON, S. M. (1928) On the embryology of a mysid crustacean *Hemimysis lamornae. Phil.Trans.R.Soc.B.***216**, 363.

MANTON, S. M. (1934) On the embryology of the crustacean *Nebalia bipes. Phil.Trans.R.Soc.B.* **223**, 163.

MANTON, S. M. (1964) Mandibular mechanisms and the evolution of Arthropods. *Phil.Trans.R.Soc.B.* **247**, 1.

MANWELL, C. (1963) The blood proteins of cyclostomes. A study in phylogenetic and ontogenetic biochemistry. *The Biology of Myxine.* Oslo (Universitetsforlaget).

MANWELL, R. D. (1961) *Introduction to protozoology.* London (Arnold).

MCLAUGHLIN, P. A. (1980) *Comparative morphology of recent Crustacea.* San Francisco (Freeman).

MEGLITSCH, P. A. (1972) *Invertebrate zoology*, 2nd ed. Oxford (University Press).

METTAM, C. (1967) Segmental musculature and parapodial movement of *Nereis diversicolor* and *Nephthys hombergi* (Annelida: Polychaeta). *J.Zool.* **153**, 245.

MILLAR, D. H. (1955) Food movement in the gut of the larval oyster. *Q.Jl.microsc.Sci.* **96**, 539.

MILLOTT, N. (1967) Echinoderm biology. *Symp.zool.Soc.Lond.* **20**.

MOORE, J. (1952) The induction of regeneration in the hydroid *Cordylophora lacustris. J.exp.Biol.* **29**, 72.

MOORE, J. P. (1959) Hirudinea In: Edmondson *Freshwater Biology* (see Edmondson (1959)).

MORTON, J. E. (1967) *Molluscs.* London (Hutchinson).

MOSCONA, A. A. (1961) How cells associate. *Scientific American*, September, 142.

NASIR, P., and ERASMUS, D. A. (1964) A key to the cercariae from British freshwater molluscs. *J.Helminth.* **38**, 345.

NEWELL, G. N. (1950) The role of the coelomic fluid in the movements of earthworms. *J.exp.Biol.* **27**, 110.

NICOL, E. A. T. (1930) The feeding mechanism, formation of the tube, and physiology of digestion in *Sabella pavonina. Trans.R.Soc.Edinb.* **46**, 537.

NICOL, J. A. C. (1950) Responses of *Branchiomma vesiculosum* (Montagu) to photic stimulation. *J.mar.biol.Ass. U.K.* **29**, 303.

NICOLL, P. A. (1954) The anatomy and behaviour of the vascular systems of *Nereis virens* and *Nereis limbata. Biol.Bull.mar.biol.Lab., Woods Hole.* **106**, 69.

NICHOLS, D. (1959) Changes in the Chalk heart-urchin *Micraster* interpreted in relation to living forms. *Phil. Trans.R.Soc.B.* **242**, 347.

NICHOLS, D. (1960) The histology and activities of the tube-feet of *Antedon bifida. Q.Jl.microsc.Sci.* **101**, 105.

NICHOLS, D. (1966) Functional morphology of the water-vascular system. In: Boolootian, R. A. (ed.). *Physiology of Echinodermata.* 219.

NICHOLS, D. (1967a) Pentamerism and the calcite skeleton in echinoderms. *Nature, Lond.* **215**, 665.

NICHOLS, D. (1967b) The rule of five in animals. *New Sci.* **35**, 546.

NICHOLS, D. (1969) *Echinoderms.* 4th Edition. London (Hutchinson).

NICHOLS, D. (1975) *The uniqueness of the echinoderms.* Oxford Biology Readers, 53. Oxford (University Press).

O'BRIEN, F. E. M. (1948) The control of humidity by saturated salt solutions. *J.scient.Instrum.* **25**, 73.

OELZE, A. (1931) Beiträge zur Anatomie von *Diastylis rathkei* Kr. *Zool.Jb.Anat.* **54**, 235.

OLDROYD, H. (1958) *Collecting, preserving and studying insects.* London (Hutchinson).

PANTIN, C. F. A. (1935a, b, c, d) The nerve net of the Actinozoa. I Facilitation, II Plan of the nerve net, III Polarity and after-discharge, IV Facilitation and the 'staircase'. *J.exp.Biol.* **12**, 119, 139, 156, 389.

PANTIN, C. F. A. (1948) *Notes on Microscopical Histological Technique for Zoologists.* Cambridge (University Press).

PANTIN, C. F. A., and PANTIN, A. M. P. (1944) The stimulus to feeding in *Anemonia sulcata. J.exp.Biol.* **20**, 6.

PAPENFUSS, E. J. (1934) Reunition of pieces of *Hydra* with special reference to the rôle of the three layers and to the differentiated parts. *Biol.Bull.mar.biol.Lab., Woods Hole.* **47**, 223.

PEARSON, J. (1908) Cancer. L.M.B.C. Memoir 16. Liverpool (University Press).

PICKWORTH, F. A. (1935) A new method of study of the brain capillaries and its application to the regional localisation of mental disorder. *J.Anat.* **69**, 62.

POTTS, W. T. W., and PARRY, G. (1964) *Osmotic and ionic regulation in animals.* London (Pergamon).

PRINGLE, J. W. S. (1957). *Insect Flight.* Cambridge (University Press).

PRINGSHEIM, E. G. (1955) Uber *Ochromonas danica* n.sp. and andere Arten der Gattung. *Arch.Mikrobiol.* **23**, 181.

REID, D. M. (1947) Talitridae (Crustacea, Amphipoda). Synopses of British Fauna No. 7, *Linn.Soc.Lond.*

ROBSON, E. A. (1953) Nematocysts of *Corynactis*: The activity of the filament during discharge. *Q.Jl.miscrosc.Sci.* **94**, 229.

ROBERTS, M. B. V. (1962a) The rapid response of *Myxicola infundibulum* (Grübe) *J.mar.biol.Ass.U.K.* **42**, 527.

ROBERTS, M. B. V. (1962b) The giant fibre reflex of the earthworm *Lumbricus terrestris L.* I The rapid response. *J.exp.Biol.* **39**, 219.

ROBERTS, M. B. V. (1962c) The giant fibre reflex of the earthworm *Lumbricus terrestris* L. II Fatigue. *J.exp.Biol.* **39**, 229.

ROUDABUSH, R. L. (1933) Phenomenon of regeneration in everted *Hydra. Biol.Bull.mar.biol.Lab.*, *Woods Hole.* **47**, 253.

ROWETT, H. G. Q. (1962) *Guide to Dissection.* London (John Murray).

RYLAND, J. S. (1962) Biology and Identification of Intertidal Polyzoa. *Field Studies* **1**, (4).

SAGER, R. and GRANICK, S. (1954) Nutritional control of sexuality in *Chlamydomonas reinhardii. J.Gen.Physiol.* **37**, 729.

SANDON, H. (1963) *Essays on protozoology.* London (Hutchinson).

SARS, G. O. (1896) Phyllocarida and Phyllopoda. *Fauna Norvegiae.* Vol. 1. Christiania.

SCHARFENBERG, U. V. (1910) Studien und Experimente uber die Eibildung und den Generationszyklus von *Daphnia magna. Int.Rev.Hydrob.Biol.Suppl.* **1**, 1.

SCHIFF, J. A. and EPSTEIN, H. T. (1965) The continuity of the chloroplast in *Euglena.* In: Locke, M. (ed.). *Reproduction: Molecular, Subcellular and Cellular.* New York (Academic Press).

SEGROVE, F. (1941) The development of the serpulid *Pomatoceros triqueter* L. *Q.Jl.microsc.Sci.* **82**, 467.

SMITH, H. G. (1938) The receptive mechanism of the background response in chromatic behaviour of Crustacea. *Proc.R.Soc.B.* **125**, 250.

SMITH, J. E. (1940) Reproductive and associated organs in *Ophiothrix fragilis. Q.Jl.microsc.Sci.* **82**, 267.

SMITH, J. E. (1966) Echinodermata. In Bullock, T. H. and Horridge, G. A. *Structure and function in the nervous system of invertebrates.* **2**, 1519.

SMITH, R. I. (1940) Studies on the effects of eyestalk removal on young crayfish (*Cambarus clarkii* Giard). *Biol. Bull.mar.biol.Lab.*, *Woods Hole.* **79**, 145.

SNODGRASS, R. E. (1925) *Anatomy and Physiology of the Honeybee.* New York (McGraw Hill).

SONNEBORN, T. N. (1950) Methods in the general biology and genetics of *Paramecium aurelia. J.exp.Zool.* **113**, 87.

SONNEBORN, T. M. (1959) Kappa and related particles in *Paramecium. Adv.Virus.Res.* **6**, 229.

STEPHENSON, T. A. (1928, 1935) *The British Sea Anemones.* Vols. 1, 2. London (Ray Soc.).

STRICKBERGER, W. (1962) *Experiments in Genetics with Drosophila.* New York (Wiley).

STRICKLAND, J. D. H., and PARSONS, T. R. (1968) *A practical handbook of seawater analysis.* Ottawa (Fisheries Res. Bd. Canada Bull. 167).

TARDENT P. (1963) Regeneration in the Hydrozoa. *Biol.Rev.* **38**, 293.

TARDENT, P. (1965) Regeneration in Hydrozoa. In: Kiortis, V. and Trampusch, H. A. L. *Regeneration in Animals and related problems.* Amsterdam (North Holland).

TATTERSALL, W. M. and O.S. (1951) The British Mysidacea. London (Ray Soc.).

TAYLOR, A. E. R. and BAKER, J. R. (1968) *The cultivation of parasites in vitro.* Oxford (Blackwell).

THOMAS, J. G. (1940) *Pomatoceros, Sabella* and *Amphitrite* L.M.B.C. Memoir 33. Liverpool (University Press).

THOMAS, J. G. (1963) Dissection of the locust. London (Witherby).

THOMPSON, T. E. (1961) The structure and mode of functioning of the Reproductive organs of *Tritonia hombergi* (Gastropoda Opisthobranchia). *Q.Jl.Microsc.Sci.* **102**, 1.

THOMPSON, T. E. (1962) Studies on the ontogeny of *Tritonia hombergi*: Cuvier (Gastropoda Opisthobranchia). *Phil.Trans.R.Soc.B.* **245**, 172.

THOMPSON, T. E. (1976) *Biology of opisthobranch molluscs.* London (Ray Society).

TILLYARD, R. J. (1917) *The biology of dragonflies.* Cambridge (University Press).

TOMPSETT, D. H. (1939) *Sepia,* L.M.B.C. Memoir 52. Liverpool (University Press).

TONOLLI, V. (1947) Il ritmo cardiaco della *Daphnia pulex* De Geer. *Mem.Ist.Ital.Idrobiol.* **3**, 415.

TREMBLEY, M. (1744) *Memoires pour servir à l'histoire d'un genre de Polypes d'eau douce, à bras en forne de cornes.* Leiden (J. and H. Verbeek).

TRUEMAN, E. R. (1966a) The mechanism of burrowing in the polychaete worm *Arenicola marina* (L.). *Biol.Bull. mar.biol.Lab,* *Woods Hole.* **131**, 369.

TRUEMAN, E. R. (1966b) Observations on the burrowing of *Arenicola marina* (L.). *J.exp.Biol.* **44**, 93

TRUEMAN, E. R., BRAND, A. R. and DAVIS, P. (1966) The dynamics of burrowing of some common littoral bivalves. *J.exp.Biol.* **44**, 469.

VICKERMAN, K. and COX, F. E. G. (1967) *The Protozoa.* London (John Murray).

VOIGT, M. (1957) Rotatoria. Die Radertiere Mitteleuropas. Borntraeger, Berlin.

WARDLE, R. A., and MCLEOD, J. A. (1952) *The Zoology of Tapeworms.* Minnesota (University Press).

WARR, J. R., MCVITTIE, A., RANDALL, Sir J. and HOPKINS, J. M. (1966) Genetic control of flagellar structure in *Chlamydomonas reinhardii. Genet.Res.* **7**, 335.

WEBB, D. A. (1940) Ionic regulation in *Carcinus maenas. Proc.R.Soc.,B.* **129**, 107.

WEIS-FOGH, T. (1956) Biology and physics of locust flight IV. *Phil.Trans.R.Soc.,B.* **239**, 553.

WELLS, G. P. (1937) Studies on the Physiology of *Arenicola marina* L. I The pace-maker role of the oesophagus and the action of adrenaline and acetylcholine. *J.exp.Biol.* **14,** 117.

WELLS, G. P. (1944) The parapodia of *Arenicola marina* L. *Proc.zool.Soc.Lond.* **114,** 100.

WELLS, G. P. (1949a) Respiratory movements of *Arenicola marina* L.; Intermittent irrigations of the tube, and intermittent aerial respiration. *J.mar.biol.Ass.U.K.* **28,** 447.

WELLS, G. P. (1949b) The behaviour of *Arenicola marina* L. in sand, and the role of spontaneous activity cycles. *J.mar.biol.Ass.U.K.* **28,** 465.

WELLS, G. P. (1950) The anatomy of the body wall and appendages in *Arenicola marina* L., *Arenicola claparedii* Levinsen and *Arenicola ecaudata* Johnston. *J.mar.biol.Ass.U.K.* **29,** 1.

WELLS, G. P. (1954) The mechanism of proboscis movement in *Arenicola*. *Q.Jl.microsc.Sci.* **95,** 251.

WELLS, G. P. (1962) Killing and preservation of *Arenicola* (*personal communication*).

WELSH, J. H., and SMITH, R. I. (1949) Laboratory Exercises in *Invertebrate Physiology*. Minneapolis (Burgess).

WENYON, C. M. (1926) *Protozoology, a manual for medical men, veterinarians and zoologists.* London (Baillière, Tindall and Cox).

WERNER, B. (1953) Ausbildungstufen der Filtrationsmechanismen bei filtrierenden Prosobranchiern. Verh. dtsch. zool. Gess., (1952), 529.

WHITEAR, M. (1953) The stomatogastric nervous system of *Arenicola*. *Q.Jl.microsc.Sci.* **94,** 293.

WIGGLESWORTH, B. V. (1950) A new method for injecting the tracheae and tracheoles of insects. *Q.Jl.microsc.Sci.* **91,** 217.

WIGGLESWORTH, V. B. (1965) *The Principles of Insect Physiology*. 6th ed. London (Methuen).

WILBUR, K. and YONGE, C. M. (ed.) Vol. 1 (1964, Vol. 2 (1966) *Physiology of Molluscs*. New York (Acad. P.).

WILLMER, E. N. (1956) Factors which influence the acquisition of flagella by the amoeba *Naegleria gruberi*. *J.exp.Biol.* **33,** 583.

WILLMER, E. N. (1963) Differentiation in *Naegleria*. *Symp.Soc.Exper.Biol.* **16,** 215.

YONGE, C. M. (1939a) The mantle cavity in the Loricata. *Q.Jl.microsc.Sci.* **81,** 367.

YONGE, C. M. (1939b) The protobranchiate Mollusca: a functional interpretation of their structure. *Phil.Trans.R. Soc.B.* **230,** 79.

YONGE, C. M. (1947) The pallial organs in the Aspidobranch Gastrapoda and their evolution throughout the Mollusca. *Phil.Trans.R.Soc.B.* **232,** 442.

YONGE, C. M. (1949) Structure and adaptations of the Tellinacea (Eulamellibranchia). *Phil.Trans.R.Soc.B.* **234,** 29.

YONGE, C. M. (1950) *The Oyster*. London (Collins).

YONGE, C. M. (1953) The monomyarian habit in lamellibranchs. *Trans.R.Soc.Edinb.* **62,** 443.

YONGE, C. M. (1926) Structure and physiology of the organs of feeding and digestion in the oyster. *J.mar.biol. Ass.U.K.* **14,** 295.

ZIMMER, C. (1941) Cumacea. Bronn's *Klassen und Ordnumg des Tierreich*. 5 (1) (4).

ZUCKERHANDL, E. (1950) Coelomic pressures in *Sipunculus nudus*. *Biol.Bull.mar.biol.Lab., Woods Hole.* **98,** 161.

Index

Invertebrates described in some detail are given in heavy type

351